Praise
The Life of Neil A. Armstrong

"To understand Armstrong on his own terms is to see a large truth of our time . . . [Hansen's] mastery of detail is put to splendid use. The narrative of the moon mission is crisp and dramatic, the science clear. He deftly takes us back into those few days of global fascination with the adventure of the three distant voyagers and the tense uncertainty about how it would turn out. . . . I finished Hansen's Apollo story with a wholly fresh sense of awe at the magnitude of NASA's achievement . . . a compelling and nuanced portrait of the astronaut."

—James Tobin, *Chicago Tribune*

"Neil Armstrong—naval aviator, research pilot, astronaut, American hero, and larger-than-life icon. He may have thought it was 'one small step for [a] man,' but it was one giant leap for the rest of us. . . . *First Man* is primed to be one of the definitive reference works on the lunar program."

—U.S. Navy Captain William Readdy,
Aviation Week & Space Technology

"Most of the astronauts' books are about the adventure. Jim Hansen's well-researched and documented book is about the adventurer. *First Man* is a compelling story of a modern-day Columbus, which provides the rare opportunity to understand the personal qualities driving explorers. Quiet, complex, and deep, Armstrong, as fuel was running out, was the right man at the right time to take America and the world to the surface of the moon."

—Eugene F. Kranz, author of *Failure Is Not an Option*

"A powerful, unrelenting biography of a man who stands as a living testimony to everyday grit and determination. . . . A must for astronaut buffs and history readers alike."

—*Publishers Weekly*, starred review

"A great read."

—*The Kansas City Star*, Noteworthy Book of 2005

"Ever since Apollo 11's 'one giant leap for mankind' in 1969 the world has wondered who Neil Armstrong really is. Now, at last, Jim Hansen has stripped away the myths and mysteries to bring us face-to-face with the man himself. This definitive portrait offers many new and fascinating details about Armstrong and his life and about the momentous and unforgettable era of exploration in which he was lucky enough—and talented enough—to play a key role."

—Andrew Chaikin, author of *A Man on the Moon*

"This impressively documented and engagingly written biography will stand the test of time."

—John Carver Edwards, *Library Journal*

"Hansen does a fine job of retelling Armstrong's childhood and remarkable career in aviation. The NASA years have been covered in many other books, but Hansen manages to keep them fresh, benefiting from Armstrong's perspective. . . . As Hansen shows, the way Armstrong chooses to carry the heavy burden of history only proves once again that he has *the right stuff.*"

—Brian Hicks, *The Post and Courier* (Charleston, SC)

"Masterfully written . . . technically accurate, scholarly yet independent and accessible. . . . Mission accomplished and a perfect touchdown."

—Leonard David, *Ad Astra*,
the magazine of the National Space Society

"Hansen's research is staggeringly impressive. . . . A work that has great appeal for anyone interested in why we explore, who we are in this aerospace age, and what it was about the United States that could enable a little kid from Wapakoneta, Ohio, to take that 'one small step' at Tranquility Base in the summer of 1969. A must read!!!"

—Richard P. Hallion, chief historian for the U.S. Air Force

"Armstrong opened his entire life to Hansen. . . . Thanks to Hansen, future historians will know more about the man than the fact he was first."

—Robert Pearlman, founder and editor
of collectSPACE.com

"[A] taut, well-told tale of our nation's race to the moon and the man who took the first step."

—Doug Allyn, *The Flint Journal*

"Let it be said at once that his book is an outstanding success. . . . Immaculately researched and packed with detail, but written in a way that will appeal to readers of all kinds. . . . This is an important book, and should be in every scientific library."

—Sir Patrick Moore, *Tes*
(Times Educational Supplement) magazine

"Jim Hansen has captured the essence of Neil Armstrong, not only as the first man on the Moon, but also as an outstanding aviator and astronaut. I was there for Neil's other major 'space step'—he recovered Gemini VIII from the ultimate end game with aggressive action, cool skill, and creative judgment seldom performed in any aviation or space endeavor. Just sixteen days after the deaths of the Gemini IX crew, he probably saved the Moon. Jim Hansen has written an exceptional and accurate account of a unique period in aerospace history and the adventures of Neil Armstrong."

—Dave Scott, Gemini VIII; Apollo 9;
Commander, Apollo 15

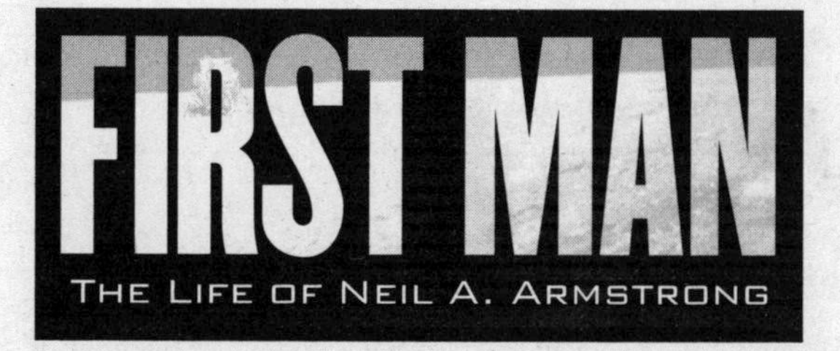

James R. Hansen

POCKET BOOKS

NEW YORK LONDON TORONTO SYDNEY NEW DELHI

Pocket Books
An Imprint of Simon & Schuster, Inc.
1230 Avenue of the Americas
New York, NY 10020

First Pocket Books paperback edition October 2018

Manufactured in the United States of America

10 9 8 7 6 5 4 3 2 1

ISBN: 978-1-9821-1047-5
ISBN: 978-1-9821-0434-4 (ebook)

For Isabelle, Mason, and Luke

Contents

Preface

If Neil Armstrong were still alive today, how would he want me to preface this new edition of his biography, coming out as it is on the eve of Apollo 11's fiftieth anniversary? I know exactly what he would say to me if I asked him that question: "Jim, it is your book. You are the author, not me. You should open the book in the way you find most appropriate."

That was pure Neil Armstrong. Once he had finally agreed to cooperate in my writing of his life story—and it took nearly three years, from 1999 into 2002, for me to secure that agreement—Neil wanted the book to be an independent, serious biography. He sat with me for fifty-five hours of interviews and agreed to read and comment on every draft chapter. But not once did he try to change or even influence my analysis or interpretation. Accordingly, he never autographed the published book, not for anyone. It was not his book, he told people: it was Jim's. I once asked him if he would sign a copy for each of my two children. He said he would think about it. I didn't ask him twice and he never brought up the subject again. It was just not his book to sign. That, too, was pure Neil.

So, how do I preface this fiftieth-anniversary edition?

• • •

My ambition is to open the book with something that is consistent with what Neil would consider significant to say at this milestone moment in the history of space exploration. Between 2018 and 2022, the world commemorates not just the first Moon landing, but ten amazing NASA missions, bred from a youthful and pioneering national space program, one carried out with remarkable speed and success as part of an epic endeavor whose name itself has become legendary: Apollo. From Apollo 8's audacious circumlunar flight in December 1968 to the final mission onto the lunar surface by intrepid astronauts on Apollo 17 in December 1972, the world watched as American astronauts ventured out from the home planet to walk on another heavenly body some quarter of a million miles away. The most notable: July 20, 1969, the day when Apollo 11, with Command Module Pilot Mike Collins, Lunar Module Pilot Buzz Aldrin, and Commander Neil Armstrong, accomplished the historic first "manned" landing on the Moon.

After much thought about how to start this book, I recalled a conversation I had with Neil in 2009—four years after the original edition of *First Man* and the year of the fortieth anniversary of Apollo 11. Our chat concerned one of the items that Neil and Buzz had purposefully left on the Moon in 1969: a tiny silicon disk about the size of a half-dollar having microscopically inscribed "goodwill messages" from leaders of 73 countries around the world on it. (The disc also carried names of the leadership of Congress and the four committees of the House and Senate responsible for legislation related to the National Aeronautics and Space Administration, as well as NASA's top management, including past administrators and deputy administrators. Also etched on the disk were four U.S. presidential statements: from then current U.S. president Richard M. Nixon and the immediate past U.S. president, Lyndon B. Johnson, plus quotes from the National Aeronautics and Space Act of 1958, signed by Dwight D. Eisenhower, and from John F. Kennedy's lunar-landing commit speech to Congress on May 25, 1961.) A man by the name of Thomas O. Paine, who was at the time the head of NASA, had corresponded with world leaders to enshrine their messages, which were photographed and reduced to 1/200 scale ultra-microfiche silicon etching. On the Moon's Sea of Tranquility, this disk, taken there by Apollo 11, still rests today inside a special aluminum case.

A year before this particular conversation with Neil, in 2008, a wonderful book had been published about the silicon disk by Tahir Rahman, a physician in the Kansas City area whose great passion, besides medicine, was space history. The title of Dr. Rahman's book was *We Came in Peace for All Mankind*, the beautiful sentiment inscribed on the plaque mounted on the ladder leg of the lunar module *Eagle*. The principal subject of Tahir's book, however, lay in the subtitle: "The Untold Story of the Apollo 11 Silicon Disk." Both Neil and I had gratefully received a copy of the book from its author, and it was our receipt of the gift copy that led to our discussion of the silicon chip and its goodwill messages.

Neil had an amazing recall of many things, and a natural forgetfulness for many others if they were matters of less concern to him. During our conversation about Tahir's book, I asked Neil which of the goodwill messages he remembered, if any, and which had impressed him the most. He specifically mentioned three of them, each of which he summarized for me and even partly paraphrased quite accurately. I didn't remember the statements (none of them so clearly, for that matter), so I wrote a few words about each one on a notepad, based on what Neil had related. The three messages he recalled came from leaders of the Ivory Coast, Belgium, and Costa Rica. When I returned home and pulled Tahir's book off my office shelf, I carefully read each message Neil had mentioned as well as the other 70 messages; Neil's choices were, in fact, three of the very best.

It is these three goodwill messages from 1969, which Apollo 11 took to the lunar surface and left there on a silicon chip fifty years ago come July 20, 2019—and which will stay memorialized on our Moon forever if left undisturbed, as they should be—that I believe Neil would like me to share in this preface:

*FROM FÉLIX HOUPHOUËT-BOIGNY. PRESIDENT OF THE IVORY COAST

At the moment when man's oldest dream is becoming a reality, I am very thankful for NASA's kind attention in offering me the services of the first human messenger to set foot on the Moon and carry the words of the Ivory Coast.

I would hope that when this passenger from the sky leaves man's imprint on lunar soil, he will feel how proud we are to belong to the generation which has accomplished this feat.

I also hope that he would tell the Moon how beautiful it is when it illuminates the nights of the Ivory Coast.

I especially wish that he would turn towards our planet Earth and cry out how insignificant the problems which torture men are, when viewed from up there.

May his work, descending from the sky, find in the Cosmos the force and light which will permit him to convince humanity of the beauty of progress in brotherhood and peace.

*From Baudouin, King of the Belgians

Now that, for the very first time, man will land on the moon, we consider this memorable event with wonder and respect.

We feel admiration and confidence towards all those who have cooperated in this performance, and especially towards the three courageous men who take with them our hopes, as well as those from all nations, who were their forerunners or who will follow them in space.

With awe we consider the power with which man has been entrusted and the duties which devolve on him.

We are deeply conscious of our responsibility with respect to the tasks which may be open to us in the universe, but also to those which remain to be fulfilled on this earth, so to bring more justice and more happiness to mankind.

May God help us to realize with this new step in world history better understanding between nations and a closer brotherhood between men.

*From José Joaquín Trejos Fernández, President of Costa Rica

I join in the wish of all Costa Ricans for the success of the historical exploit to be carried out by Apollo 11, in that it represents the scientific and technical progress attained by man in his peaceful struggle for the conquest of space and in that the crew of this ship represents human valor, will, spirit of adventure and ingenuity.

The enormous scientific and technical effort deployed in order to take the first men to the moon deserves the gratitude of mankind because from this effort will come new benefits for improving the well-being of the human race.

> *With faith we hope for better days for all mankind if there is later added to this successful endeavor—new determination for justice and liberty, as they correspond to the respect owed each human being and in favor of a major diffusion of love of one's neighbor, whose efforts we can hope will be stimulated by the spirit of humanity derived from a more clear and vivid awareness of the minuteness of this planet, which serves as our home in the cosmos.*
>
> *As representative of the Costa Rican nation, I extend my greetings to the heroes of Apollo 11 and to all those who are making this historical feat possible.*

Whether Neil Armstrong would want me to do so or not—and he likely wouldn't—I feel I should leave the final statement in this anniversary preface to Neil himself:

A month or so before the launch of Apollo 11, Armstrong, at the request of *Life* magazine, reflected, in what was surely one of his most careful bits of writing, on the meaning of the Moon landing:

> It would be presumptuous of me to pick out a single thing that history will identify as a result of this mission. But I would say that it will enlighten the human race and help us all to comprehend that we are an important part of a much bigger universe than we can normally see from the front porch. I would hope that it will help individuals, the world over, to think in a proper perspective about the various endeavors of mankind as a whole. Perhaps going to the Moon and back in itself isn't all that important. But it is a big enough step to give people a new dimension in their thinking—a sort of enlightenment.
>
> After all, the Earth itself is a spacecraft. It's an odd kind of spacecraft, since it carries its crew on the outside

instead of the inside. But it's pretty small. And it's cruising in an orbit around the Sun. It's cruising in an orbit around the center of a galaxy that's cruising in some unknown orbit, in some unknown direction and at some unspecified velocity, but with a tremendous rate of change, position, and environment.

It's hard for us to get far enough away from this scene to see what's happening. If you're in the middle of a crowd, the crowd appears to extend in every direction as far as you can see. You have to step back and look down from the Washington Monument or something like that to see that you're really pretty close to the edge of the crowd, and that the whole picture is quite a bit different from the way it looks when you are in the middle of all those people.

From our position on the Earth it is difficult to observe where the Earth is and where it's going, or what its future course might be. Hopefully, by getting a little farther away, both in the real sense and the figurative sense, we'll be able to make some people step back and reconsider their mission in the universe, to think of themselves as a group of people who constitute the crew of a spaceship going through the universe. If you're going to run a spaceship, you've got to be pretty cautious about how you use your resources, how you use your crew, and how you treat your spacecraft.

Hopefully the trips we will be making in the next couple of decades will open up our eyes a little. When you are looking at the Earth from the lunar distance, its atmosphere is just unobservable. The atmosphere is so thin, and such a minute part of the Earth, that it can't be sensed at all. That should impress everyone. The atmosphere of the Earth is a small and valuable resource. We're going to have to learn how to conserve it and use it wisely. Down here in the crowd you are aware of the

> atmosphere and it seems adequate, so you don't worry about it too much. But from a different vantage point, perhaps it is possible to understand more easily why we should be worrying.

For people who believe (wrongly) that Neil Armstrong was simply a nerdy engineer or a mere pilot of flying machines incapable of big thoughts and profound statements, all they need to do is reflect on his message quoted above to appreciate the brilliance in the man.

Neil has been dead now for six years. With that passing of time, those who knew him well understand even better how rare an individual he was, how singular his character and his accomplishments, and how much we miss him. In this retrospect, we can re-create, contemplate, evaluate, and pay tribute not just to his entire life (the first two editions of *First Man* took his biography only up to 2005, with the Preface of the second edition addressing his death only months after it happened), but to what will be his lasting legacy.

All his life, in whatever he did, Neil personified the essential qualities and core values of a superlative human being: commitment, dedication, dependability, a thirst for knowledge, self-confidence, toughness, decisiveness, honesty, innovation, loyalty, positive attitude, self-respect, respect for others, integrity, self-reliance, prudence, judiciousness, and much more. No member of the human race stepping out onto another heavenly body could possibly have represented the best of humanity more than Neil did. And no human being could have handled the bright glare of international fame or the instant transformation into a historic and cultural icon better than Neil. It was in Neil's mild and modest personality to avoid publicity and keep to the real business of the engineering profession he had chosen; he was simply not the sort of man to seek what he felt was undeserved profit from his name or reputation.

In any analysis of the quiet and unassuming way Neil lived his life after Apollo 11, the way he avoided public attention and the media for all the subsequent years, one could not fail but perceive that Neil possessed a special sensitivity that was part and parcel of his elemental character: it was as if he knew that what he had helped his country achieve in the summer of 1969—the epic landing of the first men on the Moon and their safe return to Earth—would inexorably be diminished by the blatant commercialism of our modern world, its redundant questions, all of its empty talk. At some deeply personal level, Neil understood and appreciated not just the glorious thing that had happened to him when he landed on the Moon with Buzz Aldrin while Mike Collins orbited overhead, but the glorious thing that had happened to the entire world, to all of us.

Neil had been a foremost member of the team that achieved humankind's first forays into deep space—and he always emphasized the teamwork of the 400,000 Americans instrumental to Apollo's success. He had been at the top of that pyramid, yes, but there had been nothing preordained in his becoming the commander of the first Moon landing or becoming the first man out onto the lunar surface. As he always explained, that was mostly the luck of the draw, contingent circumstances. Still, he had done what he had done, and he understood what great sacrifice, what awesome commitment, and what extraordinary human creativity it had taken to get it done. He was immensely proud of the role he had played in the first Moon landing, but it didn't turn into a circus performance for him or a money-making machine. In major respects, Neil chose to leave that particular stage of his life to the history books. It was like golfer Bobby Jones never playing competitive golf after winning the Grand Slam or Johnny Carson never again appearing on TV after leaving *The Tonight Show*. Not that Neil lived the life of a recluse after Apollo 11: that is a myth created by journalists frustrated with not getting interviews with him. After the Moon, he lived a very active life with many more accomplishments to his credit—in teaching, in research, in business and industry, in exploration. And he lived it all with honor and integrity.

For the opening epigraph to *First Man*, I selected what I felt was a profound sentence from the book, *Reflections on the Art of Living*, written by American mythologist Joseph Campbell. The sentence read: "The privilege of a lifetime is being who you are." Neil Armstrong enjoyed the privilege, and all of us should be delighted that it happened just that way for him—and for us.

James R. Hansen
March 2018

The privilege of a lifetime is being who you are.

—JOSEPH CAMPBELL, REFLECTIONS ON THE ART OF LIVING

Prologue

THE LAUNCH

After the Moon mission was over and the Apollo 11 astronauts were back on Earth, Buzz Aldrin remarked to Neil Armstrong, "Neil, we missed the whole thing."

Somewhere between 750,000 and 1 million people, the largest crowd ever for a space launch, gathered at Florida's Cape Kennedy in the days leading to Wednesday, July 16, 1969. Nearly a thousand policemen, state troopers, and waterborne state conservation patrolmen struggled through the previous night to keep an estimated 350,000 cars and boats flowing on the roads and waterways. One enterprising state auto inspector leased two miles of roadside from orange growers, charging two bucks a head for viewing privileges. For $1.50 apiece, another entrepreneur sold pseudoparchment attendance certificates with simulated Old English lettering; an additional $2.95 bought a pseudo space pen.

No football tailgate party could match the summer festival preceding the first launch for a Moon landing. Spectators firing up barbecue grills, opening coolers, peering through binoculars and telescopes, testing camera angles and lenses—people filled every strand of sand, every pier and jetty.

Sweltering in 90-degree heat by midmorning, bitten up by mosquitoes, aggravated by traffic jams or premium tourist prices, the great mass of humanity waited patiently for the mammoth Saturn V to shoot Apollo 11 toward the Moon.

In the Banana River, five miles south of the launch complex, all manner of boats choked the watercourse. On a big motor cruiser owned by North American Aviation, builder of the Apollo command module, Janet Armstrong, the wife of Apollo 11's commander, and her two boys, twelve-year-old Rick and six-year-old Mark, stood nervously awaiting the launch. Fellow astronaut Dave Scott, Neil's mate on the Gemini VIII flight in 1966, had arranged what Janet called a "*numero uno* spot." Two of Janet's friends were also on board, as were a few NASA public affairs officers and Dora Jane (Dodie) Hamblin, a journalist with exclusive coverage of the personal side of the Apollo 11 story for *Life* magazine.

Above them all, helicopters ferried successive groups of VIPs to reserved bleacher seating in the closest viewing stands a little more than three miles away from the launchpad. Of the nearly 20,000 on NASA's special guest list, about one-third actually attended, including a few hundred foreign ministers, ministers of science, military attachés, and aviation officials, as well as nineteen U.S. state governors, forty mayors, and a few hundred leaders of American business and industry. Half the members of Congress were in attendance, as were a couple of Supreme Court justices. The guest list ranged from General William Westmoreland, the U.S. army chief of staff in charge of the war in Vietnam, and Johnny Carson, the star of NBC's *Tonight Show,* to Leon Schachter, head of the Amalgamated Meat Cutters and Butcher Workers.

Vice President Spiro T. Agnew sat in the bleachers while President Richard M. Nixon watched on TV from the Oval Office. Originally, the White House had planned for Nixon to dine with the Apollo 11 astronauts the night before liftoff, but the plan changed after Dr. Charles Berry, the astronauts' chief physician, was quoted in the press warning that the president might unknowingly be harboring an incipient cold. Armstrong, Aldrin, and Mike Collins thought the medical concern was absurd; twenty or thirty people—secretaries, space suit technicians, simulator technicians—were coming into daily contact.

Two thousand reporters watched the launch from the Kennedy Space Center press site. Eight hundred and twelve came from foreign countries, 111 from Japan alone. A dozen were from the Soviet bloc.

Landing on the Moon was a shared global event that nearly all humankind felt transcended politics. British papers used two- and three-inch-high type to herald news of the launch. In Spain, the *Evening Daily Pueblo,* though critical of American foreign policy, sent twenty-five contest winners on an all-expense-paid trip to Cape Kennedy. A Dutch editorialist called his country "lunar-crazy." A Czech commentator remarked, "This is the America we love, one so totally different from the America that fights in Vietnam." The popular German paper *Bild Zeitung* noted that seven of the fifty-seven Apollo supervisors were of German origin; the paper chauvinistically concluded, "12 percent of the entire Moon output is 'made in Germany.' " Even the French considered Apollo 11 "the greatest adventure in the history of humanity." *France-Soir*'s twenty-two-page supplement sold 1.5 million copies. A French journalist marveled that interest in the Moon landing was running so high "in a country whose people are so tired of politics and world affairs that they are accused of caring only about vacations and sex." Moscow Radio led its broadcast with news of the launch. *Pravda* rated the scene at Cape Kennedy front-page news, captioning a picture of the Apollo 11 crew "these three courageous men."

Not all the press was favorable. Out of Hong Kong, three Communist newspapers attacked the mission as a cover-up for the American failure to win the Vietnam War and charged that the Moon landing was an effort to "extend imperialism into space." Others charged that the materialism of the American space program would forever ruin the wonder and beautiful ethereal qualities of the mysterious Moon, enveloped from time immemorial in legend. After human explorers violated the Moon with footprints and digging tools, who again could ever find romance in poet John Keats's question, "What is there in thee, moon, that thou shouldst move my heart so potently?" Partaking of the technological miracle of the first telecommunications satellites launched earlier in the decade, at the U.S. embassy in Seoul, 50,000 South Koreans gathered before a wall-size television screen. A crowd of Poles filled the auditorium at the American embassy in Warsaw. Trouble with AT&T's Intelsat III satellite over the Atlantic prevented a live telecast in Brazil (as it did in many parts of South America, Central America, and the Caribbean region), but Brazilians listened to accounts on radio and bought out special newspaper editions. Because of the Intelsat problem, a makeshift, round-the-world, west-to-east transmission caused a two-second lag in live coverage worldwide.

Shortly before liftoff, CBS News commentator Eric Sevareid described the scene to Walter Cronkite's audience: "Walter . . . as we sit here today . . . I think the [English] language is being altered. . . . How do you say 'high as the sky' anymore, or 'the sky is the limit'—what does that mean?"

Nowhere on the globe was the excitement as palpable as it was throughout the United States. In east Tennessee, tobacco farmers crowded around a pocket-size transistor in order to share the big moment. In the harbor at Biloxi, Mississippi, shrimpers waited on the wharf for word that Apollo 11 had lifted off. At the Air Force Academy in Colorado Springs, where 7:30 A.M. classes were postponed, fifty cadets hovered around one small TV set. In the twenty-four-hour casino at Caesars Palace in Las Vegas, the blackjack and roulette tables sat empty while gamblers stood spellbound in front of six television sets.

The multitude of eyewitnesses assembled on and around the Cape, Merritt Island, Titusville, Indian River, Cocoa Beach, Satellite Beach, Melbourne, throughout Brevard and Osceola counties, as far away as Daytona Beach and Orlando, prepared to behold one of the most awesome sights known to man. The voice of Jacksonville, Florida's Mrs. John Yow, wife of a stockbroker, quivered as she uttered, "I'm shaky, I'm tearful. It's the beginning of a new era in the life of man." Charles Walker, a student from Armstrong's own Purdue University, told a newsman from his campsite on a small inlet in Titusville, "It's like mankind has developed fire all over again. Perhaps this will be the kindling light to put men together now." In the VIP stands nearest the launch complex, R. Sargent Shriver, the U.S. ambassador to France who was married to Eunice Kennedy, a sister of the deceased president John Kennedy, who had committed the country to landing on the Moon, declared, "How beautiful it is! The red of the flames, the blue of the sky, the white fumes—those colors! Think of the guys in there getting that incredible ride. *Incroyable!*"

CBS's commentator Heywood Hale Broun, best known for his irreverent sports journalism, experienced the liftoff with several thousand people along Cocoa Beach, some fifteen miles south of the launchpad. He told Cronkite's audience of tens of millions, "At a tennis match you look back and forth. On a rocket launch you just keep going up and up, your eyes going up, your hopes going up, and finally the whole crowd like some vast many-eyed crab was staring out and up and up and all very silent. There was a small 'Aah' when the rocket first went up, but after that it was just staring and reaching. It was the poetry of hope, if you will, unspoken but seen in the kind of concentrated gestures that people had as they reached up and up with the rocket."

Even those who came to the launch to protest could not help but be deeply moved. Reverend Ralph Abernathy, successor to the late Dr. Martin Luther King Jr. as head of the Southern Christian Leadership Conference and de facto leader of the American civil rights movement, marched with four mules and about 150 members of the Poor People's Campaign for Hunger as close as they were allowed to get to the sprawling spaceport. "We are protesting America's inability to choose the proper priorities," said Hosea Williams, the SCLC's director of political education, who claimed money spent to get to the Moon could have wiped out hunger for 31 million poor people. Nonetheless, Williams stood "in admiration of the astronauts," just as Reverend Abernathy himself "succumbed to the awe-inspiring launch," declaring, "I was one of the proudest Americans as I stood on this soil. I think it's really holy ground." "There's so much that we have yet to do—the hunger in the world, the sickness in the world, the poverty in the world," former president Lyndon B. Johnson told Walter Cronkite shortly after watching the launch from his bleacher seat, wife Lady Bird at his side. "We must apply some of the great talents that we've applied to space to all these problems, and get them done, and get them done in the spirit of what's the greatest good for the greatest number."

With ten minutes left on the clock, Sevareid said on-air to Cronkite, "when the van carrying the astronauts themselves went by on this roadway just now, there was a kind of hush among the people. . . . You get a feeling that people think of these men as not just superior men but different creatures. They are like people who have gone into the other world and have returned, and you sense they bear secrets that we will never entirely know, and that they will never entirely be able to explain."

In central Ohio, a thousand miles from the viewing stands in Florida, the little burg of Wapakoneta, Armstrong's hometown, counted down. Streets were virtually empty, with nearly 6,700 residents glued to their television sets. At the center of the chaos was 912 Neil Armstrong Drive, the one-story, ranch-style home of Viola and Steve Armstrong, into which the couple had moved just a year earlier. Neil's parents had attended the Gemini VIII launch in 1966. Their son had also arranged for them to witness Apollo 10's liftoff in April. But for this flight, he advised them to stay at home, saying "the pressure might be too great" for them at the Cape. In the months leading up to the launch, Neil's mother and father had been "besieged by newsmen of every category," from England, Norway, France, Germany, and Japan. Viola recalled, "Their prying questions ('What was Neil like when he was a little boy?' 'What kind of a home life did he have?' 'Where will you be and what will you be doing during the launch?' etc., etc.) were a constant drain on my strength and nervous system. I survived this only by the grace of God. He must have been at my side constantly."

To facilitate their coverage of Apollo 11 from Wapakoneta, the three major TV networks erected a shared eighty-five-foot-high transmission tower in the driveway of the Armstrong house. The garage was turned into a pressroom with messy rows of telephones installed atop folding picnic tables, and NASA sent Tom Andrews, a protocol officer, to help the Armstrongs deal with the gaggle of reporters. Because Neil's parents still had only a black-and-white television, the TV networks gave them a large color set on which to watch the mission. On a daily basis, a local restaurant sent down a half dozen pies. A fruit company from nearby Lima delivered a large stock of bananas. A dairy from Delphos sent ice cream. The Fisher Cheese Co., Wapakoneta's largest employer, proffered its special "Moon Cheeze." Consolidated Bottling Company delivered crates of "Capped Moon Sauce," a "secret-formula" vanilla cream soda pop.

The proud mayor of Wapakoneta requested that every home and business display an American flag (and preferably also the Ohio state flag) from the morning of the launch until the moment "the boys" were safely back. Among a few locals, the media spotlight inspired a different kind of civic embellishment. Some told exaggerated stories, even outright lies, about their special connection to the astronaut. Even kids took to spinning yarns: "Listen, my dad is Neil Armstrong's barber!" or "My mom was the first girl ever to kiss Neil!" or "Hey, I chopped down Neil Armstrong's cherry tree!" Since the Armstrongs' Auglaize County phone number was public knowledge, Tom Andrews arranged to have two private phone lines run into the family's utility room, off the kitchen. Around noontime the day before the launch,

Neil called his mother and father from the Cape. Viola recalled: "His voice was cheerful. He thought they were ready for the takeoff the next morning. We asked God to watch over him."

Neil's sister and brother attended the launch. June, her husband, Dr. Jack Hoffman, and their seven children flew to Florida from their home in Menomonee Falls, Wisconsin. Dean Armstrong; his wife, Marilyn; and their three children drove down to Florida from their home in Anderson, Indiana. Viola's recall of that extraordinary morning remained sharp until her dying day: "Visitors, neighbors, and strangers gathered around to watch and listen, including my mother, Caroline; my cousin, Rose; and my pastor, Reverend Weber. Stephen and I sat side by side, wearing for good luck the Gemini VIII pins that Neil had given us.

"It seemed as if from the very moment he was born—farther back still, from the time my husband's family and my own ancestry originated back in Europe long centuries ago—that our son was somehow destined for this mission."

PART ONE

BOY PILOT

I was born and raised in Ohio, about 60 miles north of Dayton. The stories about the achievements of the Wright brothers and their invention of the airplane have been in my memory as long as I can remember. . . . Originally, my focus was on the building of airplanes, not their flying. You couldn't have success with a model that wasn't built well.

—NEIL A. ARMSTRONG TO AUTHOR,
AUG. 13, 2002

CHAPTER 1

An American Genesis

Neil Armstrong understood that neither his life story nor anyone else's began at birth. It went far back in time to the genesis of one's parentage, hundreds of years back, as far back as human memory, historical documentation, and existing genealogical records went. It cheated everyone's life story—not to mention the lifetimes, experiences, challenges, achievements, loves, and passions of parents, grandparents, great-grandparents, great-great grandparents, and beyond—to ignore one's deep family past. Neil insisted that his biography include their story.

Neil also deeply appreciated that his own family history, as with so many American families, was a story of immigrants and their courageous coming to a new land. It was an "American Genesis" as he once put it.

Neil loved America and its history. He loved what it had stood for since even before the birth of the country in its fight for independence from the mother country, England, from 1776 to 1783. For Neil, "America means opportunity. It started that way. The early settlers came to the new world for the opportunity to worship in keeping with their conscience, and to build a future on the strength of their own initiative and hard work. They discovered a new life with freedom to achieve their individual goals."

In Neil's case, the "deep family past"—that part of it which we know with certainty—goes back more than over 300 years to his earliest known Armstrong ancestors of the late-seventeenth century. The paternal line of Neil's family grew from a clan of Armstrongs who flourished from the later Middle Ages on in the notorious "Borderlands" region between Scotland and England. A small party of intrepid Armstrongs crossed the Atlantic four decades before the American Revolution. His descendants then moved steadily westward across the Appalachians, in wagons and river boats, among the boldest pioneers of the early American frontier, ultimately settling in the fertile agricultural lands of northwestern Ohio shortly after the War of 1812.

The Armstrong name began illustriously enough. Anglo-Danish in derivation, the name meant what it said, "strong of arm." Legend traced the name to a heroic progenitor by the name of Fairbairn. Viola Engel Armstrong, Neil's mother, recorded one version of the fable. "A man named Fairbairn remounted the king of Scotland after his horse had been shot from under him during battle. In reward for his service the king gave Fairbairn many acres of land on the border between Scotland and England, and from then on referred to Fairbairn as Armstrong." Offshoots of the legend say that Fairbairn was followed by Siward Beorn, or "sword warrior," also known as "sword strong arm."

By the 1400s, the Armstrong clan emerged as a powerful force in the Borders. In the sixteenth century, Armstrongs were unquestionably the Borders' most robust family of reivers—bandits and robbers. Decades' worth of flagrant expansion by the Armstrongs eventually forced the royal hand, as did their purported crimes of burning down fifty-two Scottish churches. By 1529, King James V of Scotland marshaled a force of eight thousand soldiers to tame the troublesome Armstrongs, who numbered somewhere between 12,000 and 15,000, or roughly three percent of Scotland's population. In 1530, James V marched his forces southward in search of Johnnie Armstrong of Gilnockie. Author Sir Walter Scott identified William Armstrong as a lineal descendant of Johnnie Armstrong of Gilnockie. Historians have inferred that Will was the oldest son of Christopher Armstrong (1523–1606), who was himself the oldest son of Johnnie Armstrong.

Neil Armstrong's progenitors stayed in the Borderlands until they immigrated directly to America sometime between 1736 and 1743. Adam Armstrong, born in the Borderlands in 1638 and dying there in 1696, represents Generation No. 1, ten generations before the first man on the Moon.

Adam Armstrong had two sons, one of whom was also named Adam, born in Cumbria, England, in 1685. At age twenty, Adam Armstrong II married Mary Forster. In the company of his father, Adam Abraham Armstrong III (b. 1714 or 1715) crossed the Atlantic in the mid-1730s, when he was twenty, making them the first of Neil's bloodline to immigrate to America. Father Adam died in Pennsylvania in 1749.

These Armstrongs were thus among the earliest settlers of Pennsylvania's Conococheague region. Adam Abraham Armstrong worked his land in what became Cumberland County until his death in 1779. His oldest son John (b. 1736), at age twenty-four, surveyed the mouth of Muddy Creek, 160 miles west of the Conococheague. There John and his wife, Mary, raised nine children, their second son, John (b. 1773), producing offspring leading to Neil.

Following the Revolutionary War, thousands of settlers poured into the Ohio Country. In March 1799, twenty-five-year-old John Armstrong, his wife, Rebekah, their son David; as well as John's younger brother Thomas Armstrong; his wife, Alice Crawford; and their infant, William, traveled by flatboat down Muddy Creek to Pittsburgh, then steered into the Ohio River some 250 miles down to Hockingport, just west of modern-day Parkersburg, West Virginia. The two families coursed their way up the Hocking River to Alexander Township, Ohio. Homesteading outside what became the town of Athens, Thomas and Alice raised six children. John and Rebekah eventually settled near Fort Greenville, in far western Ohio. Neil's ancestor John Armstrong (Generation No. 5) and his family witnessed negotiations for the Treaty of St. Marys, the last great assemblage of Indian nations in Ohio. In 1818, John and his family settled on the western bank of the St. Marys River. From the first harvests, the Armstrongs earned enough to secure the deed to their 150-acre property, which became the "Armstrong Farm," the oldest farm in Auglaize County.

David Armstrong, the eldest of John's children (b. 1798), and Margaret Van Nuys (1802–1831) were Neil's (unmarried) paternal great-grandparents. Margaret married Caleb Major, and David married Eleanor Scott (1802–1852), the daughter of Thomas Scott, another early St. Marys settler. Baby Stephen stayed with his mother until her premature death in March 1831, when Margaret's parents, Rachel Howell and Jacobus Van Nuys, took in their seven-year-old grandson. Stephen's father, David, died in 1833, followed by his grandfather in 1836.

Stephen Armstrong (Generation No. 7) received his grandfather Van Nuys's legacy of roughly two hundred dollars in cash and goods when he turned twenty-one in 1846. Stephen was a farm laborer, and after years working for another family, he managed to buy 197 acres and later added 218 more.

How the Civil War affected Stephen Armstrong is unknown. Stephen married Martha Watkins Badgley (1832–1907), the widow of George Badgley and mother of four. On January 16, 1867, Martha gave birth to Stephen's son, Willis Armstrong.

When Stephen Armstrong died in August 1884 at age fifty-eight, he owned well over four hundred acres, whose value was over $30,000—the equivalent today of over $700,000.

Stephen's only son, Willis, inherited most of the estate. Three years later, Willis married a local girl, Lillian Brewer (1867–1901). The couple had five children and lived in a farmhouse near River Road. In 1901, Lillie died in childbirth.

Bereaved, Willis began a part-time mail route. One stop was at the Koenig brothers' law firm. Their sister Laura worked as their secretary, and in late 1903, Willis began courting her. They married in June 1905 and lived in a house Willis bought in St. Marys. Later, they moved into an impressive Victorian on a corner of West Spring Street.

It was here that Stephen Koenig Armstrong, Neil's father, grew up. The first of two children born to Willis and Laura, the boy was welcomed on August 26, 1907, by half-sisters Bernice and Grace, and half-brothers Guy and Ray. Economic misfortune and a streak of family bad luck defined Stephen's boyhood. Willis mortgaged his farm and sank most of his money into a railroad investment scheme touted by his brother-in-law. Unfortunately so, for the investment didn't pan out, and the resulting financial disaster soured family relations, including Willis's marriage.

In 1912, Stephen's half-brother Guy died; and in 1914 the Armstrong house caught fire. Six-year-old Stephen escaped with just the clothes on his back.

In 1916, Willis, forty-nine years old and deeply in debt, quit his mail route and headed for Kansas oil fields.

In early 1919 Willis returned to Ohio. Within weeks, he moved the family back to the River Road farm, still heavily mortgaged. Soon crippled by chronic arthritis, Willis relied on Stephen to work the fields; finishing his education was a task on which Stephen's mother insisted.

Even before Stephen graduated from high school in 1925, he had decided not to pursue the farming life. Soon he fell in love with a soft-spoken young woman named Viola Louise Engel.

Stephen Armstrong's family had been living in America for well over a century when Viola's German-born grandfather, Frederick Wilhelm Kötter, sailed into Baltimore harbor in October 1864. In a family effort to circumvent enforced conscription into Prussia's military, eighteen-year-old Fritz Kötter's father sold part of his farm, located outside the village of Ladbergen in the province of Westphalia, near the Dutch border, to pay their son's passage to America.

Frederick made his way to the little burg of New Knoxville, Ohio. A U.S. state whose German immigrant population exceeded 200,000 held obvious appeal to the native of Ladbergen, Germany. Kötter's first wife died young. In the early 1870s, after purchasing eighty acres of land, Fritz married a first-generation German-American, Maria Martha Katterheinrich. They Americanized their name to Katter. The couple had six sons and one daughter, Caroline, born in 1888. Nineteen years later, on May 7, 1907, Caroline gave birth to her only child, Viola. Viola's family worshiped at the St. Paul Reformed Church, whose doctrine derived from Martin Luther's Catechism. Young Viola would become very devout and would stay that way her entire life.

On May 4, 1909, Martin Engel, a butcher by trade, died of tuberculosis at twenty-nine, with his wife and baby daughter at his side. On Viola's second birthday, he was buried in Elmgrove Cemetery. Caroline's parents cared for Viola while Caroline cooked for the wealthy McClain family. In 1911, Caroline's mother, Maria, died, and in 1916, Grandfather Katter passed away. For Caroline, loss yielded to happiness when a romance blossomed between herself and a local farmer, William Ernst Korspeter, whom she had met at the Reformed Church in St. Marys. They married in 1916. Viola matriculated at Blume High School in Wapakoneta. A slender girl with unassuming ways, Viola maintained high academic marks. A student of the piano since the age of eight, she was known for her love of music. This quality, along with inventiveness, concentration, organization, and perseverance, she passed along to her son Neil.

Yet, Viola's dearest aspiration was to devote her life to Christ as a missionary, but her parents discouraged her. Instead she earned twenty cents an hour clerking at a department store. It was then that Viola started seeing Stephen Armstrong, who had just graduated from high school. The two first spoke during a youth group meeting at St. Paul Reformed Church, and the ardor of their young love disguised their many differences—differences that became clearer and clearer over the years, to a point late in life when Viola would privately question the correctness of her marriage to such an irreligious man.

But that would not come for many years into the future. For Christmas 1928, Viola and Stephen exchanged engagement rings and were married on October 8, 1929, in the living room of the Korspeter farmhouse. For a honeymoon, the couple drove "Papa" Korspeter's automobile sixty miles for their first-ever trip to Dayton. Two weeks later, Wall Street's stock market crashed and the Great Depression began.

Stephen moved Viola into the River Road farmhouse, where she helped his mother with the housework. Stephen went to Columbus to sit for the civil service exam, and in February 1930 he received an appointment to assist Columbiana County's senior auditor. Arrangements were made to auction off the farm and to move his parents into a small house in St. Marys. In mid-May 1930 Stephen and Viola, now six months pregnant, drove 230 miles to Lisbon, near the Pennsylvania border. They were "thrilled beyond words" to have electric lights and hot and cold running water in their furnished two-room apartment.

Two weeks before her due date, August 4, Viola prepared to give birth at her parents' farmhouse. Stephen remained behind in Lisbon. On August 5, 1930, a baby boy was born. The child's jawline resembled his father's, but his nose and eyes were all Viola. Viola and Stephen called their son Neil Alden. Viola liked the alliteration, "Alden Armstrong," and the allusion to Alden from

Henry Wadsworth Longfellow's classic poem "The Courtship of Myles Standish." No one in either family had ever been christened "Neil." Perhaps they knew that "Neil" was the Scottish form of the Gaelic name Néall, which translated as "cloud," or that, in its modern form, it meant "Champion."

CHAPTER 2

Smallville

Ten days after delivering Neil, Viola rose from bed to care for the baby. The doctor did not permit her to attend her father-in-law Willis's funeral, but with Stephen at home she arranged for Neil to be baptized by Reverend Burkett, the minister who had married them. Stephen's job demanded an immediate transfer to Warren, Ohio, where he would assist a senior examiner. The Armstrong family would make *sixteen* moves over the next fourteen years, in an Ohio odyssey that wound up in Wapakoneta in 1944.

Viola found Neil to be a serene and untroubled baby, with a tendency toward shyness. Viola read to Neil constantly, instilling in him her love of books. The boy learned to read extraordinarily early, and could read street signs by age three. During his first year in elementary school, in Warren, Neil read over one hundred books. Though he started his second-grade year in the rural consolidated school in Moulton but finished at St. Marys, Neil's teacher found him reading books meant for fourth-grade pupils. He was moved up to third grade, which made him eight years old when he started fourth grade the following autumn. Nonetheless, he received top grades. Wherever the family went, Neil acclimated well and made friends easily. No children were Neil's more constant companions than his younger sister and brother. On July 6, 1933, when Neil was almost three, June Louise was born; on February 22, 1935, Dean Alan arrived.

Though they always felt loved and cherished by their parents, young June and Dean sensed that their older brother was "mother's favorite." "When it was time to plant potatoes out at our grandparents' farm, Neil was nowhere to be found. He'd be in the house, in the corner, reading a book." June recalls, "He never *did* anything wrong. He was Mr. Goody Two-shoes, if there ever was one. It was just his nature."

As older brothers go, June says, Neil "was definitely a caretaker." With brother Dean, who was five years younger, relations were more difficult: "I never violated Neil's space. It would have had to have been an invitation." Though the brothers were in the same Boy Scout troops, Neil outpaced Dean in badges earned and socialized with his older friends from school. They both loved music, but Dean enjoyed competitive sports and played on the varsity basketball team. Neil was "consumed by learning," like his mother, while Dean was more like his dad, "a fun-loving person."

Neil's unusual combination of coolness, restraint, and honesty could be read as inscrutable. But he was rarely that in the eyes of his mother. "He has a truthfulness about him," Viola said in a summer 1969 interview with *Life* magazine writer Dodie Hamblin. "He either had to be truthful and right, or he didn't think he'd engage in it. I really never heard him say a word against anybody, never." Neil was always especially reserved when discussing his father: "My father's job kept him away from our home most of the time, so I didn't think of him as being close to any of the children and did not notice if he seemed to be closer to one than another." When asked whether Neil and his father were close, June replied, "No . . ." Mother hugged the children, but Father did not. "Neil probably was never hugged by him and Neil didn't hug, either."

When Neil wrote home from college, he addressed the envelopes to "Mrs. S. K. Armstrong." His letters greeted "Dear Mom and Family." In 1943, Stephen's mother, Laura, broke her ankle. Stephen and Viola took her into their home, and she lived with them until her death in 1956. This, as well as Viola and Stephen's differences on such issues as religion and temperance, caused tensions in their marriage.

Curiously, Neil, in interviews with the author for this book, did not remember that Grandmother Laura Armstrong lived with them throughout his high school years. "Grandmother Armstrong did not come to live with us until after I left for Purdue," he related. But Neil was wrong. She had lived with them through his entire high school years. That Neil did not recall his grandmother in the house—for thirteen years!—testifies not to any general forgetfulness on his part but rather to the power of Neil's concentration even as a high school student on those aspects of his life that mattered to him most on a daily basis—his friends, his books and education, his scouting, his part-time jobs, and, most ardently, as we will see, his passion for airplanes and flying. Neil could drift off, sister June recalls. "Neil read a lot as a child and that was his escape. It wasn't an escape *from* anything; it was an escape *to* something, into a world of imagination. As a boy he felt secure enough to risk escaping, because he knew, upon returning, he would be in a nice place."

• • •

For Neil Armstrong, rural Ohio represented comfort, security, privacy, and sane human values. Leaving NASA in 1971, Armstrong would seek a return to the ordinary on a small farm back in his native state. "I have chosen to bring my family up in as normal an environment as possible," he would explain.

Armstrong's down-to-earth view was rooted in his boyhood. During these same years, cartoonist Jerry Siegel envisioned a hero named Superman who hailed from "Smallville," a town in Middle America that promoted "Truth, Justice, and the American Way."

It was not Smallville but other small towns that harbored Neil Armstrong. None registered a population in the 1930s and 1940s much above five thousand. In these genuine Smallvilles, young people—with the right kind of family and community support—grew up embracing ambition.

In addition to Neil Armstrong, that mind-set marked all seven of the original Mercury astronauts: Alan B. Shepard Jr., of East Derry, New Hampshire; Virgil I. "Gus" Grissom of Mitchell, Indiana; John H. Glenn Jr., of New Concord, Ohio; Walter M. Schirra Jr., of Oradell, New Jersey; L. Gordon Cooper Jr., of Shawnee, Oklahoma; and Donald K. "Deke" Slayton, of Sparta, Wisconsin. During M. Scott Carpenter's youth, his hometown of Boulder, Colorado, counted just over ten thousand citizens.

In the view of the Original Seven, "The Right Stuff" derived from their common upbringing. John Glenn, the first American astronaut to orbit, concurred: "Growing up in a small town gives kids something special." Children "make their own decisions" and "maybe it's no accident that people in the space program, a lot of them tended to come from small towns." For most of the history of the U.S. space program, a greater number of astronauts came from Ohio than from any other state. "The small towns, the ones that I grew up in, were slow to come out of the Depression," Neil recalls. "We were not deprived [Stephen Armstrong's annual salary was just above the 1930 national average of $2,000], but there was never a great deal of money around. On that score we had it no worse and no better than thousands of other families." To some of his boyhood friends, the fact that Neil's father had a job meant that the Armstrongs were rich.

Neil's first employment came in 1940 when he was ten—and weighing barely seventy pounds. For ten cents an hour, he cut grass at a cemetery. Later, at Neumeister's Bakery in Upper Sandusky, he stacked loaves of bread and helped make 110 dozen doughnuts a night. He also scraped the giant dough mixer clean: "I probably got the job because of my small size; I could crawl inside the mixing vats at night and clean them out. The greatest fringe benefit for me was getting to eat the ice cream and homemade chocolates."

When the family moved to Wapakoneta in 1944, Neil clerked at a grocery store and a hardware store. Later he did chores at a drugstore for forty cents an hour. His parents let him keep all his wages, but expected him to save a substantial part of it for college. Of the 294 individuals selected as astronauts between 1959 and 2003, over two hundred had been active in scouting. This included twenty-one women who served as Girl Scouts. Forty of the Boy Scouts who became astronauts made it to Eagle status. Of the twelve men to walk on the Moon, eleven participated in scouting, including Neil and his Apollo 11 crewmate Buzz Aldrin.

When the family moved to Upper Sandusky in 1941,

the town of roughly three thousand people still had no Scout troop. The Japanese attack on Pearl Harbor on December 7, 1941—breaking news of which Neil heard on the radio when his father called him inside from playing in the front yard—changed that. On the following day, when the U.S. Congress declared war, the Boy Scouts of America placed its entire resources at the service of the government. As Neil remembers, news of the war "was around us all the time, in the newspaper, on the radio. And of course there were a lot of stars in the windows of the families who had children that had gone off to fight." A new troop, Ohio's Troop 25, led by a Protestant minister, met monthly. Neil's group called itself the Wolf Patrol and elected Bud Blackford as patrol leader, Kotcho Solacoff as assistant patrol leader, and Neil as the scribe.

Troop 25 and its Wolf Patrol became, in Neil's words, "immersed in the wartime environment." Airplane recognition was a Scout forte that suited Neil perfectly. He and his friends made models that their Scout leader sent on to military and civil defense authorities so their experts could better distinguish friendly from enemy aircraft. When the minister moved away, Ed Naus, "less of a disciplinarian," stepped in, assisted by Neil's father. In their Wolf Patrol, Neil, Bud, and Kotcho entered into one of those indelible adolescent friendships that thrived on good-natured rivalry. Kotcho remembers a chem-lab prank: "I said, 'Here Neil, try some $C_{12}H_{22}O_{11}$.' To my surprise and horror, Neil grabbed a pinch full and put it in his mouth. I yelled, 'Spit it out, it's poison!' Neil said, '$C_{12}H_{22}O_{11}$ is sugar.' I said, 'I know, but I didn't think you did.' That was the last time I took for granted that I knew something that he didn't."

Over the years Wapakoneta has been consistently identified as Neil Armstrong's hometown, but it was his three years in Upper Sandusky that Neil cherished most of all. Yet, as much as the entire family enjoyed their residence from 1941 to 1944 (Neil's age eleven to fourteen) at 446 North Sandusky Avenue, circumstances forced the family to make one last move, the one to "Wapak." The major reason for the move, according to Neil, was that his father, although thirty-six years old, "thought he might get drafted." Wapakoneta, some fifty miles southwest of Upper, placed Stephen farther from his work, but, explained Neil, "Mother had her parents nearby," so if her husband got called into military service she would have support for herself and her family.

The Armstrongs bought a large two-story corner home at 601 West Benton Street. As always, Neil had no trouble adjusting to his new environs, and he immediately became active in Boy Scout Troop 14. Blume High School was six blocks from his home. His transcript shows that his best grades were always in math, science, and English. Contrary to some historical accounts that have misunderstood Blume High School's grading system, he never received any poor grades.

Always musically inclined, Neil joined the school orchestra, boys' glee club, and band. Despite his small size, he played one of the largest instruments, the baritone horn, because he liked its distinctive tone—though not many others did. On the rare Friday or Saturday evening when he played his horn in a ragtime combo, he and his fellow Wapak teen musicians, calling themselves the Mississippi Moonshiners, were lucky to make five dollars split four ways.

In high school Neil joined Hi-Y, a student organization, served on the yearbook staff, and acted in the junior class play. Elected to the student council in grades eleven and twelve, in his senior year Neil served as vice president. High school friends remember Neil as not shy but rather quiet. Neil went on very few dates in high school, although he did attend his senior prom. His dad loaned him the family's brand-new Oldsmobile for the occasion. "We double-dated with Dudley Schuler and his girlfriend Patty Cole," remembers Neil's date Alma Lou Shaw Kuffner. "Unfortunately, about three in the morning on our way back from Indian Lake, Neil fell asleep at the wheel and drove off into a ditch. A man on the way to work in Lima had to pull us out. The next morning Neil's dad found that the whole side of the car was scratched."

That May of 1946, Neil, still only sixteen years old, graduated from Blume High School. Among his seventy-eight classmates, his grades placed him eleventh in his class. Accompanying Armstrong's senior class picture in the yearbook for 1946–47 was the epigram, "He thinks, he acts, tis done." His many later successes in the control of moving vehicles would later overwhelm his stained reputation at the wheel of his father's Oldsmobile.

CHAPTER 3

Truth in the Air

Jacob Zint relished his role as Wapakoneta's Mr. Wizard. A lifelong bachelor who lived with two bachelor brothers in a sinister-looking three-story home at the corner of Pearl and Auglaize streets, just a few blocks from the Armstrong home, Zint worked as an engineering draftsman for the Westinghouse Company in Lima. On top of his garage, the scientifically minded Zint built an observatory, a domed rotunda ten feet in diameter that revolved 360 degrees on roller-skate wheels. An eight-inch reflecting telescope pointed outward at the stars and planets. Through Zint's best eyepiece, the Moon appeared to be less than a thousand miles away, rather than the actual quarter million miles' distance. It was a setup that would have pleased the eccentric sixteenth-century astronomer Tycho Brahe, one of Zint's heroes.

Jake Zint would have forever remained an obscure local oddball if not for his self-proclaimed connection to young Neil Armstrong. One evening in 1946, when the future astronaut was sixteen, Neil, his friend Bob Gustafson, and a few other members of Boy Scout Troop 14 paid a visit to Zint's home. Their purpose was to qualify for an astronomy merit badge. Because Zint, age thirty-five, did not like people coming to his place unasked, the Scoutmaster, Mr. McClintock, had painstakingly prearranged the appointment.

In Zint's estimation, the moments that followed represented a turning point in the life of young Neil Armstrong. The Moon, so Zint said, "seemed to be Neil's main interest. He would dote on it," as well as expressing "a particular interest" in "the possibility of life on other planets. We hashed it over and concluded there was no life on the Moon, but there probably was on Mars." So taken was Neil with Zint and his observatory that his visits "continued even after he went away to Purdue University." On the eve of the Moon launch, Zint claimed, Neil sent, via a visiting newsman, his old astronomical mentor a special message: "The first thing he's going to do when he steps out on the Moon is find out if it's made of green cheese."

Headline after headline during June and July 1969 featured the Zint connection to Armstrong: "Neil Dreamed of Landing on Moon Someday," "Astronomer Jacob Zint Provided Neil A. Armstrong's First Close-Up Look at Moon," "Neil Armstrong: From the Start He Aimed for the Moon," "Astronaut Realizing Teen-Age Dream," "Moon Was Dream to Shy Armstrong," and "Jacob Zint, Wapakoneta Astronomer, Says, 'Neil's Dream Has Come True.' " Many of the stories included a picture of a smiling Zint, his arms resolutely folded, standing in front of the telescope that supposedly provided Armstrong's first close-up look at the Moon.

Neil's great moment of landing on the Sea of Tranquility became Zint's own high point back in Wapakoneta: "At 2:17 A.M. on July 21, Jacob Zint hopes to have his eight-inch telescope trained on the southwest corner of the moon's Sea of Tranquility. Weather permitting, the sighting will complete an odyssey in time and space that began here 23 years ago when a small blond boy named Neil Alden Armstrong took his first peek at the Moon through Mr. Zint's lenses." Everyone wanted to know what Zint was thinking at the moment of the historic landing: "It's unbelievable, when I think of all those times Neil and I talked of what it would be like up there," he told many interested reporters. "And now he's up there."

Perversely, nothing that Jacob Zint, now deceased, ever had to say about his relationship with Armstrong was true—not a bit of it, though Zint's telescope, along with his dismantled astronomical rotunda, for many years (until shortly after the original publication of *First Man* in 2005) sat in a favored location inside Wapakoneta's Auglaize County Museum.

"To the best of my recollection," Armstrong stated in 2004 with reluctance and typical reserve, so as not to overly impugn the reputation of Wapakoneta's highly publicized amateur astronomer, "I was only at Jake Zint's observatory the one time. As for looking through Zint's telescope and having private conversations with Zint about the Moon and the universe, they never happened. Mr. Zint's story grew after I became well known. All of his stories appear to be false," though Neil never bothered to correct them or insist that Zint refrain from telling them.

Most people in 1969 had no reason to disbelieve what was being reported so widely in the newspapers. Moreover, Zint's prophetic version of Neil's "destiny" just seemed, as one journalist put it in July 1969, "Almost Too Logical to Be True."

Like Zint, Neil's favorite science teacher, John Crites, also hyped the future Moonwalker's early love of the heavens. Under a harvest Moon that was "just gorgeous,"

Crites recalled, he once inquired about young Neil's future plans. "Someday," Neil replied, pointing to the full Moon, "I'd like to meet that man up there." "That was in 1946," Crites told reporters in 1969, "when no one had a thought of going up there." "That's fiction," Neil commented tersely in an interview for *First Man*. "All my aspirations in those days were related to aircraft. Space flight would have been an unrealistic ambition."

"When Neil was approximately two or three years old," Stephen Armstrong recalled in 1969, "he coaxed his mother into buying a little airplane at the ten cents store, and there was an argument between a ten- and twenty-cent plane. Of course, his mother bought him the twenty-cent plane. From that time on, he liked airplanes because he was always zooming around in and out of the house."

Neil took his first airplane ride just before his sixth birthday when the family was living in Warren. Over the years he has heard and read so many different versions of the story that he says, "I don't know what's true. It's my belief that [the plane] was offering rides for a small fare [twenty-five cents], to tour around the city." His father remembered it this way: "One time we were headed—at least his mother thought we were headed—for Sunday school, but they had an airplane ride that was cheaper in the morning and then a price that escalated during the day. So we skipped Sunday school and took our first airplane ride."

The machine that took them up was a high-wing monoplane, the Ford Trimotor. First flown in 1928, the "Tin Goose" accommodated up to twelve passengers in wicker chairs, and cruised at a speed of about 120 miles per hour.

Sometime during his adolescence, Neil began to have a recurring dream: "I could, by holding my breath, hover over the ground. Nothing much happened. I neither flew nor fell in those dreams; I just hovered. But the indecisiveness was a little frustrating. There was never any end to the dream." Neil was never sure what they symbolized. "I can't say they were related to flying in any way. It doesn't seem there was much relationship, except for being suspended above the ground." Tongue in cheek, he would offer "I tried it later, when I was awake, and it didn't work.

"I began to focus on aviation probably at age eight or nine," Neil recalled, "inspired by what I'd read and seen about aviation and building model aircraft." An older cousin lived down the block. Once Neil "saw what he was able to do" with balsa wood and tissue paper, he became hooked.

The first model Neil remembers building was a high-wing light aircraft cabin, most likely a Taylor Cub, papered in yellow and black. "It never occurred to me to buy models *with* engines," because motors cost extra money and required gasoline—both of which were in short supply during World War II. If powered, his models were driven by twisted rubber bands.

Neil's models filled up his bedroom plus an entire corner of the basement. According to Dean, Neil built so many planes that he would fly ones that he was tired of or did not like out the upstairs window—sometimes in flames. June remembers Neil gathering "five or six at least, then he would run down the stairs and out the front door to the end of the driveway. So we're leaning out an open window upstairs, tossing out Neil's airplanes. Mother would have just died!"

Neil remembered, "I usually hung my models with string from the ceiling of my bedroom. I had put a lot of work in them and didn't want to crash them, so when I flew one of those airplanes, it was a rare occasion.

"While I was still in elementary school my intention was to be an aircraft designer. I later went into piloting because I thought a good designer ought to know the operational aspects of an airplane.

"I read a lot of the aviation magazines of the time, *Flight* and *Air Trails* and *Model Airplane News,* anything I could get my hands on." As part of the Aeromodelers Club at Purdue University, "I won a number of events or [got] second place." Neil recollected racing his "gasoline-powered 'control-line' models, flown on wires and operated at the center of a circle," to speeds well in excess of 100 miles per hour. "I absorbed a lot of new knowledge and found people, some of them World War II veterans, who had vastly more experience and intuition on how to be successful flying."

Fifteen-year-old Armstrong began saving toward the cost of flying lessons, nine dollars an hour (roughly $123.00 in 2016 dollars). Making forty cents an hour at his after-school job at Brading's Drugs, Neil worked twenty-two and a half hours to pay for a single lesson.

Early Saturday mornings Neil hitchhiked or "rode a bike with no fenders" out to a small grass airfield outside Wapakoneta. "They did what they called 'top cylinder overhauls,' " Neil recalls, "or just 'top overhauls,' for short." Once Neil turned sixteen and had his student pilot's license, he could fly the airplanes. "That's the way I built up flight time, by doing 'slow time' to coat the valves with high-octane gasoline after top cylinder overhauls." The aircraft were mostly old army planes and trainers. There was a BT-13 made by Vultee and a low-wing Fairchild PT-19. One of the newest airplanes was the Aeronca Chief, a light high-wing monoplane made in nearby Hamilton, Ohio, featuring side-by-side instead of in-line seating for two, and a control wheel instead of a control stick. A more basic version known as the Champ was Aeronca's best-selling model. It was in one of the three Champs at Wapakoneta that Neil Armstrong learned to fly.

Three veteran army pilots gave Neil flying lessons. Of the seventy students in Neil's high school class, about half of whom were boys, three learned to fly that summer of '46. Each soloed about the same time. Neil always refused, therefore, to say that his learning to fly was all that unusual.

It was unusual, though, that Neil earned his pilot's license before he got an automobile license. "He never had a girl. He didn't need a car," explained his father. "All he had to do was get out to that airport." "I believe you could solo in a glider at age fourteen," Neil said, "but in a powered airplane you had to wait till you reached your sixteenth birthday," which he celebrated on August 5, 1946. That day Neil earned his "student pilot's certificate," then made his first solo within a week or two.

The spontaneity of the event meant that the student pilot was in no position to alert family or friends. "You just heard the instructor unbuckle his belt, saw him look at you knowingly, felt his hand placed confidently on your shoulder, and thought to yourself, 'Oh, oh, here I go.' " Dean, who helped mow grass at the airfield, was on scene to observe his brother's progress. Viola was too nervous to watch her son fly, but she never tried to stop him from doing it. Part of the reason, according to June, was that Neil "never expressed any fear when he talked about it."

Armstrong had only vague recollections of his first solo, for which he got the nod from his instructor. "The first time you solo any airplane is a special day," Neil stated. "The first time ever you solo is an exceptionally special day. I'm sure there was a great deal of excitement in my mind when I got to do that first flight. I managed to make a couple takeoffs and landings successfully and bring it back to the hangar without incident." One of the positive results of making his first solo was financial. Without the need for an instructor, he only had to pay seven dollars an hour instead of nine. But the advantage was theoretical, as only ever more hours in the air would satiate his zest.

Developing his own piloting technique on a grass airfield, Neil "got in the habit of putting the airplane into a substantial slip on final approach where I would come down pretty steep so I could land on an early part of the grass runway and then have plenty of time to roll out and come to a stop." Neil also witnessed the darker side of flying. On the afternoon of July 26, 1947, twenty-year-old flying student and World War II navy veteran Carl Lange struck a power line and crashed his Champ in a hayfield. Lange died at the site from a skull fracture. His instructor survived. At the time, Neil was on the road back from Boy Scout camp. Dean remembers, "We saw the plane go down. My dad pulled off, and we all ran over there and tried to administer first aid." According to the *Lima News,* Neil, "jumping a fence, rushed to the aid of the plane occupants." The newspaper reported that Lange died in Neil's arms, but Neil said that he was not aware of the exact moment when Lange died.

Some biographical accounts have adhered to an account of Lange's fatal accident, and Neil's reaction to it, provided by Viola Armstrong. In an interview in 1969 with the Christian magazine *Guideposts*, Viola related that his experience of Lange's death shook Neil to the core. The published article "Neil Armstrong's Boyhood Crisis" said that Neil spent a solitary two days in his room reading about Jesus and pondering whether he should keep flying. Neil never remembered anything like that. According to June, "I never felt he was affected by it in any way, and it certainly did not dampen his enthusiasm for flying."

By the time of Lange's death, Armstrong had flown two cross-country solos—first to Cincinnati's Lunken Airport in a rented Aeronca. Round-trip, the flight spanned some 215 miles, each leg bookending his sitting for the navy scholarship qualifying exam. To preregister for classes at Purdue University, Neil flew to West Lafayette, Indiana, a flight of about 300 miles.

One can only imagine the astonishment of the West Lafayette airport personnel when a sixteen-year-old boy got down out of his airplane, asked for refueling, and started walking toward campus.

CHAPTER 4

Aeronautical Engineering 101

On October 14, 1947, one month after Armstrong started at Purdue University, an air force test pilot—one with whom Armstrong would later fly—broke through the mythical "sound barrier." He was Captain Charles E. "Chuck" Yeager, and the revolutionary airplane he piloted beyond Mach 1 was the rocket-powered Bell X-1. Before the military shrouded its transonic research program in secrecy, stories about the X-1's performance appeared in the *Los Angeles Times* and *Aviation Week.* Aeronautics faculty and students nationwide were discussing the meaning of "shattering the sonic wall."

For Neil, however, this new era in flight dawned bittersweet. "By the time I was old enough and became a pilot, things had changed. The great airplanes I had so revered as a boy were disappearing. I had grown up admiring what I perceived to be the chivalry of the World War I pilots—Frank Luke, Eddie Rickenbacker, Manfred von Richthofen, and Billy Bishop. But by World War II, aerial chivalry seemed to have evaporated. . . . Air warfare was becoming very impersonal. The record-setting flights—[John] Alcock and [A. W.] Brown, [Harold] Gatty, [Charles] Lindbergh, [Amelia] Earhart, and [Jimmy] Mattern—across the oceans, over the poles, and to the corners of Earth, had all been accomplished. And I resented that. All in all, for someone who was immersed in, fascinated by, and dedicated to flight, I was disappointed by the wrinkle in history that had brought me along one generation late. I had missed all the great times and adventures in flight."

As Armstrong entered college, the National Advisory Committee for Aeronautics (NACA), NASA's predecessor, along with the newly established U.S. Air Force, moved ahead ambitiously to construct new research facilities devoted to transonics, supersonics, and hypersonics (the speed regime, at around Mach 5, where the effects of aerodynamic heating became pronounced).

Armstrong's time in the aeronautical engineering program at Purdue University spanned—including a three-year stint in the military—from September 1947 to January 1955. That seven-and-a-half-year stretch saw an astonishing new era of global aeronautical development. Three months after the historic X-1 flight, the NACA activated the country's first hypersonic (capable of Mach 7) wind tunnel. A few months later, early in Armstrong's second semester, an army rocket team under Dr. Wernher von Braun launched a V-2 missile at White Sands, New Mexico, to an altitude of seventy miles. Armstrong's first full calendar year at Purdue witnessed the first flight of Convair's XF-92 airplane, with its innovative delta wing; the flight of the first civilian test pilot, Herbert H. Hoover (not related to the U.S. president), past Mach 1; the tailless X-4 aircraft's first test flights; and the publishing of an aerodynamic theory that proved critical to solving the high-speed problem of "roll coupling."

Armstrong left Purdue and reported for military duty during what would have been his spring semester of 1949. During those months, the U.S. Army established its first formal requirements for a surface-to-air antiballistic missile system; President Truman signed a bill providing a 5,000-mile guided-missile test range, which was subsequently established at Cape Canaveral, Florida; and a single-stage Russian rocket with an instrument payload of some 270 pounds flew to an altitude of sixty-eight miles. During that summer, as Armstrong took flight training at Pensacola, a V-2 rocket carried a live monkey to an altitude of eighty-three miles; the American military made its first operational use of a partial pressure suit during a piloted flight to 70,000 feet; and the first U.S. pilot ever to use an ejection seat escaped his jet-powered F2H-1 Banshee at 500 knots.

By the time he returned to his AE program in September 1952, Armstrong realized that the world of aeronautics was becoming the world of "aerospace." In 1950, the first missile launch at Cape Canaveral occurred, lifting a man-made object to the highest speed yet achieved, Mach 9. In 1951, the air force started its first ICBM program, forerunner to the Atlas program, which took the first astronauts into orbit. The following year, a centrifuge opened at the navy's medical aviation laboratory in Johnsville, Pennsylvania, that could accelerate human subjects up to speeds producing 40 g's. That same year, NACA researcher H. Julian Allen predicted that reentry-heating problems for missiles and spacecraft could be avoided by changing their nose shapes from sharp to blunt. Not just the Mercury, but also Neil's Gemini VIII and Apollo 11 spacecraft would be built around the "blunt-body" principle. During Neil's first year back in the classroom, in November 1953, NACA test pilot A. Scott Crossfield in Douglas's D-558-2 became the first person to fly at Mach 2. When Armstrong left Purdue with his degree a year later, he went to work for the NACA. In fact, Neil became a test pilot at the NACA's High-Speed Flight Station in California, where he would come to fly the experimental X-15 hypersonic aircraft seven times.

Neil Armstrong found himself immersed in this new world of the Space Age by the time he left college in January 1955.

In the early 1940s, fewer than one in four Americans completed high school, and fewer than one in twenty went to college. An eighth-grade education was the average in many rural communities. With the passage of the GI Bill in 1944, the proportion of college-goers began to rise by the early 1950s to 25 percent.

Neil was only the second person in his family to attend a university, the first being his great-uncle. College education was a precedent shared by many of the astronauts and engineers who were to become associated with the young space program. Neil was accepted to the Massachusetts Institute of Technology; however, he decided to attend Purdue University in West Lafayette, Indiana, 220 miles from Wapakoneta.

Armstrong had heard about the U.S. Naval Aviation College Program's four-year scholarships. The program, known as the Holloway Plan, required a commitment of seven years: two years of study at any school accredited by the navy, followed by three years of service, after which the student finished his last two years of college. Armstrong's medical examination recorded a weight of 144 pounds and a height of five feet nine and one-half inches; doctors categorized his general build and appearance as "athletic," his posture "good," and his body frame "medium." A heart rate of

88 beats per minute standing and 116 beats after exercise first registered Armstrong's tendency, frequently noted in his years as a test pilot and astronaut, to run a little high.

The medical form also recorded twenty hours of solo flying time in the past twelve months. At seven dollars per hour in the air, Armstrong had spent $140 (about $1,830 in 2018) on flying. That the taxi ride from Cincinnati's Lunken Airport to the downtown test site cost him seven dollars seemed "an astronomical amount of money because I could fly the airplane for seven dollars an hour."

Neil recalled "great jubilation that I had been accepted and that I had a way to go to college, paid. That was a wonderful deal."

According to Neil's navy appointment letter, dated May 14, 1947, his score of 38 points equated to 592 on Princeton's Scholarship Aptitude Test (SAT), which would likely have ranked in the top quartile among all college-bound seniors.

One month before receiving the good news from the navy, Purdue University had admitted Neil. "I just couldn't have been happier with what I was doing, going into engineering."

Purdue's AE program was more hands-on and less theoretical in focus than MIT's. In their first semester, students in the new School of Aeronautics learned to weld, machine, and heat-treat metals, and do sand casting. Six days a week Neil had three hours of classes each morning and three hours of lab each afternoon.

Instead of joining Naval ROTC, Neil fulfilled the program's requirements by playing in the university band, which functioned as a military band. The first semester he lived in a Lafayette boardinghouse; after that, he rented a room in a house closer to campus. Neil's cumulative GPA for his first year was 4.65, equivalent to a low B.

A rare surviving letter written by Armstrong near the end of his second semester details his college routine:

> *Sunday P.M.*
>
> *Dear Mom & Family,*
>
> *Thanks for the laundry, letters, & girl Scout Cookies. The other fellows saw them & when I got home last night, they were nearly all gone.*
>
> *You don't have to worry about my getting a job this summer. I am going to summer school. It's an order. I have my schedule for classes all-ready. It is:*
>
> *Differential Calculus.*
> *8–10 A.M.—Mon., Tue, Wed. Thu. Fri.*
> *Physics*
> *10–12 A.M.—" " " " "*
> *Physics Lab*
> *1–3 P.M.—Tue. Thu.*
>
> *That's all, but it's a load. The two labs are the only afternoon classes and there are no Saturday classes. I should be able to get home on weekends some times pretty easily. There will be a lot of homework in the summer even though there aren't many classes.*
>
> *Today, we went to Indianapolis to the first model airplane contest. My control lines broke on the first official flight so I didn't have a chance to win anything.*
>
> *I think I am doing better in my studies lately. I enjoy analytics and I understand a little of the chemistry we are studying.*
>
> *I'll send the laundry again & hope you'll send Les's blanket back. There are six weeks of school left yet.* (6)

I won't be out until the middle of June. I saw a show tonight that was the best I had seen in a long time. It is called "Sitting Pretty" with Clifton Webb, Maureen O'Hara, and Robert Young. I especially recommend it to you & Dad. It is a comedy. I'm running out of paper, so I'll stop.

Love, Neil

In the fall of 1948, Neil was informed that he would begin military service early, after three regular semesters plus a term of summer school.

When Neil left for navy flight training in February 1949 after four semesters at Purdue, he was only eighteen and a half. When he returned to the university in September 1952, he had just turned twenty-two. "I was really getting old," he related, laughing. "When I went back to the university, kids looked so young!"

After significant exposure to operational flying and handling high-performance jets, Armstrong thought "maybe there was a way to find a combination where I can do both airplane design and piloting." An internship for the summer of 1954 at the Naval Flight Test Center at Patuxent River in Maryland cemented his career goal. At Purdue Armstrong was now taking specialized courses in his major, and in no engineering course did he receive a grade lower than 5 (out of a possible 6). In the same fall 1953 term, Armstrong taught a section of General Engineering, Aircraft Layout and Detail Design, a course he had aced. Just as Neil's academic life improved during his second stay at Purdue, so did his social life. He pledged a fraternity, Phi Delta Theta, and lived in the frat house. Armstrong sang in Phil Delt's musical program in the school's variety show. The following spring he became the fraternity's musical director. For the "Varsity Varieties" Neil wrote and codirected two short musicals, "Snow White and the Seven Dwarves" and "The Land of Egelloc" ("college" spelled backward). His showmanship might have affected his grades, as he received several Cs and withdrew from Introductory Nuclear Physics.

No doubt eighteen-year-old Janet Shearon, the first love of his life, further distracted Neil from his academic work. He met Janet, who was studying home economics, at a party cohosted by Janet's sorority and Neil's fraternity. The second time they spoke was early one morning when she was on her way to a home economics lab and he was delivering the school newspaper. Neil also drove a tomato truck for a local cannery, and in the summertime sold kitchen knives door-to-door.

He also had intermittent weekend responsibilities as an officer in the U.S. Naval Reserve, carpooling with his Purdue navy buddies to the Naval Air Station in Glenview, Illinois, north of Chicago, to fly F9F-6 jets. In civilian garb, he flew among fellow veteran military pilots as part of the Purdue Aero Flying Club, which he chaired during the 1953–54 academic year. Lafayette's Aretz Airport housed the club's few small planes: an Aeronca and a couple two- and four-place Pipers.

One weekend in 1954, Armstrong suffered a minor accident following an air meet in Ohio. Neil thought he would fly a club Aeronca to Wapakoneta, but a rough landing in a local farmer's field caused "damage sufficient to prevent flying it back, so I took off the wings and returned it disassembled to West Lafayette on my grandfather's trailer." Armstrong finished his last coursework in early January 1955. He did not attend commencement exercises, instead returning to Wapakoneta to prepare for his job in Cleveland at NACA Lewis. Purdue sent by mail the diploma awarding his B.S. degree in aeronautical engineering. His final grade point average of 4.8 on a 6.0 scale represented a highly respectable performance in a very demanding field stretched out over nearly seven years. After returning from the navy, his GPA averaged 5.0, which included the equivalent of A or B grades in twenty-six out of a total of thirty-four courses.

For the rest of his life, engineering would be Armstrong's primary professional identity. Even during his years as a test pilot and as an astronaut, Neil considered himself first and foremost an aeronautical engineer, one whose ambition to write an engineering textbook set him apart from virtually all of his fellow fliers: "I am, and ever will be, a white-socks, pocket-protector, nerdy engineer—born under the second law of thermodynamics, steeped in the steam tables, in love with free-body diagrams, transformed by Laplace, and propelled by compressible flow. As an engineer, I take a substantial amount of pride in the accomplishments of my profession."

It would, indeed, be engineering—more than science—that would accomplish the Moon landing, and an engineer who would be the first to set foot on another world.

PART TWO

NAVAL AVIATOR

I'll always remember him for the way he could talk about flying, no bragging, no great statements, just a man who was cool, calm, intelligent, and one of the best fliers I ever knew.

—PETER J. KARNOSKI, ARMSTRONG'S ROOMMATE FOR BASIC TRAINING, CLASS 5–49, NAVAL AIR TRAINING COMMAND, NAS PENSACOLA

CHAPTER 5

Wings of Gold

If Neil Armstrong had not become a naval aviator, he would not have been the first man to walk on the Moon.

The first American to fly into space, Alan B. Shepard Jr., was a U.S. Navy aviator. So was the commander of the first Apollo flight, Walter M. Schirra. Of the one dozen human beings privileged to walk on the Moon, seven of them wore, or had worn, the navy's wings of gold. Most remarkably, six of the seven commanders chosen to pilot Apollo spacecraft down to lunar landings were naval aviators. This included not only the first man to walk on the Moon but also the last man to leave its surface to date, Eugene A. Cernan, in Apollo 17. In between Armstrong and Cernan, fellow navy pilots Charles "Pete" Conrad Jr. (Apollo 12), Alan Shepard (Apollo 14), and John W. Young (Apollo 16) flew Apollo spacecraft down to the lunar surface. Navy captain James A. Lovell Jr. would also have done so if not for the near-tragic mishap on Apollo 13's outbound flight. Only David R. Scott (commander of Apollo 15) did his military flying with the U.S. Air Force.

In 1955, the Eisenhower administration selected navy's Vanguard to become the first U.S. satellite. (As early as 1946, the navy commenced feasibility studies of global command and control of the U.S. fleet via Earth satellite vehicles. The Naval Research Laboratory and Office of Naval Research's Viking rockets set a number of altitude records, including a flight in May 1949 that reached an altitude of 51.5 miles.) The Vanguard program, a joint effort with the National Academy of Sciences, lagged behind the Soviet Sputniks, first launched in the fall of 1957. That December, the Vanguard program suffered the nationally televised humiliation of a launchpad explosion at Cape Canaveral. No one was killed in what was dubbed "Flopnik" and "Kaputnik," but the disaster reinforced President Eisenhower's decision, made after Sputnik 2, to green-light the alternative U.S. army satellite program headed by Dr. Wernher von Braun. On the last day of January 1958, von Braun's team launched the nation's first satellite, Explorer I, on the first try. Vanguard remained grounded until March 1958.

Still, much about the emerging U.S. space program would be defined by "the Navy Way." The service's TRANSIT satellites, first launched in April 1960, established satellites' effectiveness as navigational aids. Three of the seven original Mercury astronauts were naval aviators (Shepard, Carpenter, and Schirra) and one (Glenn) flew as a marine. Five of the "New Nine," the second group of astronauts, of which Armstrong was a member, were navy pilots (Armstrong, Conrad, Lovell, Stafford, and Young), as were many ensuing groups.

The training of a naval aviator culminated in landing an airplane on the deck of an aircraft carrier. Between February 1949, when Neil reported to the commander of the 4,000-acre naval air training base in Pensacola, Florida, and August 1950, when, just two weeks after his twentieth birthday, he ceremoniously received his navy wings of gold, Armstrong passed the test.

Orders arrived on January 26, 1949, for Neil, as well as Purdue classmates Donald A. Gardner, Thomas R. "Tommy" Thompson, Peter J. "Pete" Karnoski, and Bruce E. Clingan, to start flight training.

Neil went by train from Wapakoneta to Cincinnati, where the group convened, joined by two other Holloway students (from Miami University of Ohio), David S. Stephenson and Merle L. Anderson, for the 720-mile rail journey to Pensacola. On February 24, 1949, eight days after passing medical exams at the naval air station, they pledged their oath as midshipmen, the lowest grade of officer in the U.S. Navy.

The navy designated the Pre-Flight training group to which Armstrong (serial number C505129) and his six friends were assigned as Class 5-49, the fifth class to begin training at NAS Pensacola in 1949. New classes formed about every two weeks that year, for a total of nearly two thousand trainees. Yet during World War II, as many as 1,100 cadets *per month* were beginning Pre-Flight. In 1945 alone, 8,880 men completed flight training with the U.S. Navy.

Forty midshipmen belonged to Class 5-49, which included roughly the same number of naval cadets, "NavCads," enlisted men selected for navy flight training. "Pre-Flight" ground school lasted four months.

For sixteen weeks in the classroom, Armstrong and his mates took intensive courses in Aerial Navigation, Communications, Engineering, Aerology (i.e., meteorology), and Principles of Flight. They studied aerodynamics and the principles of aircraft engines. They learned how to send Morse code and understand the basic tenets of weather forecasting. The navy also required eighty-seven hours of Physical Training, and thirteen hours of Gunnery, among other courses. Armstrong and the others were drilled by marines, taught flight basics by marines, and disciplined by marines.

Training included a mile-long swim in the base's pool and a torture drill called the Multi-Phase Ditching Trainer (or "Dilbert Dunker"). The fully clothed candidate was fitted with a parachute, then strapped into a simulated cockpit sent on rails into a swimming pool. His task was to unharness, knock out the flipped canopy, exit the sinking plane, and swim up to the surface before running out of breath. Many required the help of frogmen to survive the Dunker, but Armstrong handled the trial with ease.

Class 5–49 finished its sixteen weeks of Pre-Flight on June 18, 1949. Armstrong's marks averaged 3.27 on a 4.0 scale, which ranked him near the top 10 percent of the class.

Six days after completing Pre-Flight training, Class 5–49 moved to Whiting Field for Stage A of flight training. The largest of NAS Pensacola's auxiliary airfields, Whiting consisted of North and South airfields located about a mile apart, each equipped with four 6,000-foot-long paved runways.

Armstrong's instructor was Lee R. P. "Chipper" Rivers, "a very good instructor, quite authoritarian but fun-loving." Stage A included 20 hops, of which A-19 became the "safe for solo" check flight, and A-20 the first solo in the North American SNJ, the most famous of all World War II trainers, with retractable landing gear and a radial engine of 600 horsepower. "The SNJ was a big step up from both the Aeroncas and Luscombes for me," Armstrong explained. Its greater "finesse and control force" flew "very much like the F6F Hellcat that was the predominant navy fighter in World War II." All in all, the SNJ was an "ideal training plane."

Neil's first hop in the SNJ occurred on July 6, 1949.

Over the next several lessons, Armstrong worked to improve on his deficiencies—most notably, landing. He made his fifteenth hop on August 23 with a different instructor who rated the attempt an overall "Unsatisfactory," specifically for his approaches ("overshot wind-line every time"). Although Armstrong's problems with altitude and speed control and judging his approaches to landings continued, Rivers gave Neil many more average than below-average marks. After the eighteenth hop, he judged Neil "safe for solo." After a check flight on Wednesday, September 7, Armstrong made his first navy solo. Afterward, a couple of Neil's mates observed navy tradition by cutting off the lower half of his tie, and Neil gave Chipper Rivers a bottle of his favorite whiskey. Stage B of Basic Training—maneuvers—began the next day after Armstrong's first solo flight. Neil made seventeen flights in nineteen days.

On September 27, 1949, Neil's check flight instructor made the final evaluation: "Student obviously knew all work and was able to fly most of it average to above. Towards last of period he got so nervous it began to show up in his work. Should be able to continue on in program and make an average pilot."

Stage C—aerobatics—began the following week at nearby Corry Field. Armstrong proved "above average" from the start at "inverted stall, wingover rolls, & loops." Stage D had Armstrong and his fellows "flying" in the Link Trainer. Dating to the late 1920s, the machine (which would stall and spin if maneuvered incorrectly) was equipped with the stick, throttle, and rudder pedals of a single-engine fighter, as well as a layout of standard navigation instruments.

But the real test came "under the hood," in the rear seat of the SNJ. In Partial Panel flying, the instructor could turn off the gyro horizon and directional gyro. In this training Armstrong intuited by logic to trust only the instruments, an ability he would later apply to piloting a spacecraft through the vacuum of space.

Armstrong's D stage was marred by "weak transitions," but for his ten Stage D instrument flights, Neil received nothing but "ups."

His next five flights (completed from November 15 to 18) involved the Radio Range Phase of D stage, where his instructors continued to focus on Neil's poor altitude control.

He made his two mandatory night flights (Stage E) on Friday, November 4, both of them resulting in "ups." By Thanksgiving 1949, Armstrong had completed the first five stages of Basic Training. He had made forty hops with 39.6 hours in dual instruction and 19.4 hours solo.

Saufley Field, an outlying landing field (OLF) northwest of Pensacola off Perdido Bay, was the site for formation flying (Stage F), conducted simultaneously with Primary Combat (Stage H) and Cross-Country Navigation (Stage I), for which he received very favorable evaluations in the end of January 1950. Though ground strafing and dive-bombing proved challenging for him, Neil's marksmanship was superior.

Those students who made it to Carrier Qualifications entered the crucible in which naval aviators were truly made. In late February 1950, Armstrong began final Basic Training at Corry Field. Field Carrier Landing Practice (FCLP) occurred on a 600-foot runway painted onto an OLF twenty-three miles west of Pensacola. The place was known as "Bloody Barin" for the large number of accidents that had occurred there during World War II.

Neil's class of ten spent the next three weeks learning to follow "a landing signal officer completely," Armstrong explains. "The LSO had a paddle in each hand and he would, just by the arrangements of the paddles, tell you that you're a little high or a little low, a little fast, or you need to turn a little more. If the LSO deems you cannot complete a successful landing or a safe landing, he will wave his paddles at you—a so-called wave-off—and you're immediately commanded to add full power and 'go around' and try again." Following the K-12 check flight, Neil was "field qualified" and ready to make his first landing at sea.

On March 2, 1950, Armstrong headed out over the Gulf of Mexico to make Stage L's required six landings on the USS *Cabot,* a light carrier steaming a short distance off Pensacola. "The SNJ was a relatively low-speed airplane," and, he remembered, "even if you had thirty knots across the deck, you could take off easily, without a catapult." Landing, of course, was the major challenge. Naval wisdom holds that "a good carrier landing is one from which you can walk away. A great carrier landing is one after which you can use the aircraft again."

Armstrong likened his first carrier landing to his first solo back in Wapakoneta, another "very emotional achievement" in his flying life. "It is certainly a highly precise kind of flying. It works because you, in a very precise manner, get the airplane through that very small window that will allow it to land successfully on a very short flight deck."

He related that he received no wave-offs and qualified in carrier landings. His marks were nine "Average" and two "Below Average," one for "fast starts" and another for "poor lineup" in final approach. This ended Basic Training and punched tickets to Advanced Training.

"I requested fighters, and fortunately was assigned

to fighters" at NAS Corpus Christi. "The fighter pilots always said that only the very best men got to be fighter pilots," Neil admitted, laughing. "My own guess is that a large part of it had to do with what needs the navy had at the time you graduated." Neil added, "I was assigned the F8F-1 Bearcat as my advanced training aircraft, which I was delighted with because it was a very high-performance airplane." First flown in 1944, the Bearcat, with its full "bubble" canopy, was the last propeller-driven fighter aircraft built by Grumman for the navy. Many consider it the finest piston-engine fighter in service with the U.S. Navy at the end of World War II. A small plane with an outstanding power-to-weight ratio, the F8F-1 offered both great agility and great speed, up to 434 miles per hour (377 knots). Compared to anything that Armstrong had ever flown, the Bearcat was a hot rod, with fantastic acceleration and climbing ability.

At Cabaniss Field, one of Corpus Christi's six outlying auxiliary bases, Neil began his indoctrination with VF Advanced Training Unit No. 2 on March 28, 1950. In the three months ending June 21, 1950, he made thirty-nine flights and logged over seventy hours in the air, all but one hour of it solo. His last five flights at Cabaniss showed considerable improvement.

By mid-July 1950, Neil was back at NAS Pensacola preparing to make his next six required qualification landings, this time in the cockpit of an F8F Bearcat. On his fifteenth FCLP hop on August 10, Armstrong was deemed "Field Qualified."

His next day's destination was the Gulf of Mexico and the USS *Wright* (CVL-49). Armstrong experienced an extraordinarily good day at sea—not a single "Below Average" mark. According to Neil, holding back a smile, "The Bearcat was able to take off in a very short distance. They wanted you still steering down the runway when you passed all the senior officers who were standing on the bridge." This was not just a matter of military decorum. "As soon as your wheels lifted off, then you're going to be susceptible to wind tending to drift you one way or the other," possibly right into the bridge.

On August 16, 1950, five days after Armstrong aced his carrier qualifications in the F8F, Naval Air Training Command Headquarters at NAS Pensacola informed the midshipman by letter that he had "successfully completed the full course of the prescribed syllabus of training for Naval Aviators" and was "hereby designated a Naval Aviator (Heavier-than-Air)."

Graduation took place a week later, on August 23. Neil's mother and sister drove 825 miles to attend. His father was unable to attend because he was testifying in a court case.

After a short leave, Midshipman Neil Armstrong reported to ComAirPac, that is, air command of the Pacific Fleet—"for duty involving flying." Armstrong explained: "Typically you're going to ask for an assignment that is similar to your recent training. In my case, it was fighters, so I was going to ask for a fighter squadron, and the choice would be East Coast or West Coast. I'd never been to the West Coast and thought it would be nice to see that part of the country."

Reaching California in early September 1950, Armstrong served ten weeks with Fleet Aircraft Service Squadron (FASRON) 7—based at NAS San Diego (NAS North Island as of 1955). From October 27 to November 4, 1950, Armstrong trained at the close-air support school run by the Marine Corps at its amphibious base on the south strand of Coronado Island down from NAS San Diego. Fifty miles north at marine Camp Pendleton, mock aerial combat drills set "offense" pilots (Neil continued to fly an F8F-2) to find and attack "enemy" ground targets and disrupt the "defense."

On November 27, 1950, ComAirPac ordered Armstrong, along with Pre-Flight buddy and FASRON 7 mate Herb Graham, to "proceed immediately and report to the Commanding Officer of Fighter Squadron 51." This was a veteran squadron just returning stateside aboard the USS *Valley Forge*. It was the first of three cruises to the Far East that VF-51 would make during the Korean War.

Armstrong wanted to fly jets in VF-51, the first all-jet squadron in the United States Navy. As Herb Graham told it: "When VF-51 was being formed in 1950, jets and jet-trained pilots were scarce and VF-51 was to be an all-jet squadron flying F9F-2s. Neil was in that assignment pool and he was an excellent young pilot. It was a dream spot."

VF-51's commanding officer was Lieutenant Commander Ernest "Ernie" Beauchamp. A flight instructor at NAS Pensacola before the attack on Pearl Harbor, Beauchamp flew Grumman F-6F Hellcats in World War II with VF-8, a key player in the victory in the Philippines. Aboard USS *Bunker Hill* in mid-1944, Beauchamp's squadron in six months' time took down 156 Japanese aircraft, producing thirteen different aces, pilots who had destroyed five or more enemy aircraft in air-to-air engagement. But Ernie was more than an outstanding fighter pilot. He had a brilliant mind for fighter tactics. In the spring of 1945, Beauchamp took command of Fighter Squadron VF-1, aboard the USS *Midway,* but the war in the Pacific ended before the squadron could deploy.

Beauchamp stayed in the navy after the war, retaining his squadron command before taking a staff post with the Deputy Chief of Naval Operations for Air in the Navy Department in Washington. On June 25, 1950, the very day the Korean War started, Lieutenant Commander Beauchamp left his desk job to assume command of VF-51, which became as close to a handpicked squadron as the navy ever got.

Serving temporarily at NAS North Island as officer-in-charge of a brand-new jet transition unit (JTU) for flyers from reserve F4U squadrons that had been recalled to active duty, Beauchamp not only saw the records but also observed the performance of a large number of pilots. "Only two or three pilots in the [then] currently deployed VF-51 would be available for a second tour," Beauchamp explained in 2002. Beauchamp managed to secure the assignments of four of his veteran aviators. Lieutenant Richard M. Wenzell (who became VF-51's operations officer), Lieutenant William A. Mackey, Lieutenant Daniel V. Marshall, and LCDR Bernard Sevilla.

Still, Beauchamp was short on pilots. On the word of "Wam" Mackey, Beauchamp recruited four additional aviators from Whiting Field: JTU instructors LTJG Robert E. Rostine and LTJG John Moore; along with JTU graduates and class 5-49 members LTJG Thomas B. Hayward (future Chief of Naval Operations) and LTJG Ross K. Bramwell. Beauchamp told his chosen nine officers to wield "a fine mesh screen" for eleven additional "nuggets," or new aviators who had not yet received an assignment.

No one person has ever claimed credit for identifying Armstrong as a person to bring into VF-51. Filling out the group with Armstrong and Graham were, in alphabetical order, ENS James J. Ashford, LTJG William W.

Bowers, LTJG Leonard R. Cheshire, ENS Hershel L. Gott, ENS Herbert A. Graham, ENS Robert J. Kaps, ENS Kenneth E. Kramer, ENS Donald C. McNaught, ENS Glen H. Rickelton, LTJG George E. Russell, and LTJG Harold C. Schwan. Carrying on with VF-51 from the *Valley Forge* cruise were LTJG Francis N. Jones and LTJG Wiley A. Scott. The squadron was fortunate to retain experienced chiefs and first-class petty officers. The VF-51 nuggets still faced a competitive selection process. One potential disadvantage for Armstrong was that, at the time of his assignment to the squadron in late November 1950, he had not yet flown a jet.

Some say the transition from props to jets was "like switching from a high-powered race car with a 'four on the floor' stick to a faster one with 'automatic' drive." Others say the changeover was more problematic. On Friday, January 5, 1951, Armstrong first took off in a Grumman F9F-2B. The "spectacular" first flight in the Panther lasted a little over an hour, and was another "one of those magic moments" in Neil's career as a pilot. "That was very exciting to me, to be in the front lines of the new jet fighters."

Though Neil was only a few months past his twentieth birthday, his fellow aviators held him in high regard. Wam Mackey characterized Neil as "very serious and very dedicated. He was a fine young pilot—a very solid aviator, very reliable." But it was Beauchamp himself who most needed to be impressed, and he was. Because the squadron possessed so few jets (only six planes for twenty-four pilots), flying was, in Armstrong's words, "a bit scarce"—about three flights per week per aviator through the first two and a half months of 1951. By mid-March, the winter fog had lifted and VF-51 had a full complement of aircraft allowing each pilot to fly between five and seven hops per week, periodically maintaining their instrument proficiency by flying "under the hood" in old twin-engine Beechcraft SNB trainers.

As the squadron trained, the specter of the enemy loomed. "We felt that we would possibly be fighting swept-wing MiG-15s," recounts Herb Graham. "They cruised above our top speed and could climb at a higher speed than we could dive. It was similar to the start of World War II when the navy F4F Wildcat fighters were faced with the much higher-performance Japanese Zero." Having read the combat action report covering the Panther's encounters with the "superior performance" MiG back in late 1950, Beauchamp felt "grave concern" that if MiGs were "manned by pilots as aggressive and well trained as ours, that our own pilot and plane losses would have been great."

According to Armstrong, "We didn't know to what extent we would be offensive, in the sense that we would be dropping bombs or shooting guns, to what extent we might be defending the fleet against Chinese or Russian incoming aircraft, or to what extent it might be air-to-air or air-to-ground. I was very young, very green."

Contemplating the threat of facing the Russian MiG, and struggling to keep their focus through a grueling training schedule, unmarried pilots lived in the bachelor officers' quarters at North Island. If Armstrong's age and youthful looks did not single him out, his hobbies did. Besides being an avid reader, he remained passionate about building models. Training once again built toward the finality of carrier "quals" in the Panther, this time on the recently modernized 27,100-ton carrier USS *Essex* (CV-9). Neil had previously made twelve carrier landings—six in the SNJ and six in the F8F. The older pilots had many more carrier landings in prop planes, but they had no more experience in jet landings than Neil did. "The speeds tended to be higher on the jet,"

Neil noted. "We were flying at slightly over a hundred knots typically in a pattern, which was maybe twenty knots faster than we'd been flying with the Bearcat."

"I happened to be a day fighter pilot," Armstrong was glad to say. "We had night fighters on the ship I was on, and I thought they were crazy." He qualified on June 7, 1951, roughly two months before his twenty-first birthday and just two days after his preset "date of rank" for promotion to ensign. On final approach, with his powers of concentration intent on the paddles of the landing signal officer, Armstrong reduced his speed to just above a stall, about 105 knots. In an instant, the ramp of the *Essex* flashed below him, the jet dropped abruptly, and its all-important tailhook blessedly snagged one of the arresting wires. With only 150 feet left before the airplane hit the protective barriers, Armstrong's F9F Panther jerked to a stop, having gone in a heartbeat from about 105 knots to teeth-rattling zero. Armstrong faced the LSO and his little green record book seven more times that day.

Following the exhilaration of his first carrier landing in a jet, Armstrong experienced the thrill of a "cat shot," cannonading airborne by one of the navy's powerful H8 hydraulic catapults. At this point, following eight successful carrier landings, Beauchamp must have finalized his choice of Armstrong as one of VF-51's officers beginning the cruise on the *Essex.* A month earlier, Beauchamp had assigned him to serve as both the squadron's assistant education officer and its assistant air intelligence officer. Based on Beauchamp's input, Captain Austin W. Wheelock, commanding officer of the *Essex,* noted in Armstrong's officer's fitness report dated June 30, 1951: "Ensign ARMSTRONG is an intelligent, courteous, and military appearing officer. As a naval aviator, he is average to above average and is improving steadily. He is recommended for promotion when due." Armstrong's combined 215 hours in the SNJ, 102 hours in the F8F, 33 hours in the SNB, and 155 hours in the F9F made for a total of 505 hours in the air since Neil joined the navy.

On Monday, June 25, 1951, Fighter Squadron 51 received its orders. Three days later, at 1430 on June 28, the *Essex* upped anchor. As she approached the Hawaiian Islands on July 3, most of the carrier's aircraft flew ahead to Oahu's southwestern tip.

At NAS Barbers Point, the squadron's aircraft were first equipped with heavy bomb racks. Ken Kramer remembers: "We had expected that we would be fighting MiGs, and we had practiced our dogfighting tactics probably more than any other squadron before us. Instead, we became a ground attack squadron," "a big letdown for us" as naval aviators.

Yet, the navy's decision to add the bomb racks was sound. The FJ-1 lacked carrier suitability; among other reasons, the plane kept losing its tailhook in its landing gear. VF-51 became a fighter-bomber unit because in the eastern half of Korea there were simply no MiGs to engage.

The training in Hawaii lasted from July 4 through 31. By August 23, 1951, the *Essex* was fifteen days out of Pearl Harbor, and already on station some seventy miles off the northeast coast of Korea near the harbor at Wonsan. Joining Fighter Squadron 51 aboard the *Essex* were one squadron of F4U Corsairs (VF-53), one squadron of AD Skyraiders (VA-54), and one squadron of F2H-2 Banshee jets (VF-172). Also embarked were four VC detachments: VC-61 with F9F-2P photo planes; VC-3; VC-11; and VC-35 ("VC" designating a "composite squadron" trained in night attack and defense, air early warning, and antisubmarine warfare).

The replacement at Pearl Harbor of the Banshee squadron for VF-52 (and its F9F-2s) was an unhappy surprise for the Panther pilots, who took no delight in the notion they might play second fiddle to the Banshees.

They need not have worried.

CHAPTER 6

Fighter Squadron 51

The men of VF-51 were more excited than scared about the prospects of combat, feeling that they were embarking on one of the great adventures of their lives.

A bad omen for what was to come, typhoon Marge battered the *Essex* for two straight days, rolling the ship just ten degrees shy of its capsize point. On August 22, the *Essex* joined Task Force 77 about seventy miles off Wonsan. Looking out the large bay door of the hangar deck, Armstrong saw his first American carrier battle group. The carrier *Bon Homme Richard;* the battleship *New Jersey;* two cruisers, the *Helena* and *Toledo;* and some fifteen to twenty destroyers numbered among some two dozen warships that would swell in the following months to four carriers and three cruisers in simultaneous action.

Air Group 5's first stretch of combat operations commenced on August 24, when CVG-5 launched seventy-six sorties against "targets of opportunity." It was not Armstrong's turn to fly that first day. Nor did he participate on the twenty-fifth in a massive air raid on the railyards at Rashin near the Soviet border—the first time navy fighters escorted air force bombers over hostile territory. According to Armstrong, "The four-plane division was the mainstay of the operation." A division consisted of two sections of two airplanes each. In flight, the sections stayed separated by a quarter- to a half-mile.

Beauchamp divided his twenty-four pilots into six divisions scheduled to receive approximately the same number of hops. The head of the sixth division was John Carpenter. Carpenter's section leader was John Moore. The junior officers flew as wingmen. At the start of the cruise, Armstrong usually flew as Carpenter's wingman, in the division with John Moore. Later, Neil flew mostly with Wam Mackey. Like the other pilots, Armstrong also flew a number of photo escorts, which were not done in divisions.

The fact that CAG Marshall Beebe always asked for the squadron's youngest aviator as his wingman did not go unnoticed. Beebe's aggressive approach to combat flying may have put Neil even more into harm's way. Beebe, the World War II double ace with 14 kills, seemed fearless. "Marsh" had a well-deserved reputation for staying "feet dry"—overland—for too long, maximizing air time over enemy targets but leaving the planes with barely enough fuel to make it back to their carrier. On a couple of occasions Beebe managed to get authorization for his jets to fly up into MiG Alley in the uppermost regions of North Korea, though the requisite defensive maneuvering against MiGs would have used up too much fuel to afford safe return to their carrier. Armstrong remembered when "I would have appreciated a couple of hundred more pounds of fuel in the landing pattern."

Armstrong's first action over North Korea came on August 29, when he escorted a photoreconnaissance plane above the 40th parallel over the port of Songjin, then flew a routine combat air patrol over the fleet.

Three of the next four days he flew armed reconnaissance over Wonsan, Pu-Chong, and again up to Songjin. A few VF-51 aircraft encountered small-arms ground fire on the twenty-ninth, but the squadron's first taste of potent AA fire did not come until September 2. Beauchamp's divisions' main objective was disruption of the transport system that fed the North Korean and Chinese armies. "We did that by blowing up trains and bridges and tanks," explained Armstrong, "and just being as contrary as we could."

In its first ten days of action, Air Group 5 experienced a nasty rash of casualties. Covering the week ending Sunday, September 2, the ship's combat action report noted, "Not a day had gone by but at least one plane had been hit by AA." The next week almost ended Neil Armstrong's life.

On September 3, 1951, Armstrong suited up for his seventh combat mission. Donning a naval aviator's two-part "poopy suit" drew comparisons to putting on a straitjacket. The call to "Flight Quarters!" commenced a noisy, frenetic choreography on deck. The "plane captain" started the jet engines even before the pilots arrived to make their assisted climb into the cockpit, where the captain connected the shoulder and lap straps and arranged the parachute harness. Following a check of his oxygen mask and the status of his life raft and radio, the aviator was primed for the ship's powerful H8 catapults. His cat shot would be Armstrong's twenty-eighth in three months.

Armstrong's task was to fly an armed reconnaissance mission into a hot zone that U.S. naval intelligence called "Green Six." Located west of Wonsan, Green Six was the code name for a narrow valley road that led to the interior border of South Korea.

The principal targets for September 3, 1951, were

freight yards and a bridge. According to Rick Rickelton, who was flying wing for Mackey, "We really ran into a terrific concentration of AA; fairly heavy stuff. I think I could have walked on it." Flak hit Lieutenant Frank Sistrunk's AD Skyraider while Sistrunk was bombing the bridge. The Skyraider crashed, and Sistrunk became Air Group 5's fourth casualty during the *Essex* cruise.

Armstrong, flying as John Carpenter's wingman, made a number of attacking runs that day. On one bomb run, at approximately three hundred and fifty miles per hour, Neil sliced through a cable, presumably a North Korean–devised booby trap for low-flying attack aircraft. Close to six feet of Neil's right wing was shorn. He barely managed to fly to friendly territory, where his only option was to eject. Carpenter stayed with him until Neil ejected as planned in the vicinity of an airfield near Pohang, designated K-3, located far down the coast of South Korea and operated by the U.S. Marines. The term "punching out" does not do justice to the "kick in the butt" of the Panther's British-made Stanley Model 22G ejection seat, which was survivable at anything over five hundred feet when not compromised by any sort of "sink rate." Armstrong's was Fighter Squadron 51's first-ever ejection-seat bailout. The jump was also Armstrong's first.

Neil "intended to come down in the water," but misjudging the wind, he floated inland and landed in a rice paddy. Aside from a cracked tailbone, Neil was virtually unhurt. No sooner had he picked himself off the ground when a jeep drove up from K-3. Inside the jeep—Neil could barely believe his eyes—was one of his roommates from flight school, Goodell Warren. "Goodie" was now a marine lieutenant operating out of Pohang airfield. Warren told Armstrong that the explosions he was hearing out beyond the coastline came from North

Koreans laying mines in the bay. If Neil's parachute had stayed on course, he might very well have splashed down in the deadly minefield.

Late in the afternoon of September 4, Armstrong returned to the *Essex* aboard a mail and personnel transfer craft nicknamed a "codfish." According to Ken Dannenberg, VF-51's intelligence officer, "Naturally we had to rough him up a bit." As per ejection procedure, Neil had removed and dropped his helmet, which broke when it hit the ground. "Neil had that broken helmet in his hand and a smile on his face," Dannenberg recalled. "You know, Neil, you're going to have to pay the government for that helmet." Kidding aside, Armstrong "received a lot of favorable notice for his cool handling of the situation," Herb Graham remembered.

In letters home, Neil virtually never mentioned combat, and certainly not what happened to him that day. All he did was make a note in his logbook for September 3, 1951: "Bailed out over Pohang." Next to it he drew a little picture of an open parachute with a tiny figure of a man hanging from it. The airplane was the first Panther lost to Fighter Squadron 51. But that was not the reason there was no celebration on the *Essex* the night Armstrong returned. Earlier that day, two of his squadron mates, James Ashford and Ross Bramwell, had been killed in action. Twenty-four-year-old Bramwell lost control of his aircraft after getting hit by enemy flak. Neil flew in the same division as the twenty-five-year-old Ashford and might have been in ops with him if not for his ejection the day before. During a reconnaissance mission in the region between Simp'yong and Yangdok, northwest of Wonsan, Ashford's jet, heavily loaded with ordnance, failed to pull out while making a rocket run on a truck, then flew into the ground and exploded. "What a price to pay for a goddamn truck!" was a lament that the entire squadron shared.

According to Beebe's combat action report, through September 4, 1951, "The Air Group had destroyed seven bridges, ninety railroad cars, twenty-five trucks, twenty-five oxcarts, two hundred and fifty troops, and damaged about twice as many of each, the price being the lives of five pilots, one air-crewman and ten aircraft." On September 5, the entire task force took a day off from combat to replenish, giving them a chance to reflect. "They never missed an opportunity to shoot at you," Armstrong would relate. "We saw all kinds of guns, all kinds of sizes, and some were radar-controlled and some were not. They had those long-barreled 85s that could reach up a long way. There was always a lot of concern about getting hit. I had a lot of bullet holes in the airplanes I flew, but usually got them back." Over the next nine days, Neil flew four combat air patrols, one photo, and four armed reconnaissance missions.

The biggest disaster of the entire *Essex* cruise happened not in the air over North Korea but on the carrier deck. On September 16, 1951, an F2H Banshee from VF-172 came in for an emergency landing. LTJG John K. Keller fought to bring his Banshee home following a midair collision. At the head of his Panther division, Ernie Beauchamp had just entered the *Essex* landing pattern. The Skipper was turning crosswind for final approach when he heard Keller calling for a "straight in." Beauchamp put on power, picked up his wheels and flaps, and cleared the landing approach, as did the other three planes in his division, flown by Rostine, Kaps, and Gott. A series of mistakes escalated into catastrophe. Keller forgot to lower his tailhook for landing, and the hook spotter and the LSO mistakenly thought it was down. The oversight brought the eight-ton Banshee slamming into the deck at nearly 130 knots. The jet plane jumped all of the heavy crash barriers, then tumbled headlong into an array of aircraft just moved to make room for the returning aircraft. Some of those pilots and plane captains had yet to exit their planes. The mushrooming explosion of parked planes—some fully fueled with almost a thousand gallons of high-octane gas—was tremendous. The *Essex*'s forward flight deck was a ball of fire. The only choice for Beauchamp's division was to fly over and land on the *Boxer,* where they stayed overnight. The consequences of the crash were obscene. Four men burned to death. Engulfed in the gaseous envelope of the flames, five others leaped into the ocean seventy feet below, only to face burning aviation gasoline on the surface. A tractor shoved the offending Banshee overboard, with its dead young pilot still inside, and did the same to a few other burning airplanes. By the time the conflagration was extinguished several hours later, seven men had died. Sixteen were seriously injured. Eight jets had been turned to cinders. Fortunately, the Skyraiders, loaded as they were with fuel plus a 5,000-pound bomb load, were parked safely over to the other side.

Armstrong was serving as the squadron duty officer that day. Rules prescribed that he would stay at his position in the ready room. Consequently, he did not see the fire and took no part in the firefighting. For the next three days, the men of the *Essex* mourned. With the loss of Armstrong's plane, the deaths and serious injuries, and the fiery destruction of four additional Panthers, a demoralized Fighter Squadron 51 counted only nine serviceable aircraft, down from sixteen, and twenty-one pilots, down from twenty-four.

It was a somber *Essex* crew that gathered for a memorial service while in route to Yokosuka on September 20. The service honored the memory of the thirteen men in CVG-5 killed since the cruise began. Armstrong considered himself lucky. He survived his September 3 flight by the skin of his teeth. Furthermore, had he not served as squadron duty officer the day the Banshee crashed, Armstrong would likely have been on the deck taxiing one of the Panthers.

Arriving at Yokosuka in the early evening of September 21, 1951, Neil Armstrong experienced his first overseas "rest and relaxation." The U.S. Navy had taken over a number of resort hotels on the east side of Japan, the most beautiful and luxurious of the "R&R camps" being the Fujiya Hotel, in the cool shadow of magnificent Mount Fuji. Armstrong more than once enjoyed the wonderful food, drink, and service, all for very little charge. On the resort's golf course Neil also decided to try his hand at the game, which he later came to love. The *Essex* stayed in port for ten days. On October 1, 1951, it headed to the northeast coast of Korea to rejoin Task Force 77.

During this second combat period Neil would fly ten missions. One of them turned into an experience he would never forget but chose almost never to talk about, not even with his squadron mates.

Passing over a ridge of low mountains in his Panther jet during a dawn combat patrol in mid-October 1951, Neil saw laid out before him rows and rows of North Korean soldiers, unarmed, doing their daily calisthenics outside their field barracks. He could have mowed them down with machine-gun fire, but he chose to take his finger off the trigger and fly on. As Neil would relate to the author many years later (after the original version of *First Man* was published in 2005), "It looked like they were having a rough enough time doing their morning exercises." No one else in his fighter squadron ever heard the story, because Neil never told it, but, when informed about it following Neil's death in 2012, they accepted it without hesitation as true. They themselves would have all fired their guns, they admitted, but there was "something too honorable in Neil for him to kill men who were in no position to defend themselves."

But Armstrong flew bravely, fought with courage, and always carried out his orders. On October 22, 1951, Neil's division found two trains for the ADs and Corsairs to destroy and then hit several supply points. On the twenty-sixth, his division hit bridges and busted rails in the region of Pukch'ong. On the thirtieth, Neil was part of an attack that flew quite far north, well above the 40th parallel. The day before, he flew about as far west as he ever got, during a fighter sweep in the area of Sinanju, a part of MiG Alley.

Fighter Squadron 51 suffered nary a casualty during this second tour. Overall, the entire air group lost only three pilots and the aircraft that carried them, a great improvement over the initial weeks of the first operational period. During the month of October, the squadron expended 49,299 twenty-millimeter rounds and dropped 631 general-purpose hundred-pound bombs. Neil personally fired an estimated seven thousand rounds, dropped forty-eight bombs, and fired thirty rockets during the initial two-and-a-half-month combat period. During his twenty-six flights, of which nine were combat air patrols, he accumulated more than forty-one and a half hours of air time.

Following another refurbishing of the ship in Yokosuka lasting from October 31 to November 12, 1951, the men returned to action, again off Wonsan Bay. With the onset of winter, carrier activities in the Sea of Japan turned miserable. During November and December 1951, the Screaming Eagles unloaded 135,560 pounds of bombs. Strafing remained the most effective weapon for VF-51, with 43,087 rounds fired, an average of 2,051 rounds per pilot. In December 1951, prior to leaving again for refurbishing in Yokosuka on the thirteenth, Armstrong took to the air eight times. On December 2, at high altitude and over water, the engine in Armstrong's Panther jet quit on him. Flameouts were a serious problem plaguing gas-turbine engines. Neil's flameout was caused by a fuel control mechanism being stuck at a low-altitude setting due to salt corrosion. Advancing the throttle at the higher altitudes required by CAP missions had injected too much fuel into the mix, extinguishing the jet's flame. Fortunately, the jet relit and Armstrong finished his flight without further trouble.

During its third tour in the Sea of Japan, VF-51 had some close calls but suffered no fatalities. On December 14 the *Essex* arrived back in Yokosuka, where it would spend Christmas 1951. On the day after Christmas, the ship was bound for yet another combat tour in Korea. *Essex*'s fourth tour proved to be by far the nastiest, most strenuous, and longest lasting of the entire cruise. For thirty-eight days, until February 1, 1952, the pilots of Air Group 5 flew a total of 2,070 sorties. Armstrong himself flew twenty-three missions, with a total time in the air of over thirty-five hours. Twenty-three cat shots, twenty-three carrier landings, all in one month, all in combat conditions: this was Neil's experience. Relying on half-frozen catapults and bone-cold aircraft carrying icy guns, Armstrong and his mates performed an unenviable job.

On January 4, 1952, which ended the first week of the *Essex*'s fourth tour, the men of CVG-5 got happy news. At the end of January they were to spend two weeks in port in Yokosuka and then head back to the States.

But before that could happen, Ensign Rick Rickelton's Panther was hit by flak, nose-dived into the ground, and exploded. That night in his journal squadron mate Bob Kaps wrote: "Hope the Lord can see through this mess; don't think I can. There has to be a reason for prolonging this business but I just don't see it."

With Rickelton gone, Wam Mackey's division needed another wingman. The job fell to Neil. For the rest of the cruise, he flew primarily with Mackey, Chet Cheshire, and Ken Kramer. Two days after Rickelton's death, as Mackey remembered, "The admiral came down to the wardroom and said, 'I've got some bad news: such and such a ship has had problems and is going to be delayed in relieving us, and we are going to have to come back one more time.'"

The focus again would be bridges deep inside North Korea. From the beginning of the war, bridges had been the principal targets in the interdiction campaign. According to official Pentagon wartime statistics, navy planes destroyed 2,005 North Korean bridges out of a total of 2,832 that U.S. military forces destroyed in all.

Over time the navy learned—at great cost—that the key to effective bridge strikes was coordinating the props and the jets into a single unified and well-timed assault. Marshall Beebe and the squadron commanders of Air Group 5 hatched the basic plan on the *Essex* in the latter months of 1951. Jets, with their higher and steeper "drop-down approach" to a target and their faster escape speed, had a significantly better chance of penetrating a bridge's defenses. Yet the jets were not the best instruments for actually taking out a bridge. That took two-thousand-pound bombs, which jets could not carry. The job of the jets was to quell the antiaircraft fire. Then came the Corsairs, which also bombed and strafed the AA positions, followed finally by the Skyraiders deploying the heavy ordnance. Typically, at least twenty-four aircraft would be involved in a major bridge strike: eight jets, eight Corsairs, and eight Skyraiders. The successful new tactic was quickly adopted throughout Task Force 77, with one alteration. To prevent the dust created by the jets' airbursts from concealing the props' bomb targets, the air groups directed the jets to move their suppression points of aim farther from the bridges.

Although the new tactics lessened air group casualties, there was no way to avoid losing some men. Three other pilots died after Rickelton. The death that hit Armstrong and the rest of VF-51 the hardest was their own, LTJG Leonard R. Cheshire, on January 26, 1951.

Like Rickelton, "Chet" Cheshire was from New Mexico—Albuquerque, to be exact. Cheshire had been married just before he left for Korea. After the war was over, he planned to become a teacher. Neil and Chet slept right across the aisle from each other, on lower bunks. The two men—the squadron's youngest member and the other the squadron's oldest junior officer—became close friends. Mackey's division was making its second run on a camouflaged train sitting in the Kowan area, just adjacent to Wonsan Bay when Cheshire's plane was hit by the fatal AA.

That evening over the ship's PA system, the chaplain said a prayer, as he always did, for the men who lost their lives. Since the ship had left Hawaii for Korea, he had said this prayer for Beebe's men twenty-eight times.

At 1330 hours on February 1, 1952, the *Essex* left Task Force 77 for Yokosuka, ending its fourth combat tour number when it was thirty-seven grueling days out of port. Flying well over two thousand sorties during that stretch (441 of them done by VF-51), Air Group 5 had fired nearly four hundred thousand rounds of ammunition, dropped almost ten thousand bombs, shot off approximately 750 rockets, and hit the enemy with just under three thousand pounds of napalm. This resulted in 1,374 railroad track cuts, the destruction of thirty-four bridges and damage to forty-seven more, along with a multitude of damaged or destroyed war matériel and infrastructure. In accomplishing these attacks, CVG-5 lost five men, two of them from VF-51, and the services of more than a dozen aircraft.

Armstrong's fifth and final tour of combat started on February 18, 1952. Mercifully, it lasted only two weeks. Neil was in the air every one of the days that flying was done, for a total of thirteen flights. On the morning of the 25th, Neil finished the work of the night hecklers in a morning attack that destroyed both locomotives and forty cars on a long train. Neil's last flight in the Korean War came on March 5, 1952. On that day, the pilots of VF-51 transferred their planes to the *Valley Forge*. Overall, Neil flew every one of VF-51's F9Fs at least once, except the few that were lost early.

Armstrong flew a total of seventy-eight missions for over 121 hours in the air. Thirty of them were CAP, fifteen were photo escort, and one was for gunnery training. In the other thirty-two, he flew recco, fighter sweeps, rail cuts, and flak suppression.

On March 11, 1952, the *Essex* departed for Hawaii. Finally on March 25 came the glorious sight of the California shoreline. As did his fellow aviators, Armstrong arrived home with a chest full of war medals. Like most of his crewmates, Neil typically downplayed his achievements, saying, "They handed out medals there like gold stars at Sunday school." His first award, the Air Medal, came in recognition of his first twenty combat flights; his second, a Gold Star, in recognition of his next twenty. With his mates, Neil also received the Korean Service Medal and Engagement Star.

PART THREE

RESEARCH PILOT

In the end the accuracy of the results really depends upon the flyer, who must be prepared to exercise a care and patience unnecessary in ordinary flying. Get careful flyers whose judgment and reliability you can trust and your task is comparatively easy; get careless flyers and it is impossible.

—CAPTAIN HENRY T. TIZARD,
TESTING SQUADRON,
BRITISH ROYAL FLYING CORPS, 1917

CHAPTER 7

Above the High Desert

With Armstrong's contract with the navy expired, technically he was free to return to college. However, since Squadron 51 was still in combat, "My options were either to extend my time in the service or swim home, so I extended." On February 1, 1952, while he was still serving aboard the *Essex,* the navy terminated his regular commission and reappointed him as an ensign in the U.S. Naval Reserve.

Arriving back in the States with his shipmates on March 25, 1952, Neil spent the next five months based ashore in Southern California ferrying aircraft in and out of Naval Air Station San Diego for Air Transport Squadron 32. He left the navy on August 23, 1952, in the month of his twenty-second birthday; during his last flight up to the San Francisco Bay area, he made a celebratory, unauthorized pass in his aircraft *under* the western span of the Bay Bridge, whose clearance below was 220 feet. Promoted to lieutenant junior grade in May 1953, he remained in the U.S. Naval Reserve until he resigned his commission in 1960. Back in school at Purdue, Neil flew regularly with Naval Reserve Aviation Squadron 724 at NAS Glenview, outside Chicago. When he became a test pilot for the National Advisory

Committee for Aeronautics at Edwards Air Force Base at Muroc Dry Lake, northeast of Los Angeles, Neil would do his reserve flying with VF-773 at NAS Los Alamitos, near Long Beach.

Graduating from Purdue in January 1955, Armstrong entertained several job options. He could have stayed in the navy. He interviewed for positions with Trans World Airlines (TWA) and Douglas Aircraft Company. Neil also briefly considered graduate work in aeronautical engineering. If he had taken the job offered by Douglas, Neil would have become a production test pilot, test-flying each new aircraft of a given type.

Another option, and the one he took, was to become an experimental test pilot, as epitomized by the fledgling Society of Experimental Test Pilots, which had committed at its establishment in 1955 "to assist in the development of superior aircraft." Specifically, Neil's ambition was to be a "research pilot." A special class of experimental test pilot, the research pilot strove to advance the science and technology of flight across a broad front. Employment opportunities for research pilots existed primarily at private research organizations or with the federal government, most prominently the National Advisory Committee for Aeronautics. From boyhood, Armstrong had regularly followed the results of ongoing NACA research in *Aviation Week* and other aviation magazines, and NACA reports were part of the curriculum in his aeronautical engineering classes at Purdue. In the summer before his last semester, Armstrong presented his credentials to the NACA. Specifically, he applied to be a test pilot at the NACA's High-Speed Flight Station at Edwards AFB, the facility where the X-planes were being flown in their assault on the mythical "sound barrier." As Edwards had no openings, the NACA circulated his application to all of its research centers. Irving Pinkel, an engineer from the NACA's Lewis Flight Propulsion Laboratory in Cleveland, Ohio, "asked if he could come down and talk with me." Pinkel headed the physics division at Lewis; his brother, Benjamin, was in charge of the thermodynamics research division. Sometime during the fall of 1954, Irving interviewed Neil. Pinkel could not offer much money, but promised him the excitement of aeronautical research.

Armstrong accepted the post at Lewis Laboratory. It did not hurt that the job kept him in Ohio, because by this time Neil was hoping to marry his college sweetheart, Janet Shearon, a Midwestern girl and home economics major from suburban Chicago.

Assigned originally to the lab's Free-Flight Propulsion Section, Armstrong's official job title was Aeronautical Research Pilot, responsible for "Piloting of aircraft for research projects and for transportation, and engineering in free flight rocket missile section." His first test flight at Lewis came on March 1, 1955. For civil service purposes, the NACA labeled Armstrong a "research scientist." Yet, as with most NACA employees, his work served the organization's legislated mission, "the scientific study of the problems of flight, with a view to their practical solutions."

The chief test pilot at Lewis was William V. "Eb" Gough Jr. Like Armstrong, Eb Gough had earned an engineering degree and became a naval aviator facing Japanese Zeroes in World War II, reaching the rank of lieutenant commander. When the war ended, Gough became a test pilot for the NACA. Mel Gough, Eb's older brother, had been chief of NACA Langley's Flight Research Division since 1943. The Langley group featured a half-dozen talented engineer-pilot hybrids, including John P. "Jack" Reeder, Robert A. Champine,

John M. Elliot, John Harper, and James V. Whitten. Neil came to regard Reeder as "the best test pilot I ever knew."

When Armstrong joined the NACA in February 1955, most of its research pilots were trained engineers. Yet most of the NACA's flight research took place at Langley, the High-Speed Flight Station, or the Ames Aeronautical Laboratory in Northern California. Armstrong found himself at NACA Lewis to be one of just four test pilots, with Eb Gough, William Swann, and Joseph S. Algranti, future chief of the Aircraft Operations Division at the Manned Spacecraft Center in Houston.

Armstrong stayed at Lewis for less than five months, investigating new anti-icing systems for aircraft. He also worked in his first space-related flight program, studying high-Mach-number heat transfer. In early tests, various air-launched models descended at speeds reaching Mach 1.8. On March 17, 1953, a T40 rocket air-launched by a Lewis test pilot achieved the hypersonic speed of Mach 5.18, the first time that the "NACA flew successfully an instrumented vehicle to greater than Mach 5." On May 6, 1955, Algranti and Armstrong flew the forty-fifth test in this series. The pilots steered their P-82, North American's Twin Mustang, over the Atlantic Ocean beyond the NACA's Pilotless Aircraft Research Station at Wallops Island, off Virginia's Eastern Shore. Attached to the belly of the P-82 was a solid-rocket model designated ERM-5. A conventional ballistic shape with a sharp nose, slender body, and tail fins, the ERM-5 was equipped with a T-40RKT rocket motor that had been developed by the Jet Propulsion Laboratory in Pasadena. Reaching the optimum altitude, Algranti released the model. The ERM-5 reached a hypersonic speed of Mach 5.02 and an acceleration rate of 34 g.

Armstrong "did a lot of analyzing data; designing components for advanced versions of the rockets, and doing calculations, and drawings for them." The proactive identity of the engineering test pilot that was fostered by the NACA—and by its successor, NASA—fit Armstrong perfectly. Neil had always felt that even though the NACA position was the lowest-paying job he was offered coming out of college, "it was the right one."

"The only product of the NACA was research reports and papers," Neil explained. "So when you prepared something for publication, you had to face the technical and grammatical 'Inquisition.'. . . The system was so precise, so demanding."

Neil's last test flight in Cleveland occurred on June 30, 1955. A week or so earlier, Abe Silverstein, Lewis's deputy director, had called him. "I walked over to his office," recalled Armstrong, "and he said he had gotten a letter from Edwards and would I still like to transfer out there." Work at Cleveland had been interesting, but Edwards was a test pilot's Shangri-La, the place where the sound barrier was broken in October 1947, and where the newest and most revolutionary experimental aircraft—the X-1A, X-1E, X-3, X-5, Douglas D-558-2, YRF-84F, F-100A, and YF-102—were being piloted to speeds of Mach 2 and beyond.

In early July 1955, after a brief visit with his family in Wapakoneta, Armstrong took off for Southern California. Neil had purchased his first car, a 1952 Oldsmobile, the same make his father owned, for $2,000. Dean Armstrong had come out to California following Neil's return from Korea. The brothers took the Olds sightseeing from Mexico to Canada before heading home. Another cross-country trip took Neil to his new job at Edwards in July 1955. On the way, Armstrong planned to make one important stop in Wisconsin, to visit Janet Shearon.

Neil and Janet met as students at Purdue University the year Neil returned from Korea. He was a twenty-two-year-old junior, and she was eighteen and a freshman. What attracted Neil was Janet's poise and bearing, her smarts, her good looks, and her lively personality. Born on March 23, 1934, Janet Elizabeth Shearon was the daughter of Dr. Clarence Shearon and his wife, Louise. Dr. Shearon was chief of surgery at St. Luke's Hospital, and taught at Northwestern University's medical school in Evanston, Illinois.

The Shearon family lived a comfortable upper-middle-class life in an affluent suburb of Chicago. Interestingly, Dr. Shearon owned and flew his own airplane, a Piper Cub. In November 1945, when Janet was eleven, her father died suddenly of a heart attack. Although her father's career as a physician had kept him away from home a lot, Janet dearly loved him. On the verge of her teenage years, the loss of her father was devastating. Janet did not always get along well with her mother, who, like Janet, was quite strong-willed. So her father loomed large in Janet's mind as her hero, the one person who appreciated her worth, including her skills as a competitive swimmer. Graduating from high school in 1952, Janet attended Purdue University and majored in Home Economics. As part of her busy college life, Janet swam in the intramural program and joined the women's synchronized swim team. She also joined Alpha Chi Omega sorority. One of Janet's good friends in college turned out to be the man who would become the last Apollo astronaut to leave the surface of the Moon, Apollo 17 commander Eugene Cernan. Gene and Janet met through Cernan's fraternity brother in Phi Gamma Delta, William Smith, who had gone to high school with Janet. It wasn't that Neil met Janet through Cernan or his Fiji friend, but rather vice versa.

Neil met Janet informally one day while walking on campus. Never very social or active in dating, "Neil knew me for three years before he ever asked me for a date," recalled Janet. "That wouldn't be so bad except that, after we were married, his roommate told me that the first time Neil saw me he came home and told the roommate that I was the girl he was going to marry. Neil isn't one to rush into anything." Janet, on the other hand, was dynamic and self-confident. According to Neil's brother Dean, who himself started at Purdue in 1953, and who recollected that he met Janet before Neil did, Janet was "as strong as horseradish. She looks you in the eye. Her body language is dramatic. She crosses her arms to say, 'And what do you mean by that!?' "

Janet and Neil got engaged during her junior year in 1955 after Neil had graduated and was working in Cleveland. The courtship was unusual in that there really wasn't any. The betrothed were virtual strangers to each other. "We never really dated," Janet explained. "My philosophy was, 'Well, I'll have years to get to know him.' I thought he was a very steadfast person. He was good-looking. He had a good sense of humor. He was fun to be with. He was older. He had a better sense of maturity than a lot of the boys I dated, and I had dated a lot of boys on campus." Brother Dean Armstrong remembered, "It shocked the heck out of me that they were engaged, because I had no idea that he was serious with her. Maybe opposites attract." According to

Cernan, "Neil and Jan must have found something in common. Jan was a classy girl, and I could see her being attracted to someone who was not trying to impress her. She probably had to drag it out of him."

Their wedding took place at the Congregational Church in Wilmette, Illinois, on January 28, 1956. Dean served as the best man and their sister June was one of Janet's attendants. The newlyweds honeymooned in Acapulco.

The couple took an apartment in Westwood, so Janet could take classes toward her college degree at UCLA. Neil returned to his bachelor's quarters on North Base at Edwards and commuted to Westwood on the weekends, a round-trip of over 180 miles. According to Neil, "That was for one semester. Then we moved up to the Antelope Valley and rented a house in an alfalfa field." In late 1957, they bought mountainside property in Juniper Hills that came with a small cabin. The move meant that Janet never earned her degree, something she always regretted.

The six-hundred-square-foot cabin overlooking Antelope Valley was rustic. Its floor was bare wood. There were no bedrooms per se, just a room with four bunks. The cabin had a tiny bath and a small kitchen, but only primitive plumbing and no electricity. Even after Neil finished installing the wiring, Janet did all of the cooking on a hot plate. They enjoyed neither hot water nor a bathtub. Neil hung a hose out over a tree branch to pass as a shower. Janet bathed baby Ricky—née Eric Allen, born on June 30, 1957—outside in a plastic tub. Only slowly, after a lot of remodeling, did the cabin really become livable. Yet the remote setting up in the San Gabriel Mountains was gorgeous and provided total relaxation away from everything. After son Ricky came daughter Karen Anne, born April 13, 1959. The Armstrongs' third and last child, Mark Stephen, was born April 8, 1963, after the family moved to Houston in the fall of 1962.

Neil's job at Edwards was, as Janet once said, "some fifty miles but only one stop sign away." Neil carpooled with fellow High-Speed Flight Station employees who lived in nearby towns. Being a test pilot made him one of the worst carpoolers. "He wasn't very reliable," remembered Betty Scott Love, one of the "human computers" at the High-Speed Flight Station, who together with other women employed there by the NACA did the tedious mathematical work of converting all the flight data into meaningful engineering units. (Betty Scott Love was the HSFS's counterpart to the likes of Katherine Johnson and other women computers who worked at NACA/NASA Langley Research Center and whose story—in terms of the segregated computing group at Langley—was depicted in the 2016 film *Hidden Figures*.)

Armstrong himself had an interesting collection of motorcars, some of which he used for carpooling. Soon after he moved to California, he traded his 1952 Oldsmobile toward a new Hillman convertible, a snappy European import. "Then a fellow at the High-Speed Flight Station had a '47 Dodge," Neil explains. "He threw a rod on the way to work, so he sold me the car for fifty dollars 'as is.' I hauled the car up to the cabin and rebuilt the engine."

"I don't know if you could say that Neil drove like he flew or flew like he drove," Betty Love recalled. "He always sat back in the driver's seat like he was in an easy chair and crossed his left leg over his right knee." Once while contemplating the mathematics of the snow level in the San Gabriels, Neil crossed the center line and ran a truck into a ditch. "It happened to be an air police!" laughed Betty Love. "Neil showed him his

ID and the MP, instead of bawling him out, saluted and told him to get on his way." At Edwards Neil's driving of an automobile became the stuff of legends. Ultimately, one story goes, "nobody wanted to ride with Neil." Even Janet wasn't comfortable with him driving. Apparently driving a car in the earthbound two dimensions simply did not engage his mind in the way flying an airplane did.

Armstrong reported to work at the High-Speed Flight Station on July 11, 1955. His formal job title was Aeronautical Research Scientist (Pilot). With the phenomenal growth of American airpower during World War II, the army's airfield on Muroc Dry Lake ("Muroc Field") mushroomed in size and purpose, the site's succession of aviation "firsts" topped by Bell X-1's push through the sound barrier in 1947. The newly established U.S. Air Force that year took over the army operation, later renaming it Edwards Air Force Base in honor of Glen W. Edwards, an air force captain. It was at Edwards that the first supersonic fighter to enter U.S. military service, North American's YF-100A, debuted in May 1953. Although Edwards AFB and the NACA's High-Speed Flight Station were officially independent entities, most people referred colloquially to both facilities simply as "Edwards."

Originally, the NACA employed only twenty-seven individuals in its Muroc Flight Test Unit. Its entire operation was hemmed in on a few dozen acres at "South Base." In 1951, Congress appropriated an additional 120 acres and $4 million. Via a new concrete parking apron and taxiway, this dual-hangar facility, which opened in June 1954, gave the HSFS ready access to the enormous runway, fifteen thousand feet long and three hundred feet wide, on the west side of the dry lake bed.

HSFS chief Walter C. Williams led the first detachment of NACA personnel from Langley to Pinecastle, Florida, and from there on to Muroc in 1946 for the purpose of flying the X-1. He ran the NACA's desert flight research operation until joining the Space Task Group in September 1959 to develop launch operations and oversee the building of a worldwide tracking network. As one of the top men in Project Mercury, Williams served as the director of flight operations for the first three Mercury flights, those made by Shepard, Grissom, and Glenn in 1961 and 1962. Armstrong's Flight Branch was part of the Flight Operations Division. Total HSFS staff numbered 275, a fraction of the nearly nine thousand at Edwards AFB.

Flight Operations Division reported to Joseph R. Vensel, a former research pilot. Vensel's authority extended to all aircraft maintenance, inspection, and operations engineering. Ops Engineering required Vensel to be knowledgeable about aircraft design, because research aircraft often needed new wings, tails, appendages, or other alterations built on-site in NACA shops. Adjacent to his office were the desks of all his test pilots.

Under Vensel was Neil's immediate boss, the head of Flight Branch, chief test pilot Joseph A. Walker. Walker earned a degree in physics in 1942. Entering the Army Air Corps, he flew P-38 fighters in North Africa during World War II, earning the Distinguished Flying Cross and the Air Medal. In March 1945, Walker became a test pilot for the NACA in Cleveland, contributing to the laboratory's aircraft-icing research. Walker came to Edwards in 1951. His promotion to chief test pilot came just months before Neil arrived. A close friendship would grow between the two men, and between their wives.

The twenty-four-year-old Armstrong once again ranked the most junior pilot. Joe Walker's ten years of research pilot experience included an estimated 250 flights at Edwards, well over one hundred in experimental aircraft, including the Bell X-1, Douglas D-558-1 and D-558-2, Douglas X-3, and Northrop X-4. Walker had made seventy-eight test flights in the Bell X-5, America's first high-performance variable-geometry ("swing wing") aircraft.

Even more experienced than Walker was HSFS test pilot Scott Crossfield. It was Crossfield whom Armstrong came to replace, yet, recalled Armstrong, "We were side by side in the office for nearly a year. He had announced that he was going to be the pilot on the X-15 program." At thirty-four years old in 1955, Crossfield was already a legend. A naval aviator, Crossfield had earned a degree in aeronautical engineering from the University of Washington, one of the Guggenheim Fund–supported AE programs. Joining the NACA as a research pilot at Muroc in June 1950, Scott flew hundreds of research flights, including eighty-seven in the rocket-powered X-1 and sixty-five in the jet-powered (straight-wing) D-558-1 and rocket-powered (swept-wing) D-558-2 aircraft. In November 1953, Crossfield became the first person ever to fly at Mach 2—faster than 1,320 miles per hour—in the D-558-2 Skyrocket.

Fellow HSFS test pilots in July 1955 were Stanley P. Butchart and John B. McKay. Both were naval aviators during World War II. Butchart had served in the same torpedo-plane squadron with future U.S. president George H. Bush: VT-51. In 1950, both Butchart and McKay received degrees in aeronautical engineering, at the University of Washington and Virginia Polytechnic Institute, respectively. Butchart came to the HSFS as a research pilot in May 1951. McKay assumed pilot status in July 1952. Both men flew a variety of research aircraft, including the D-558 and X-5. Butchart became the station's principal multiengine pilot. Hundreds of times he flew a B-29 Superfortress up over 30,000 feet in order to air-launch a research aircraft.

Stan Butchart first met Armstrong through Eb Gough in March 1955 at NACA Langley. Neil still had on his old navy flight jacket and Butchart thought, "Boy, this kid is not even out of high school yet! He looked so young." Gough told Butchart that Edwards was really where Armstrong wanted to be. Looking at Armstrong's résumé, Butchart figured that "somebody had to pick him up quick." Walker and Vensel agreed and tabbed him for Crossfield's slot.

Armstrong started flying the first day he arrived at Edwards, in a P-51 Mustang, one of America's most significant and most beloved military airplanes. "It was quite elegant," Armstrong said. "Just didn't have the performance of my F8F Panther Jet."

"I was in a learning mode for the first few weeks," Armstrong recalled, flying almost every day, either in the P-51 (with an F-51 designation) or in the NACA's R4D, a military version of the celebrated Douglas DC-3 transport. "As they became more confident in my abilities, and as I became more experienced, they gave me more and more jobs."

Though in position to chase D-558-2 Skyrocket and X-1A launches that ultimately aborted, on August 3, Armstrong saw his first actual drop while flying chase in the F-51 on Crossfield's D-558-2 flight investigation of stability and structural loads at supersonic speeds. Later that month, Armstrong also checked out in the YRF-84F, the prototype of Republic Aviation's swept-wing jet fighter (maximum speed 670 mph), and first crewed on the B-29. Armstrong's first launch assist on a research aircraft came on August 24, 1955, again with Crossfield piloting the Skyrocket.

"Generally, the person in the left seat was in command of the drop," Neil explained. "The person in the right seat did most of the flying. Over the years I flew in both positions probably an equal number of times." Without question, this was challenging flying. "We were usually taxing the performance limit of the aircraft because there was a lot of excess drag due to having the [research] aircraft slung beneath the B-29's belly. We also wanted to get as high as we could for the launches," typically up in the 30,000- to 35,000-foot region, which would take an hour and a half or more. After that, "it was a matter of getting into the proper position."

In air-launching lurked unanticipated dangers. On August 8, 1955, just an instant before Joe Walker was to be dropped in the X-1A, an explosion within its rocket engine rocked the B-29. "I thought we'd hit another airplane," remembered pilot Stan Butchart, "and in those days there wasn't anybody else up there above twenty thousand feet!" Alarmed by the big bang, Walker immediately scrambled up and out of the X-1A and into the bomb bay of his mother ship. The X-1A was too damaged to fly, and the B-29 could not risk landing with it still hooked to its underside. Butchart had no choice but to jettison the research aircraft into the desert. The machine exploded on impact, ending the X-1A program.

Armstrong saw the whole thing. Butchart remembered, "Armstrong was flying off our wing in the F-51. So we gave him a good introduction to how the game went." The cause of the accident proved to be a simple leather gasket that sealed the propellant plumbing joints. When saturated with liquid oxygen, the leather was so unstable that a shock of any magnitude caused the gasket to blow. Unfortunately, a number of accidents occurred before engineers identified the problem and fixed it.

Eight months after arriving at Edwards, Armstrong experienced one of his own closest shaves ever. On March 22, 1956, in the NACA's launch B-29 modified and designated P2B-1S, Armstrong was flying in the right seat with Butchart in command to his left, along with five crew members. Their job was to take the number two D-558-2 research airplane up to an altitude of a little over thirty thousand feet and then drop it so HSFS research pilot Jack McKay could take it through a flight investigation of its vertical tail loads.

Approaching thirty thousand feet, one of the B-29 engines quit. Passing the controls over to Neil, Butchart turned around to consult with flight engineer Joseph L. Tipton. With no power, the propeller blade on number-four engine windmilled in the air stream.

"I wasn't too concerned about it, really," recalled Butchart. "B-29 engines are not all that dependable." On his control panel, Butchart had four "feathering" buttons designed to shut down, or "feather," the rotation of a propeller up to three times. Feathering his far starboard engine, he expected the propeller to come to a standstill. Instead, just as the prop came close to stopping, it started spinning again. With Neil flying the plane, the propeller came back up to full speed, and then *exceeded* the rpm's of the other props.

Armstrong and Butchart faced a critical choice: "try to slow down and hope we can keep the rpm of the propeller under control" or "speed up and get rid of the rocket plane underneath."

Butchart hit the feather button twice more, to no avail. In the meantime McKay down in the cockpit of the Skyrocket called up, "Hey, Butch, you can't drop me! My Grover loader valve just broke." Given that the rebel propeller could fly loose at any moment, Butchart announced: "Jack, I've got to drop you!"

Already, Butchart had motioned to Armstrong to nose down the B-29. If the speed at launch was anything less than 210 mph, the Skyrocket would come out in a stall—falling but not *flying.* But the runaway prop spun even faster, increasing its likelihood of busting loose.

Butchart put his hand on the emergency release lever and pulled. Nothing happened. He pulled two or three times. Nothing. Then he reached up and hit the two toggles that armed the "pickle switch" (conventionally used to drop bombs), which the NACA had adapted to drop its research airplanes. The D-558-2 fell away sharply from the B-29, and the prop let go.

The blades flew off in every direction, one of them slicing through its bomb bay where test pilot Jack McKay had been sitting a few seconds earlier, and hit its number-two engine on the other side.

Getting the B-29 down for landing was not going to be easy. The starboard number-three engine was still running, but its instrument readings had shut down. The pilots shut that engine down. Number one had not been damaged, but it had to be shut down too, because of the torque it caused with neither of the starboard engines running. Butchart and Armstrong had to fly the B-29 down from thirty thousand feet with only one engine.

Butchart tried to take over the flying from Armstrong, but his wheel was loose and floppy. He looked over and said, "Neil, you got control?" and Neil answered, "Yeah, a little bit." Both pilots had rudder and longitudinal control, but Butchart did not have pitch control, nor did he have any control of roll because his cables to the ailerons were shot. What controls Armstrong had were dicey.

"So we just made a slow, circling descent, tried never to get to a very large bank angle, and were successfully able to make a straight-in landing onto the lake bed," Armstrong remembered. According to Butchart, during the descent "Neil kept saying, 'Get your gear down! . . .' and I said, 'Wait a minute. I have to make sure I can make that lake!' because there was no way of going around and I couldn't use too much power even on [number] two because we couldn't hold the rudder down. We were both standing on rudder. . . . So it was pretty tense coming down."

With typical understatement, Armstrong summed up the experience: "We were very fortunate. It could have turned ugly."

McKay in the Skyrocket also landed safely.

Over the course of his seven-year career at Edwards, Armstrong piloted or copiloted a launch plane more than one hundred times. He dropped or flew chase for every type of NACA/NASA research airplane then flown at Edwards. Virtually every day that conditions were suitable, the young test pilot took to the air. From the time he came to Edwards in July 1955 to the time he left to join the astronaut corps at the end of September 1962, Armstrong made well over nine hundred total flights, an average of over ten flights per month.

Flight Operations Division logbooks indicate approximately 2,600 hours total flight time, roughly fifteen and one half weeks of twenty-four-hour days in the cockpit of some of the country's most advanced, high-performance, and risk-laden experimental aircraft. Most of his flights came in jets. More than 350 of his flights took place in one of the famous "Century" series fighters: the North American F-100 Super Sabre, the world's first fighter capable of sustained supersonic speeds in level flight; McDonnell F-101 Voodoo; Convair F-102 Delta Dagger; Lockheed F-104 Starfighter;

Republic F-105 Thunderchief; and Convair F-106 Delta Dart.

The first time Armstrong broke the sound barrier came in October 1955, an F-100A flight investigation of longitudinal stability and control characteristics involving various wing slots and slats in different leading-edge configurations.

In June 1956, Armstrong started flying the F-102, newly supersonic thanks to NACA aerodynamicist Richard T. Whitcomb's recent development of the "area rule," by which the drag of a wing and the drag of the body of an aircraft must be considered as a mutually interactive aerodynamic system. "I flew the YF-102, which was the pre-area-rule F-102," Armstrong remembered. "Kind of a dog of an airplane," it was "not a lot of fun to fly," and "I don't think I could ever get it supersonic." Pinching the waist of its fuselage improved the F-102's speed and overall performance even with approximately the same engine thrust. However, it suffered from very high drag due to lift. In the NACA's F-102s, Armstrong "did a lot of landing work, because we more than anyone else at that point in time were flying the rocket airplanes and having to make unpowered landings." Armstrong also flew dead-stick landings in the F-102 as well as the F-104.

About a third of the nine-hundred-plus flights piloted by Armstrong at Edwards were true "research" flights. The other two-thirds involved familiarization flights, chase, piloting air launches, or flying transport. Considering the two-year period from 1957 to 1958 as a representative sample shows that Armstrong flew the greatest number of flights in the R4D/DC-3 followed by the F-100A, F-104, B-29, F-100C, and B-47. Besides the F-51 Mustang and the aforementioned Century series fighters, Armstrong logged time in the venerable T-33 "T-Bird," a two-seater derivative of the F-80 Shooting Star fighter; North American's F-86E Sabre; McDonnell's F4H Phantom; Douglas's F5D-1 Skylancer; and Boeing's KC-135 Stratotanker. Armstrong pushed past Mach 2 in Bell's X-1B and X-5, and went hypersonic in the North American X-15. He also piloted a unique experimental vehicle called the Paresev.

The flights Armstrong made lasted on average less than one hour apiece, particularly the research flights. Typically, fewer than ten flights in any year lasted more than two hours, and only four or five lasted more than three hours. Many of these longer flights took place in the R4D/DC-3 on transport missions to other NACA laboratories, to aircraft manufacturers, or to military bases, or involved taking the B-29 up to high altitude for air-launch operations.

"Our principal responsibility was engineering work," Armstrong explained. "It was program development, looking at the problems of flight. It was a wonderful time period, and it was very satisfying work, particularly when you found a solution."

Almost everyone who has ever rated Armstrong as a pilot, including his commanders back in the navy, made a connection between his piloting skills and his engineering background and talents. Flight Research Center colleague Milt Thompson wrote that Neil was "the most technically capable of the early X-15 pilots." William H. Dana, who as a NASA research pilot flew in some of the most significant aeronautical programs ever carried out at what became NASA Dryden Flight Research Center, emphasized how "bright" Armstrong was about the aircraft he flew: "He understood what contributed to a flight condition. He had a mind that absorbed things like a sponge and a memory that remembered them like a photograph. That set him apart from mere mortals."

As impressive as Armstrong's abilities were to pilot-engineers, aeronautical engineers who did not fly appreciated Armstrong as a pilot even more. At Edwards, Neil often worked with Gene J. Matranga, a 1954 graduate in mechanical engineering from Louisiana State University. "Neil ran circles around many test pilots, engineering-wise," Matranga declared. "The other guys who flew seat of the pants knew instinctively what to do, but they didn't always know why. Neil knew why. As long as he could convince himself that something was going to be successful," Armstrong's "openness to doing things," in Matranga's opinion, compared favorably to a "pretty hard and fast reluctance on the part of many pilots" to surrender any of their authority to nonfliers. "Neil did not have that bias."

Ultimately, there can be no doubt that Armstrong's experience and talents as a professional engineer served the cause of his flying career extremely well. Those who handpicked him in 1962 for the second class of astronauts, without question, favored Neil's engineering qualifications.

A telling admission came from Christopher C. Kraft Jr., a NACA flight researcher and one of the founding fathers of the American space program: "I was prejudiced for the fact that this guy's been a NACA test pilot. He was above the capability of the other test pilots we had in the loop because he'd been through the daily contact with flight engineers, of which I was one."

According to Kraft, key people on the astronaut selection board, notably NACA veterans Robert R. Gilruth, Walter Williams, and Dick Day, felt even more

CHAPTER 8

At the Edge of Space

The rarefied conditions into which Armstrong "zoomed" in his sleek fighter jet were far closer to those on the Martian surface than anything down on Earth. Streaking upward past 45,000 feet, he passed the biological threshold at which a person could survive without the protection of a space suit. When his near-vertical climb reached 90,000 feet, atmospheric pressure fell to a scant 6 millibars, about 1 percent of the pressure at sea level. Outside his cockpit, the temperature dipped to 60 degrees below zero F.

This *was* space. The only way to control his plane at the top of its ballistic arc was to invoke Newton's Third Law and expel some steam via jets of hydrogen peroxide. A pilot in a near vacuum could maneuver his airplane in pitch, yaw, and roll just as manned spacecraft would later do. With all the energy from the zoom dissipating, Armstrong's jet came close to a virtual standstill, sitting on its tail. For over half a minute at the top of his climb, he experienced a feeling of weightlessness. At about 70,000 feet, Neil had shut down the engine to prevent it from exceeding its temperature limit. The cockpit's ingenious auxiliary pressurization system released a squirt of compressed gas.

partisan in Armstrong's favor, especially Williams and Day. Both men were themselves engineers rooted in the NACA's engineering research culture. Both came to NASA's Manned Spacecraft Center after spending years in flight research with NACA/NASA at Edwards, where they had come to know and admire young Armstrong. "Neil was about as good as you could come by in evaluating a man from a test-pilot-performance capability," Kraft stated. The only real uncertainty came down to whether *he,* Neil, personally *wanted* to become an astronaut.

For why choose to become an astronaut when Armstrong was already so deeply and so creatively involved in the biggest, most technically challenging flight programs ever attempted? Two of these programs—the X-15 and Dyna-Soar—had as their goal not just flying *piloted winged* vehicles at *hypersonic* speeds, but flying them *transatmospherically,* into and back from space.

The engine's *not* running at the top of the arc was critically important to the goal of the flight test. If not shut down, the engine would have introduced yaw motions challenging Neil's capacity to control the aircraft.

Streaking down nose-first into the atmosphere, enough air molecules eventually passed through the jet's intake ducts to allow Armstrong to restart his engine, and, at a speed of about Mach 1.8, begin his recovery from the unpowered dive. From that point on, with luck, the rest of the flight was routine all the way down to the runway. If Neil did not get an engine restart, he could make a dead-stick landing. If necessary, in the moments after touchdown, he could pull a lanyard to deploy a drag chute housed just below the plane's vertical stabilizer to decrease his landing roll-out distance.

In this fashion, Neil Armstrong and his fellow NASA test pilots at Edwards—at the controls of a long pointy jet plane nicknamed "The Missile with a Man"—made the country's first dramatic excursions to the edge of space. They did so for research purposes more than half a year before Commander Alan B. Shepard became the first American astronaut to fly in space.

These facts fly in the face of popular lore. Thanks to author Tom Wolfe's 1979 bestseller *The Right Stuff,* and the 1983 Hollywood film adaptation, most people believe that the man who first flew *in an airplane* to the edge of space was U.S. Air Force test pilot Captain Chuck Yeager. However, much that has been written about Yeager and his December 1963 flight is factually inaccurate. Most important, Yeager and the U.S. Air Force Test Pilot School at Edwards were not responsible for "developing the first techniques for maneuvering in outer space," as some air force publications have claimed; instead, NACA/NASA, with the F-104 and previously with the X-1B, led the way in this critical new technology of the emerging "Space Age." (The X-1B flights occurred in 1957 and 1958, but they were not effective in terms of reaction-control research.) And Yeager was not even close to being the first pilot to zoom into the high stratosphere. Some NASA test pilots began to make zooms to ninety thousand feet as early as the fall of 1960. And in the rocket-assisted NF-104A, air force pilots performed zooms into the upper stratosphere before Yeager.

In addition, well before December 1963, a far more remarkable and historically significant flying machine had pushed the envelope considerably further than any zooming F-104. This machine was the X-15, the fastest and highest-flying manned winged vehicle ever built—and one that Chuck Yeager never flew. Conceived by the NACA in the early 1950s and built by North American Aviation under the sponsorship of the air force, the navy, and the NACA, the X-15 was constructed not just to explore the hypersonic flight regime existing above Mach 5, but also to study the possibilities of flying a winged vehicle outside the sensible atmosphere (the region where aerodynamic control surfaces will function). First flown in June 1959, the rocket-powered X-15 was a veritable "aerospace plane." By the end of 1961, the year President Kennedy committed the nation to the Moon landing, the X-15 attained its primary design goals of flying to a speed in excess of Mach 6 (over four thousand mph) and to an altitude of over two hundred thousand feet (or nearly thirty-eight miles high). In 1962, a year that saw the Mercury flights of astronauts Glenn, Carpenter, and Schirra, air force pilot Robert White, in a pressure suit similar to the Mercury space suit, flew the X-15 more than fifty miles high (264,000 feet), the altitude that technically qualified him as an "astronaut" according to a policy invented by the U.S. Air Force (and never endorsed by NASA). The total number of X-15 pilots who earned "astronaut wings" according to the air force definition was eight. That was one more than the original group of Mercury astronauts, only six of whom made it into space (and only four into orbit) as part of the Mercury program. (Mercury astronaut Deke Slayton did eventually fly into space, in 1975, as part of the Apollo-Soyuz Test Program.)

Following over thirty zooms in the F-104, Neil Armstrong would fly the X-15 seven times before joining the second class of American astronauts in September 1962. Neil never made it above the fifty-mile mark, but on April 20, 1962, in his sixth X-15 flight, he did reach 207,500 feet, just under forty miles high.

In retrospect, the movement of aeronautics from subsonic to transonic, then to supersonic and on to hypersonic (and beyond that to "hypervelocity"), seems inevitable. As the emerging Cold War crystallized into an atomic face-off between the United States and the Soviet Union, the sharpest focus for hypersonic enthusiasm lay in the development of an intercontinental ballistic missile (ICBM) armed with nuclear warheads. Yet for those enthusiasts for whom aeronautics still meant piloted, winged *airplanes,* the ambition was to design a rocket-powered vehicle to take men and cargo on "hyperfast" flights across global distances, on trajectories that, at their apex, flew out into space.

Rocket-powered experimental research airplanes were air-dropped into flight. Armstrong piloted his first on August 15, 1957, the first check-out flight of the modified X-1B, zooming to about sixty thousand feet.

Although it was the highest altitude that Armstrong had yet flown, at only 11.4 miles the dynamic pressure simply was not low enough to test the reaction controls.

In landing the aircraft, his nose landing gear "failed." According to Neil's official report, he "inadvertently touched down at 170 KIAS [Knots Indicated Airspeed], nose wheel first." "It didn't really *fail,*" Neil admits. "I broke it. I was landing on the lake bed, and it was fairly normal. But at touchdown the airplane began to porpoise and, after several cycles of the porpoising, the nose wheel bracketry failed. I felt devastated, of course, but that was improved a little when I found out that was the thirteenth or fourteenth time [due to the coupling of the geometry] that had happened [with the X-1 series]."

His second flight in the X-1B, on January 16, 1958, was aborted due to systems problems. The ten-year-old X-1B flew only one more time, on January 23, when Armstrong and Stan Butchart air-dropped pilot Jack McKay for a zoom to fifty-five thousand feet, one that did not slow enough at the top to check out reaction controls. Immediately after McKay's flight, mechanics found irreparable cracks in the rocket motor's liquid oxygen tank, ending the entire X-1B program.

Supersonic jets differed from their slower predecessors in the design of their relatively shorter swept-back wings, denser shapes, and a much greater mass concentration around their fuselages. Unexpectedly, this altered geometry brought on some serious aerodynamic difficulties known as "roll coupling" (also called "inertial coupling" or "roll divergence").

As Armstrong reported to work at the HSFS in the summer of 1955, no problem was receiving more attention than roll coupling. Not only was the problem endangering the F-100, it had also threatened the D-558-2,

X-2, and the NACA's newest research airplane, the Douglas X-3. A long, slender, dart-shaped aircraft, the X-3 Stiletto experienced coupling instability during abrupt roll maneuvers that caused it to go wildly out of control. Built for Mach 2, the X-3 was barely able to reach Mach 1.2 because it never received the higher-rated-thrust turbojets intended for it. The NACA retired the plane in May 1956 after only twenty flights. So all the attention turned to the F-100. Quickly, a fix was found—the addition of a much larger tail. Then, flying its own modified F-100C, the NACA tested a new automatic control technique—one that used pitch damping as a means of lessening the divergence of the yaw—to resolve the roll coupling problem more generally. Armstrong checked out in the airplane on October 7, 1955, and piloted many of the flights for that program during the next two years.

This partially automatic flight-control system that Armstrong helped to develop for the F-100 was one of the first to incorporate "feedback compensation." In essence, the idea was for the control surfaces on the aircraft (ailerons, rudder, elevator, and such) to communicate as part of an integrated, self-regulating system. Starting in April 1960, Neil consulted about this technology with engineers at the Minneapolis-Honeywell Corporation. After Honeywell installed the prototype system—called MH-96—on an F-101 Voodoo in early 1961, Neil traveled to Minnesota in March 1961 to fly it. Based largely on his favorable written reports, NASA decided to install the MH-96 on the final X-15 (X-15-3), which was scheduled to be test flown for the first time late in 1961. Given his role in the MH-96's development, NASA assigned Armstrong to pilot the first flight. In Minneapolis as at Edwards, Neil explained, "We used airplanes like the mathematician might use a computer, as a tool to find answers in aerodynamics."

The NACA's High-Speed Flight Station virtually invented the flight simulator for research purposes. In 1952, the NACA had convinced the air force to buy an analog computer, which could be reprogrammed by HSFS engineers into a flight simulator. By the time Armstrong arrived at Edwards in 1955, flight simulators had been making important contributions to a number of research programs, notably the X-1B and the X-2, the latter of which the NACA was supposed to receive after the air force finished testing it. Unfortunately, a tragedy with the X-2 stopped that from happening. In his first flight in the X-2, air force test pilot Melburn G. "Mel" Apt lost control due to roll coupling, the vehicle entering into a wild series of diverging rolls. Apt tried frantically to regain control of the aircraft but could not. His only option was to eject via the plane's escape capsule. Though the capsule's drogue parachute opened, its larger parachute did not. Apt attempted to bail out of the pod but there wasn't enough time. The capsule smacked into the hardpan of the Edwards bombing range, the rest of the X-2 crashing five miles away. The flight made Mel Apt the pilot of the fastest aircraft that had ever flown, over three times the speed of sound, but his death completely overwhelmed all thought about that. But not about what had happened to him, to his aircraft, and why. Subsequently, as Neil remembered, prospective X-15 pilots would be shown—more than once—the on-board film of Apt's fatal flight, taken by a stop-frame camera mounted behind Apt in the cockpit.

The Apt tragedy deepened the NACA's commitment to the development of its research simulators. In the Sim Lab, Neil learned "that there were many ways to induce errors into the programming. Often the outputs to the instruments were improperly mechanized so the instrument would not accurately represent the airplane motions. I found this to be true much later in Houston, and always took the time with a new simulator to check the accuracy of its response." Armstrong may have spent more time in simulators than any other pilot then at Edwards, gaining experience where he actually saw and felt the results, "constantly picking up new information and potentially valuable techniques."

Armstrong also became one of the first NACA/NASA test pilots to endure the torture of the navy's Johnsville, Pennsylvania, centrifuge. (The National Aeronautics and Space Administration came to life on October 1, 1958, the product of the National Aeronautics and Space Act of 1958, signed into law by President Eisenhower on July 29, 1958, with the NACA officially becoming defunct but in actuality serving as the nucleus of the new NASA.) The purpose of "riding the wheel" was to see "whether the *g* field that you had to go through in a rocket-launch profile would adversely affect your ability to do the precision job of flying into orbit." Armstrong explained the purpose of the research: "We hypothesized that it would be possible to pilot an aircraft into orbit—that a vertically launched rocket could be manually flown into orbit without the need for an autopilot or any sort of remote control."

A team of seven pilots took part in the experiment: Armstrong, Stan Butchart, and Forrest "Pete" Petersen from the NASA Flight Research Center (formerly the HSFS); two other NASA pilots, one from Langley and one from Ames; and two air force pilots. Lying on their backs and strapped into molded seats contoured to fit the form of the individual pilot in his pressure suit, Armstrong and his mates were put through the wringer. Every possible force and stress and every possible flight condition was brought to bear on the pilots as they whirled dizzily at the end of the fifty-foot-long arm. At the highest speed and angle of the wheel, they experienced acceleration rates as high as fifteen g's. Only a couple of the pilots handled g-forces that high, and Armstrong was one of them. Gene Waltman, one of the FRC technicians on the scene, remembered Armstrong saying that at fifteen g's so much blood left his head that he could only see one of the instruments in the simulated cockpit. Neil recalled, "We persuaded ourselves that it was, indeed, a doable task, operating the controls of a launch vehicle or aircraft accelerating at those high rates." With FRC engineers Ed Holleman and Bill Andrews, Armstrong coauthored a NASA report announcing the surprising results. Many people in the aerospace community questioned the finding that g-forces up to about eight g's actually had very little effect on a pilot's ability to operate flight controls until it was proven to be true in the X-15 and Mercury programs. Armstrong later went back to Johnsville to fly X-15 entry trajectories with various flight control system settings.

But the key component of X-15 flight preparation was the electronic simulator. Two main X-15 simulators were built. Both of them were analog machines, because digital computers were still far too slow to do anything in "real time." North American erected the simulator called the "XD" on company property on what is now the south side of Los Angeles International Airport. Armstrong visited several times to experience the simulation of all six degrees of freedom. Flying down in an R4D, Day remembers Neil regularly asking for an ILS (instruments) approach into Los Angeles airport. "We did several flights down, basically entries. We would go up to 2,500 or 3,000 feet and we would do entries at different angles of attack and then plot angle of attack versus maximum dynamic pressure. It turned out to be a straight line, which was a special equation. And Neil learned that in case he had trouble."

Under Dick Day's direction, NASA built at Edwards an X-15 simulator that replicated the X-15 cockpit. According to Armstrong, the machine was "probably the best simulator that had ever been built up to that time, in terms of its accuracy and dependability." In preparation for each one of his seven X-15 flights, he spent fifty to sixty hours in the simulator.

"The actual X-15 flights were only ten minutes long, and generally in the simulator you didn't have the ability to do the landing," Neil explained. "You'd just do the in-flight, and they were only a couple minutes long. We would put together a little team—the pilot, one of the research engineers, and one of the guys from the computer group—and say, 'Here's what we want to do,' and they'd take what data we had and put it in and find out what we could learn from it. You could kind of begin to understand a problem."

The X-15 program came together quickly. Barely a year after building began in September 1957, the first one rolled out of the factory. Six months later, in March 1959, the X-15 made its first captive flight and, three months after that, its first glide flight. On September 17, 1959, less than four years since the project's inception, Scott Crossfield took it on its first powered flight. Wind tunnel tests indicated that the X-15 at low speed possessed a very low lift-drag ratio (L/D), that is, one producing very little aerodynamic lift. Once its rocket burned itself out, the X-15 would come down fast and steep. Normal power-off landing techniques were inadequate. Beginning in the summer of 1958, Armstrong flew L/D approaches testing "various and sundry combinations of speed brakes and flaps" well into 1961.

Everybody involved in the X-15 program seemed to hold an opinion about the best landing approach. Armstrong and other NASA pilots proposed a version which they believed offered the greatest flexibility. According to project engineer Gene Matranga, "Our technique involved a 360-degree spiraling descent starting at about 40,000 feet" right above the desired touchdown point on the runway. From that "high key" position, the pilot moved into a 35-degree bank (usually to the left) while maintaining an airspeed of 285 to 345 miles per hour. At roughly twenty thousand feet, after some 180 degrees of the spiral had been completed, the X-15 reached the "low key." At this point, the aircraft was headed in the opposite direction of the landing runway and was about four miles abeam of the touchdown point. From the low key, the turn continued through the other 180 degrees until the X-15 lined up with the runway at about a five-mile distance. The rate of descent through the spiral averaged over two miles per minute, which meant it took on average about three minutes to go from high key to that point where the X-15 was ready to head straight in for landing.

To determine where the flare should begin, Armstrong and Walker were forced to resort to the imprecise explanation of "I feel it." In this case, Matranga understood: "We tried to work mathematical models for determining the starting point, and it just could not be done. It was just something that the pilots, with their own experience, knew intuitively, and it could, from flight to flight, vary pretty significantly." After the back of Crossfield's plane broke in a rough landing, North American adopted the spiral technique that Armstrong and his mates worked out. The technique developed by

NASA became standard. In fact, the basic technique developed at the Flight Research Center worked well later in the so-called lifting body program, and it also worked well for the Space Shuttle. Armstrong coauthored two papers on the F-104 low L/D landing investigations, as well as a number of other technical papers on various topics.

Crossfield flew the X-15 thirteen times before North American turned it over to NASA–air force–navy partnership. Armstrong watched as many of those flights as he could. Two of Crossfield's flights were in the number-one airplane, the rest in number two. The highest speed he reached in any of them was Mach 2.9, the highest altitude 88,116 feet, and the farthest distance 114.4 miles.

Armstrong did not fly the X-15 for the first time until November 30, 1960. Prior to that, he did fly chase on two occasions. In all, Neil flew chase for the X-15 on six occasions. Often Neil was in the Edwards control center, on the microphone with the pilot, and monitoring the radar and telemetry. The last time he flew chase as an Edwards employee was on June 29, 1962, when NASA colleague Jack McKay flew the number-two airplane nearly to Mach 5. For the majority of X-15 flights, four chase planes were employed; in the longer-range flights, a fifth was added.

On November 30, 1960, Neil sat in the cockpit of the number one airplane high over Rosamond Dry Lake anxiously waiting to be launched in an X-15 for the first time. At the controls of the B-52 drop plane were Major Robert Cole and Major Fitzhugh Fulton. Flying the chase planes for Neil were Joe Walker, Lieutenant Commander Forrest S. Petersen, and Captain William R. Looney. Overall, it was the twenty-ninth flight in the X-15 program, the seventeenth involving the X-15-1, and the seventh made by a NASA pilot.

With Neil at the controls for the first time, the purpose of flight number 1-18-31 was simply pilot familiarization, but nothing was ever very simple about flying the X-15. He had been in the X-15 simulators for hundreds of hours, but the real thing was very different. "When you're dressed up in that pressure suit, and you get the hatch closed down on you, you find that it is a very, very confined world in there. The windshield fits over you so snugly that it's very difficult to see inside the cockpit." Looking out of the windshield, Neil saw nothing at all of the aircraft he was flying. "There's a lot of tension when you're in that situation even though you know it's been done before. Everybody else has been able to handle it, so you ought to be able to."

At forty-five thousand feet, Fitzhugh in the B-52 started the same sort of countdown that would be used later in space shots: "Ten seconds, launch light is on. Five, four, three, two, one, launch." Armstrong had been air-launched before, in the X-1B, but the X-15 came off much more dramatically, with more of a clank. Then came the challenge of getting the rocket motor started, *right away.*

The engine powering Neil's X-15 was the XLR-11, built by Reaction Motors. The XLR-11 comprised two rocket motors, an upper and a lower. Each motor had four chambers and each chamber gave 1,500 pounds of thrust, a total of twelve thousand pounds of thrust. But chamber number three would not light, reducing the total thrust to 10,500 pounds. Even if up to four chambers had not been operating, the vehicle still could have been flown, though it would have had to stay close to base and immediately prepare for landing. Fellow test pilot Jack McKay told Neil to "go ahead and proceed with the original flight plan."

Other than the number three chamber failing to light, Armstrong's first X-15 flight went without incident. After the aircraft came level at 37,300 feet, Neil put it into an eight-degree climb that took him to an altitude of 48,840 feet before "pushover," or nosing back downward. His maximum speed was only 1,155 mph, or Mach 1.75. But Walker and the rest were pleased with what they saw from Armstrong that day. Armstrong's second X-15 flight, and his first for research purposes, came ten days later, on December 9, 1960, also in the number-one airplane. Flight number 1-19-32 first tested the X-15's newly installed "ball nose." Until this flight, the X-15, typical of all research aircraft up to this time, had a front-mounted boom with vanes to sense airspeed, altitude, angle of attack, and angle of sideslip in a free aerodynamic flow field. At such high altitudes and high speeds, the X-15 would melt its nose boom, destroying measurement data.

The ingenious solution was to design a sphere that could be mounted on the front of the aircraft. The sphere would be subject to the highest temperatures on the airplane, but it could be cooled from the inside by liquid nitrogen. Dropped by the B-52 at the standard 45,000 feet, the X-15 climbed to 50,095 feet at a speed of Mach 1.8. Burnout of the rocket came immediately after Neil extended the aircraft's speed brakes. The ball nose worked so well that it would be used throughout the remainder of the X-15 program. Neil's own performance was, again, solid.

It would be over a year before Armstrong would make another X-15 flight. Throughout 1961, Armstrong continued to work on the new automatic flight-control system for X-15-3, the aircraft in which he would make his third through sixth X-15 flights starting in December 1961. Until then, there would not be nearly so much test flying for Neil as in previous years. But there would be more travel than ever to Minneapolis-Honeywell and to Seattle, where he consulted for NASA on the air force's new X-20 space plane program, known as Dyna-Soar.

CHAPTER 9

The Worst Loss

In late spring 1961, the Armstrong family was temporarily in Seattle because Neil was working at Boeing, the contractor on Project Dyna-Soar, a joint NASA–air force program to develop a manned hypersonic boost-glide, designated X-20, which some advocates thought might beat Project Mercury's ballistic capsule into space. But Alan Shepard's suborbital flight just a few weeks earlier, on May 5, 1961, put an end to that dream—yet Dyna-Soar lived on. Visiting the public park on Lake Washington had become a regular weekend outing for the Armstrongs in Seattle. Almost four, Ricky enjoyed the swings, as did Karen, age two.

While leaving the public park on June 4, Karen tripped and fell. She wound up with a knot on her head, had a little nose bleed, and that night her eyes were crossed. Neil and Janet feared that "Muffie," as Neil called his daughter, might have suffered a concussion. A Seattle pediatrician told Janet to have Karen checked out thoroughly back in California, where the family was headed at the end of the week. Karen's regular pediatrician in Lancaster sent her on to an ophthalmologist, who told Janet to see how the child did at home and to bring her back in a week. A mother of one of

Janet's swimming students was a registered nurse, who was alarmed to observe that Karen seemed to be getting progressively worse. She kept tripping and her eyes were almost constantly crossed. The nurse told Janet to hospitalize Karen for a comprehensive series of tests.

Janet made the arrangements, because Neil, upon returning from Seattle, had left to do some work at Minneapolis-Honeywell. "He didn't know anything about it, so I finally called him and told him I was hospitalizing her." That day Karen's eyes began to roll and she could not talk plainly anymore. The little girl went through a battery of tests at the Daniel Freeman Memorial Hospital in Inglewood, culminating in an encephalogram, which required a spinal tap and an injection of air into Muffie's spinal canal. The results along with X-rays showed that Karen had a glioma of the pons, a malignant tumor growing within the middle part of her brain stem. Even today, the prognosis for brain-stem gliomas is remarkably poor: a majority of children still die within a year of diagnosis.

"They immediately started X-ray treatment on her to try and reduce the size of the tumor," Janet recalled. "In the process, she completely lost all her balance. She could not walk; she could not stand. She was the sweetest thing. She never, ever complained." At the hospital, "I was with her around the clock, or Neil was there. He took a week off and we stayed in a motel down there. One of us kept Ricky and one of us stayed at the hospital." The initial week of radiation was followed by six weeks of outpatient treatment. "During this time, she learned to crawl again and eventually she learned to walk again," Janet noted. "Then it got so I could bring her home on weekends and took her back to L.A. through the week." Over the seven-week period, the hospital gave Karen the maximum 2,300 roentgens of

X-ray. For the next month and a half, Karen's condition improved. The radiation temporarily arrested the tumor.

But before long, Karen's symptoms were back—the difficulty with coordination and walking, the crossed eyes and double vision, the inability to speak clearly, the sagging of one side of the face. Returning to the hospital in Inglewood, Neil and Janet knew there was only one possible treatment remaining: cobalt. They decided to try the cobalt, involving a gamma ray beam that irradiated deeper into the brain but which killed not only cancer cells but also healthy tissue. Doctors told them the cobalt was her only chance. But Karen's weakened body could not take it. The doctors at Freeman Memorial were very straightforward. Rather than hospitalizing her again, everyone involved realized she would be happier at home. The family even traveled to Ohio for the holidays. "She made it through Christmas," Janet recounted. "She couldn't walk by this time—she could crawl—but she was still able to enjoy Christmas. It seems like the day Christmas was over, she just went downhill. It just overcame her."

In the last weeks of their little girl's life, Neil and Janet leaned on their friends Joe Walker and his wife, Grace, for support, especially Janet. "Jan brought Karen around a number of times," Grace recalled. "We put her in the high chair and tried to feed her Jell-O or pudding. She would try to eat it and then she would just throw up." Grace remembered that Neil also drove over once with Muffie: "Just a short visit on a Sunday. Neil wanted to show her our new baby girl, who was about three months old. I wanted to do something overt—putting hands on Karen or saying a prayer, or something—but I didn't feel Neil would consider it acceptable. I felt Neil came to me because he wanted somehow to encourage Karen and to hold on to hope. You could see that he loved his little girl very deeply."

On January 28, 1962, Karen died at home in the family's Juniper Hills cabin after an agonizing six-month battle with the brain tumor. The week leading up to her death must have been particularly difficult on Janet as Neil was away on job-related travel. The day she died was Neil and Janet's sixth wedding anniversary. Final rites were held on Wednesday, January 31, with burial in the children's sanctuary at Joshua Memorial Park in Lancaster. In honor of Neil's daughter, the Flight Research Center grounded all test planes the day of her funeral. Grace Walker remembered Neil being very stoic and showing little emotion, in contrast to Janet, who was visibly shaken. Grace thought about hugging Neil but stopped herself: "I think he always felt like that wasn't the thing to do. He was very tight emotionally."

People who knew Armstrong well indicated that Neil never once brought up the subject of his daughter's illness and death. In fact, several of his closest working associates stated that they did not know that Neil ever had a daughter.

Neil was back in the office on February 5, and back in the air the day after that. Armstrong took no other time off from the job until a family trip to Ohio in mid-May, though there was another month-long stay in Seattle from February 26 to March 20, when Neil was again consulting on the Dyna-Soar program. "It hurt Janet a lot," recalled Grace Walker, "that Neil went right back to work." Janet was "a very directed and self-sufficient person, but she desperately needed her husband to help her. Neil used his work as an excuse. He got as far away from the emotional thing as he could. I know he hurt terribly over Karen. That was just his way of dealing with it." According to Grace, "Jan was angry for a very long time: angry at God and, I think, at Neil, too." Unwilling to get into difficult verbal exchanges, "Neil left Janet in limbo."

There is no question that Karen's death shattered Neil to the core. "It was a terrible time," his sister June recalled. "I thought his heart would break. Somehow he felt responsible for her death, not in a physical way, but in terms of 'Is there some gene in my body that made the difference?'" June remembered a telling incident from the spring after Karen's death, when Neil took his family to Wapakoneta for a short vacation and informal family reunion: "A baby sheep had died at the Korspeter farm. The men went out to the barn to attend to the dead lamb. My husband, Jack, later told me that Neil could not go into the barn. Neil waited outside while the other men took care of the animal." However, back in California such intense feelings of sadness and loss did not keep Neil from regularly visiting Muffie's grave.

Later in life, during Armstrong's most celebrated days as an astronaut, there would be some curious personal moments that hearkened back to the loss of Karen. The most extraordinary of these came during the Apollo 11 crew's post-flight visit to London, England, in October 1969. Under the headline "2-Year-Old Girl Bussed by Neil," the story began by explaining that Neil, Buzz Aldrin, and Mike Collins were about to set out for Buckingham Palace and an audience with Queen Elizabeth and Prince Philip. "But it was a tiny girl who came to see the spacemen only to be nearly crushed against a barrier who won the heart of the slender, blue-eyed Armstrong, the first man to set foot on the Moon. A policeman had picked up Wendy Jane Smith, two, when she was shoved against a barricade in front of the U.S. Embassy. Armstrong caught her eye and quickly stepped forward and kissed her while a crowd of more than three hundred cheered."

Was there an intensely personal—perhaps subconscious—relationship between Karen's death at the end of January 1962 and Neil's decision to submit his name for astronaut selection just a few months later? "I never asked him," June confessed. "I couldn't." Yet it was clear to June that, through his becoming an astronaut, Neil turned it all around, at least for himself: "The death of his little girl caused him to invest those energies into something very positive, and that's when he started into the space program."

CHAPTER 10

Higher Resolve

Armstrong never consciously related his decision to become an astronaut to his daughter's death: "It was a hard decision for me to make, to leave what I was doing, which I liked very much, to go to Houston. But by 1962 Mercury was on its way, the future programs were well designed, and the lunar mission was going to become a reality. I decided that if I wanted to get out of the atmospheric fringes and into deep space work, that was the way to go."

On October 4, 1957, the Soviet Union had launched into orbit the world's first artificial satellite, Sputnik 1. This stunning technological achievement gave a new sense of urgency to the American aerospace community, and led to the formal abolition of the NACA and its amelioration by NASA. NASA's first priority was to place a man in space through a program known as Mercury. Of all the pilots who became the first astronauts—Gordon Cooper, Gus Grissom, and Deke Slayton from the air force; Scott Carpenter, Wally Schirra, and Alan Shepard from the navy; and John Glenn from the marines—Neil knew only Schirra well, from working with him on the navy's preliminary evaluation of the McDonnell XF-4H, which later became the F-4. Even after the first suborbital Mercury flights in 1961, Armstrong thought "we were far more involved in spaceflight research than the Mercury people.

"I always felt that the risks we had in the space side of the program were probably less than we had back in flying at Edwards or the general flight-test community. The reason is that we were exploring the frontiers, we were out at the edges of the flight envelope all the time, testing limits. That isn't to say that we didn't expect risks in the space program. But we felt pretty comfortable because we had so much technical backup and we didn't go nearly as close to the limits as much as we did back in the old flight-test days."

A significantly higher rate of fatalities in the world of flight test supports Armstrong's contention. Not a single American astronaut was lost in an actual space flight until the loss of the seven members of the Space Shuttle *Challenger* crew in 1986. In contrast, in 1948 at Edwards alone, thirteen test pilots were killed. In 1952, sixty-two pilots died there in the span of thirty-six weeks. Armstrong might well have chosen to remain in the challenging world of test flying. Neil's final X-15 flight occurred on July 26, 1962, but there were 135 more X-15 flights to follow in the next six years, before the program ended in October 1968. In November 1960, NASA named Armstrong a member of the air force/NASA Dyna-Soar "pilot consultant group." Although the air force eventually complicated Dyna-Soar by trying to make it operational, its original intent was for research. Its objective was demonstrating controlled *lifting* reentry, a technique that created enough aerodynamic lift to give a transatmospheric vehicle the cross-range necessary to maneuver down to established runways, as the Space Shuttle would later do. Lifting reentry provided a flexibility that the nonlifting, blunt-body ballistic capsules sorely lacked. Because it pushed technology so fast and so hard in so many areas, Dyna-Soar served as a critical focal point for a wide range of future-oriented aerospace R&D.

Although NASA engineers at Dryden had considered the possibility of air-launching the X-20 from a B-52 or B-70 mother ship, NASA and the air force decided to loft the boost-glider into orbit on top of a Titan III. This raised the problem of how to rescue the X-20 and its crew if some emergency, like a fire or booster failure, occurred on the launchpad. (Such a nightmare scenario almost happened in the Gemini program, when, in December 1965, Wally Schirra in Gemini VI-A came awfully close to yanking the seat ejection ring between his legs and blowing himself and fellow astronaut Tom Stafford up and off their Titan.) Because Dyna-Soar was a winged vehicle capable of real flying, a pilot inside the X-20, once blasted clear of his Titan booster, could perhaps fly the vehicle down safely to a runway landing.

Armstrong conceived of a way to test the rescue concept. The small escape rocket being planned for Dyna-Soar shot the X-20 up several thousand feet, and it occurred to Armstrong that "maybe we could duplicate that. So I set about finding out if we could, and seeing if we could get an airplane for it."

The F5D Skylancer was an experimental fighter built by Douglas that the navy had decided not to produce. Only four of the aircraft were ever built, with two of the prototypes given to NASA in late 1960. Armstrong flew one of the F5Ds on September 26, 1960, during a visit to NASA Ames. Neil realized immediately that the F5D could serve particularly well in a study of Dyna-Soar abort procedures because its wing planform was a good match for the X-20's slender delta-shaped wing. Armstrong knew it took a plane like the F5D whose gear could extend out fully and safely at high speed, over 300 knots (345 mph). Armstrong began flight tests in the F5D in July 1961, just shortly after Karen's illness was diagnosed. While he and Janet were initiating what became the little girl's first round of X-ray treatment, Neil occupied his mind with the problem of figuring out what kind of separation flight path and landing approach would best bring the X-20 down safely. Between July 7 and November 1, 1961, Armstrong made no fewer than ten test flights in the F5D. By early October, he had developed an effective maneuver for the abort. Neil simulated the act of being shot away by the escape rocket by making a steep vertical climb in the F5D to seven thousand feet. At that point, he pulled on his control column until the "X-20" lay on its back. Rolling the craft upright, he initiated the low L/D approach. Landing came on a specially marked area on Rogers Dry Lake, a parcel that simulated the ten-thousand-foot landing strip at Cape Canaveral.

Late in the summer of 1961, NASA installed a Cinerama camera into the nose of the F5D to film the abort procedure. On October 3, 1961, Armstrong demonstrated the Dyna-Soar rescue during a special visit of Vice President Lyndon B. Johnson to Edwards. Much of this flying came during the troubled times following the diagnosis of Karen's tumor.

It was six weeks after Karen's death, on March 15, 1962, that the air force and NASA jointly named Armstrong as one of the six "pilot-engineers" for Dyna-Soar. The only other NASA pilot named was Milt Thompson, so the selection was quite an honor. The other four pilots were air force designees. At thirty-one, Armstrong was the youngest of the group. If a small fleet of X-20s actually got built, the sextet would be the prime contenders for first flying the X-20 when it came on line, then scheduled for 1964.

As Armstrong looked into his professional future following his daughter's death, he saw three choices: "I could have kept flying the X-15. I was also working on the Dyna-Soar. That was still an on-paper airplane, but it was a possibility. Then there was this other project down at Houston, the Apollo program. . . . Apollo was just so overpoweringly exciting that I decided to give up these other opportunities to pursue it, even though I knew it may never happen."

Armstrong admitted that the growing excitement surrounding Project Mercury had something to do with his decision. On February 20, 1962, three weeks after Karen's funeral, Mercury astronaut John H. Glenn orbited the Earth three times in *Friendship 7.* No celebration since that for Armstrong's hero Charles A. Lindbergh in 1927 matched the national outpouring in Glenn's honor. If ever there was a time to entice a pilot out of his airplane and into a spacecraft, this was it. "Astronaut Glenn" appeared on the cover of countless newspapers and magazines in the winter and spring of 1962, including *Life*. Armstrong deliberated for around five months over his decision to apply for astronaut selection. All the while, he continued to grieve for his daughter—and he continued to fly.

Armstrong claimed there were no noticeable ill effects on his work at Edwards, but in the months immediately following Karen's death, he had a few flying mishaps there. Twice during Karen's illness, Armstrong piloted the X-15 rocket plane, his third and fourth of seven X-15 flights overall. Both flights came off without any hitch, at least not in Armstrong's performance.

Preparing for an X-15 flight took high intensity from

everyone involved, but no one felt the pressure like the pilot. Furthermore, the X-15 flight with Armstrong at the helm was to be the first run of the number-three aircraft. Already the X-15-3—or more precisely, the plane's powerful new XLR-99 rocket engine—had a checkered history, including an explosion on its test stand in June 1960. The investigators determined that a frozen regulator, a faulty relief valve, and a rapid buildup of back pressure caused the center structure of the X-15's ammonia tank to ram and smash open the control system's hydrogen peroxide sphere. Not until the entire pressurizing and pressure release systems were thoroughly analyzed, redesigned, and tested could another pilot step into the X-15 cockpit. By the time Armstrong got into the plane, the first flight of X-15-3 had been delayed by sixteen months, at a cost of $4 million.

Following a new round of engine ground testing, Neil had every reason to think that the X-15-3's problems had been solved.

Rebuilding the aircraft gave North American the opportunity to update the newest X-15 airplane's research equipment and outfit it with the MH-96 "black box" that Armstrong had been helping Minneapolis-Honeywell develop for the program. Flight testing the innovative adaptive control system became the primary purpose of Neil's December 1961 flight.

The flight was scheduled to launch on December 19, but was aborted when instrumentation involving the X-15's ball nose did not read correctly, so it got pushed back to the next day. The flight was not without problems. Immediately upon being dropped from a B-52 over Silver Lake, all three axes of the new stability augmentation system on the MH-96 disengaged and "a severe right roll occurred with accompanying yaw and pitch excursions."

Armstrong recalled that the failure did not cause him much trouble: "One of the aspects of the MH-96 was its reliability. It was a system designed to run for 76,000 hours between failures. That was a medium-speed flight. I think it was faster than I had ever flown before, though," up to a speed of Mach 3.76, or 3,670 mph.

Armstrong landed the X-15 after less than ten and a half minutes, having flown a distance of 150.9 miles. The highest altitude reached was 81,000 feet. Armstrong brought the X-15-3 down gingerly onto Rogers Dry Lake. Armstrong's next X-15 flight came on January 17, 1962, a week and a half before Karen's death. The flight was again to evaluate the MH-96 system and was the first time for Armstrong past Mach 5. It was also the first time he ever flew above 100,000 feet. In fact, he surpassed both marks, with a speed of Mach 5.51 and an altitude of 133,500 feet. Launched by a B-52, the X-15-3 traveled a distance of 223.5 miles in a little less than eleven minutes before touching down safely.

His X-15 flight on January 17, 1962, was Armstrong's last flight of any kind until a week after Karen's funeral, when on February 6 he took an F5D up over Edwards for a low L/D approach. In the entire month of February, Neil flew only three other days, on the twelfth, thirteenth, and sixteenth. Armstrong worked on Dyna-Soar in Seattle from February 26 to March 20.

Upon returning to work at Edwards on Monday, March 23, Armstrong immediately began to prepare for his next X-15 flight. Most of his flying involved "touch-and-go" landings in an F-104, which amounted to practice landings of the X-15. Because of various problems with the planes, not until April 5 did Armstrong make the flight. Then, just as he was being dropped at altitude north of Death Valley, his rocket engine did not ignite. In an X-15, there was only time enough for one relight.

The remaining time until touchdown was required to complete the jettison of the propellants. If a second relight was attempted, the X-15 would still have some propellants in the tanks at landing, which, in Armstrong's words, "was not desirable." He recalled what "sure seem[ed] like a long time the second time for that engine to light up."

Accelerating to a top speed of Mach 4.12, Neil thundered up to 180,000 feet. It was the first time he had reached a high enough altitude to fully integrate the MH-96 reaction controls. The test flight spanned 181.7 miles in a little over eleven minutes before landing.

The airplane still had not been flown to the point of testing the MH-96 system limit, or "g limiter," in part designed by Armstrong, to prevent the pilot from exceeding 5 g's, and he "felt the obligation to demonstrate every component and aspect of the MH-96."

It was this commitment that led to Armstrong's making what some came to feel was his biggest pilot error in the X-15 program.

Flight 3-4-8 occurred on April 20. Armstrong remembered, "It was the highest I'd ever gone"—to 207,500 feet, an altitude that remained his highest until Gemini VIII. "The views were spectacular. The system ran pretty well up there. The reaction control systems were operating satisfactorily 'across the top.' It kept a good attitude reference. Everything worked well. It was well outside the atmosphere so that we were flying completely on reaction controls. Aerodynamic controls were completely ineffective, like flying in a vacuum." Coming down from peak altitude, part of the flight plan was to check out the g limiter. Armstrong explained, "I thought I got the g's high enough, but it was not kicking in. That was my job, to check out that system."

Armstrong let the X-15 nose up just a little, causing

it to balloon to a high enough altitude—roughly 140,000 feet—where "the airplane returned to the wings-level attitude with essentially no sideslip. At about fifteen or sixteen degrees angle of attack and four g, I elected to leave the angle of attack in that mode and I was hoping that I would see the g limiting in action. We had seen g limiting on the simulator operation at levels approximately four g so I left it at this four-g level for quite a long time hoping that this g limiting might show up. It did not, and apparently this is where we got into the ballooning situation."

Over the radio, "NASA 1" told Neil rather emphatically, "We show you ballooning, not turning. Hard left turn, Neil!" "Of course I'm trying to turn," Neil explained, "but nothing's happening. I'm just on a ballistic path and I get over on to a very steep bank angle trying to pull down into the atmosphere. But the aerodynamics are not doing anything. The plane's going to go where it's going to go. It's on a ballistic path. I rolled over and tried to drop back into the atmosphere, but the aircraft wasn't going down because there was no air to bite into.

"I had no reason to suspect that ballooning would cause any trouble, because I had fiddled around with this lots of times in the simulator and never, never had any kind of problem with bouncing out like that."

Eventually the X-15 fell back down into the atmosphere where Armstrong was able to start making the turn. But by that time, Neil recalled, the airplane had gone "sailing merrily by the field"—at a speed of Mach 3! By the time he rolled into a bank, pulled up the angle of attack, and started to turn back in a northeasterly direction toward Edwards, Armstrong found himself approaching Pasadena. Subsequent Edwards lore had Neil flying as far south as the Rose Bowl, but Neil was forty-five miles south of Edwards though still above

100,000 feet. Downward visibility was very limited, so Neil did not know how far south he was, but he was nowhere close to the Rose Bowl.

"It wasn't clear at the time I made the turn whether I would be able to get back to Edwards. That wasn't a great concern to me because there were other dry lakes available. My easiest choice was to land at a lake called El Mirage, and I could easily get there. The only other alternative at that point would have been Palmdale municipal airport, and I didn't want to get into their traffic pattern." So Armstrong committed himself to trying for Edwards: "After I got on the . . . northbound track for Edwards, it was clear that I was going to be able to try to go in. I'd have to make a 'straight in.' "

Armstrong's X-15 flight of April 20, 1962, established X-15 program records for the longest endurance (12:28:07) and for longest distance (350 miles, ground track). Local Edwards lore had Neil trekking right down amid the Joshua trees as he made his landing on the southern tip of Rogers Dry Lake; in fact, the jest was that the Joshua trees were passing *above* Neil. Fellow NASA test pilot Bruce Peterson was situated on the north lake bed waiting to send up locator flares. "Neil was supposed to land on runway 18 on the north lake bed," Peterson related. "Then I heard on the radio that he was going to go to the south lake, so I got in my vehicle and I must have been doing a hundred miles per hour, racing down that lake bed to see if I could get to the south lake bed and throw some flares. I watched him come in and I knew he was close to the edge of the lake."

People who were not even at Edwards on April 20 came to believe that Armstrong made it back by the skin of his teeth. NASA pilot Bill Dana, who was to fly the X-15 sixteen times, had taken an F-104 to Albuquerque, New Mexico, that day, "but I sure heard about it when I got back!" Air Force test pilot Pete Knight did not see any of the flight, either, but "I heard about it when fellow pilots started teasing Neil for his 'record cross-county flight.' We thought it was funny at the time, to bounce back up and get into the thin air where you can't turn. It's not too bright." Major Bob White, who was flying chase for Armstrong in an F-100, admitted that he "kind of giggled over it a little bit" and "never did discuss the overshoot with Neil because it might have been a little embarrassing." When the report of the flight went to Washington, what Neil did sounded like a "screw-up" to NASA officials who read it, including Brainerd Holmes, director of the Office of Manned Space Flight. "I just assumed that was because Holmes really didn't understand," Neil felt. "He didn't have any technical knowledge of the problem involved."

Armstrong would later explain, "It might have been well advised for me to think, 'Well, if the g-limiting isn't kicking in, I'm not going to push it. I'll leave that to the next flight and try it again.'" Typically, Armstrong regarded his infamous "overshoot to Pasadena" as "a learning thing."

Just four days after his X-15 overshoot, Armstrong was involved in a second incident, indicating that Karen's death may have temporarily affected his job performance. On April 24, Armstrong and Chuck Yeager made their only-ever flight together.

The X-15 flight plan necessitated emergency landing sites all along the trajectory. One of the farthest flung was Smith Ranch Dry Lake, located some 380 miles due north of Edwards.

Conditions on a dry lake bed needed to be carefully

checked out, especially during the wet winter season. Teams of inspectors would walk the lake bed, dropping six-inch-diameter lead balls from a height of five feet. Measuring the diameter of the depressions made by the balls and comparing them to measurements that had been made on a firm, usable lake bed, the inspectors determined whether the ground would support the fifteen-ton weight of the X-15.

The winter of 1962 was a particularly wet one in the western desert. Many roads leading to and from Edwards were closed, and very little flying took place.

On Monday, April 23, NASA's Joe Walker took an F-104 up to Smith Ranch Dry Lake to check it out for possible emergency use by X-15-1, which Walker was scheduled to fly down from Mud Lake. The NASA R4D Gooneybird, flown by Jack McKay and Bruce Peterson, reported that day that Smith Ranch might be sufficiently dry to support a landing.

Paul Bikle, the head of the FRC, wanted to be absolutely sure of the condition of Smith Ranch Dry Lake for Walker's flight. On the twenty-fourth, after White's X-15-2 flight was canceled due to clouds, Bikle made a phone call to Colonel Chuck Yeager, the new commander of the Aerospace Research Pilots School at Edwards—and who also, incidentally, had been copiloting the launch B-52 just that morning. According to Yeager, he told Bikle the lake bed was too wet, but that he would attempt a landing if Armstrong flew and as long as he wasn't responsible for anything that happened. Armstrong sat in front, Yeager in back of the T-33. On the sunny and warm afternoon, both men wore just flying suits and gloves.

"We went up there and looked it over," Armstrong recalled, "and it looked like it was damp on the west side but pretty dry on the east side. So I said to Chuck,

'Let's do a touch-and-go and see how it goes.'" The touch-and-go took place with absolutely no trouble. Neil landed, ran the wheels over the surface, added power, and took off. The problem for Armstrong came next, when Yeager told him, "Let's go back and try it again, and slow down a little more."

"Okay, we'll do that," Neil agreed. "So we landed a second time and cut the power back and slowed down, and then I could feel it starting to soften a little bit under the wheels so I added some throttle, and then it settled some more, and I added some more throttle. Finally, we were at a full stop, full throttle, and we started to sink in." Armstrong related, "Chuck started to chuckle. Slowly he got to laughing harder. When we came to a full stop, he was just doubled over with laughter."

As Armstrong and Yeager got out of the T-33, an air force pickup truck immediately drove up to them. "The driver came out and he had a chain," Armstrong remembered. "So we put it around the nose gear and hooked it up to the truck and tried to pull the airplane out of the mud, unsuccessfully. We couldn't do it, so we just sat there on the wing." Neil took eight-millimeter film of the plane stuck in the mud with an inexpensive movie camera. The mishap took place at about 3:30 in the afternoon. With the sun dropping behind the high mountains to the west, the temperature fell quickly. For men wearing only thin flying suits, it soon grew cold. "Any ideas?" Yeager claimed he asked Armstrong, with Neil grimly shaking his head no. Sometime after 4:00 P.M., they heard the sound of NASA's Gooneybird approaching. Because Edwards had not heard from the T-33, NASA radioed McKay and Dana to fly over to Smith Ranch and take a look. Bill Dana recalled the "ribbing" that Neil took from Yeager, and that Neil "did not rise to the bait." It was clear to both Dana and

McKay, as it certainly was to Armstrong, that "Yeager took delight in Neil's embarrassment."

In his autobiography and in interviews, Yeager expressed harsh sentiments toward Armstrong, culminating in this opinion: "Neil Armstrong may have been the first man on the Moon, but he was the last guy at Edwards to take any advice from a military pilot." To which Neil only responded wryly, "On this occasion at Smith Ranch, I did take his advice!"

Back at the High-Speed Flight Station, Armstrong seemed to be suffering through a prolonged streak of bad luck. On Monday, May 21, Neil returned to work following his family vacation in Ohio.

Joe Vensel told Armstrong to fly up and inspect Delamar Lake, about ninety miles north of Las Vegas. After a half-hour flight in an F-104, Neil set up an approach that would allow him to practice his dead-stick landings. "I did it just like we always did," he recalled. "We made our flare and came down steeply just like the X-15, simulating putting the gear down in the middle of the flare, touching down, and then adding power and taking off. On this occasion I was doing that, but I was looking into the sun and the glare was very difficult."

Very few of the intermediate lake beds benefited from painted stripes and other markings, as did the regular runways on the big dry lakes near Edwards. From one lake bed to another, the texture of a surface could vary dramatically as could the surface cracks in its clay crust. Every experienced desert pilot knew that landing on a dry lake was like trying to judge height above glassy water. Two factors contributed to the "accident." Armstrong failed to judge his height precisely enough; and he didn't realize that, when he extended his landing gear during the flare, the gear did not extend fully and lock in place, causing the fuselage to smack into the lake bed.

"So I lost hydraulic pressure," Armstrong explained. "I wanted to leave the gear down; I couldn't pick it up, anyway. I couldn't make it back to Edwards on the fuel I had. I decided to go to Nellis Air Force Base, near Las Vegas, which was a lot closer."

His radio antenna gone, Armstrong could not communicate: "So I had to make a no-radio approach where you go over the field and waggle your wings, and the people in the tower, they're supposed to see you and realize that you're making a no-radio approach."

What Armstrong did not know was the loss of hydraulic pressure had triggered the release of his emergency arresting hook. If Armstrong had known his arresting hook was down, his landing at Nellis AFB would have come off trouble-free; after all, he was a naval aviator with loads of experience making tailhook landings. The Nellis arresting gear consisted of a steel cable attached to a long length of ship's anchor chain, each link weighing over thirty pounds.

"There was a good jolt when I hit it," Armstrong related, "and it was completely unexpected because I didn't even think about the hook being down, because I couldn't see exactly what my situation was." Down the runway for hundreds of feet, this way and that, links of heavy, broken anchor chain went careening like desert tumbleweeds. The F-104 stopped dead in its tracks.

It took the air force thirty minutes to clear the runway and considerably longer to rig a makeshift, interim arresting gear. Driven to the building where the base operations officer was on duty, Armstrong took off his gear, explained what happened to the perturbed base ops officer, and mustered his nerve to telephone back to NASA to report his accident. By then everyone at NASA had been fearing the worst. The Edwards control tower had no information. A few minutes later the tower reported that Neil had encountered a problem, but had landed safely at Nellis. Soon, NASA test pilot Milt Thompson went to pick him up in the only two-seat aircraft available, an F-104B. However, a strong crosswind caught the plane, forcing Thompson to plunk down hard enough to blow the left main tire. A fire truck and base ops vehicle quickly joined Thompson's crippled airplane as he parked it off the center taxiway. The only person who felt worse than Thompson at this moment was Armstrong as he watched the base ops officer shut down the runway for the second time that afternoon.

NASA now had two stranded pilots. It had no choice but to send a third plane to Nellis. Unfortunately, the only available plane was a T-33, another two-seater. As Bill Dana headed in, it looked like he was going to overshoot the runway. "Oh no, not again!" lamented the base ops officer, while Neil hid his head in his arm and Thompson watched "transfixed." Fortunately, Dana got the airplane stopped in time. "Please don't send another NASA airplane!" the air force officer begged. "I'll personally find one of you transportation back to Edwards."

True to his word, when an air force C-47 happened to be passing through Nellis on its way to Los Angeles, the ops officer expedited the refueling to haul Thompson away. For years thereafter, the base ops officer related "the tale of the three hot-shot NASA test pilots" that ruined his runways.

The day after the debacle in Nevada, Armstrong left on a two-week trip to Seattle, returning on June 4. His first flight following the Nellis Affair came on June 7, when he piloted an F-104 in the company of Bill Dana.

Armstrong by this time had decided to apply for astronaut selection. NASA formally announced that applications would be accepted for a new group of astronauts on April 18, 1962, two days before Armstrong's X-15 overshoot. Very possibly, Neil knew nothing about NASA's announcement until April 27. On that day, the FRC's in-house newsletter ran a story entitled "NASA Will Select More Astronauts," specifying an additional five to ten slots. The new pilots would participate in support operations for Project Mercury, and then join the Mercury astronauts in piloting the two-man Gemini spacecraft.

The requirements for selection could not have suited Armstrong better if had they been written for him specifically. The successful applicant had to be an experienced jet test pilot—preferably one presently engaged in flying high-performance aircraft. He must have attained experimental flight status through military service, the aircraft industry, or NASA. He had to hold a college degree in the physical or biological sciences or in engineering. He needed to be a U.S. citizen who was under thirty-five years of age and six feet or less in height. His parent organization, in this case NASA's Flight Research Center, had to recommend him for the job.

The director of the Manned Spacecraft Center in Houston, Robert R. Gilruth, would be accepting applications until June 1, 1962. Pilots meeting the qualifications were to be interviewed in July. Those who passed a battery of written examinations on their engineering and scientific knowledge were then to be thoroughly examined by a group of medical specialists. The training program for the new astronauts was to include work with design and development engineers, simulator flying, centrifuge training, additional scientific training, and flights in high-performance aircraft. Virtually the entire training syllabus involved activities that Armstrong had already done.

From May 9 to 11, 1962, Armstrong was in Seattle to attend the Second Annual Conference on the Peaceful Uses of Space, an event cosponsored by NASA and several other aerospace societies to explore the potential international applications of space science and technology. Armstrong, Joe Walker, Forrest Petersen, and Bob White, all members of the "100,000 Foot Club," gave a presentation on "The X-15 Flight Program." Other speakers at the conference included NASA Administrator James E. Webb, Vice President Lyndon B. Johnson, and other notables. Attendance at this conference and at the conjoining Seattle World's Fair could not help but impress Armstrong. The star attraction on the second day of the Fair was astronaut John Glenn, fresh off his orbital flight for Project Mercury. "Throngs of awestruck admirers" lined Seattle's streets to catch a glimpse of him.

One member of the selection panel for the second group of astronauts was Dick Day, an FRC flight simulation expert with whom Armstrong worked closely; in February 1962, Day had transferred from Edwards to Houston to become assistant director of the Flight Crew Operations Division at the Manned Spacecraft Center. Day oversaw all astronaut training programs and served as the astronaut selection panel's ad hoc secretary. According to Day, Armstrong's application for astronaut selection missed the June 1 deadline. "There were several people from Edwards who had gone on to Houston. His former boss Walt Williams, for one. Walt had gone on to be the operations director in Houston for the Space Task Group. He wanted Neil to apply, and I wanted Neil to apply. Neil's application came in late, definitely, by about a week. But he had done so many things so well at Edwards. He was so far and away the best qualified, more than any other, certainly as compared to the first group of astronauts. We wanted him in."

When it came in, Day slipped it into the pile with all the other applications prior to the selection panel's first meeting. Virtually everyone at Edwards thought that Armstrong was a great choice to become an astronaut, especially when it was announced in early June 1962 that he was to receive the prestigious Octave Chanute Award. Presented by the Institute of the Aerospace Sciences, the Chanute Award went to the pilot that the IAS deemed had contributed the most to the aerospace sciences during the previous year. According to Dick Day, Paul Bikle, the director of the Flight Research Center at Edwards and Day's former boss, did not think so positively of Armstrong. Bikle chose not to recommend Neil for astronaut selection because, in his mind, Neil's immediate past record in the air raised some serious concerns about his performance. In late May 1962, Bikle had even pulled the plug on Neil's scheduled test flight in the U.K. of Handley-Page's new HP-115 supersonic research airplane. Dick Day was also aware of that.

Another key person at the Manned Spacecraft Center in Houston was Christopher Columbus Kraft Jr., the original director of NASA's manned spaceflight operations at Mission Control in Houston. Upon graduating with an aeronautical engineering degree from Virginia Tech in 1944, Kraft had gone to work in the Stability and Control Branch of NACA Langley's Flight Research Division, where he rubbed elbows with such talented flight test engineers as Bob Gilruth, Charles Donlan, and Walt Williams, men who in the summer of 1958, following Sputnik, took Kraft with them into the Space Task Group, which planned and administered Project Mercury. No one understood the mentality of test pilots or astronauts better than these four men. Though

Kraft did not serve on the selection panel for the second group of astronauts, he had a lot to do with defining the selection criteria. "Charles Donlan was in charge of that," Kraft recalled, "and he talked to me about it because he valued the association I was having with the first seven astronauts. I emphasized that we should go talk to the people who know the candidates, know their character, and know their capabilities. People like Gilruth, Williams, and myself were looking for qualified test pilots."

Kraft "hardly knew" Armstrong out at Edwards. "I didn't know about his daughter's death. I did know he had had a few accidents—what pilot hasn't. But I never associated them with any psychological event. What I knew was that Walt Williams thought Armstrong was first-rate. When we met him, Gilruth and I and everybody else felt the same way.

"He would make a fine astronaut."

CHAPTER 11

I've Got a Secret

While at work in his office at Edwards in early September 1962, Neil got a phone call from Deke Slayton, head of the astronaut office at the Manned Spacecraft Center, then still under construction on Clear Lake, southeast of Houston.

Deke came right to the point: "Hi, Neil, this is Deke. Are you still interested in the astronaut group?"

"Yes, sir," replied Armstrong.

"Well, you have the job, then. We're going to get started right away, so adjust your schedule and get down here by the sixteenth." Slayton told Armstrong that he could tell his wife, but otherwise to keep the news quiet.

Neil's parents did not get the news about their son becoming an astronaut until sometime that weekend when they received a phone call from a NASA public relations officer who was helping CBS set up the couple's Monday evening appearance on the television show *I've Got a Secret.* (After panelist Betsy Palmer guessed the secret—that the Armstrongs' son that day had just been named an astronaut—host Garry Moore offered the prophetic statement, "Now, how would you feel, Mrs. Armstrong, if it turned out—and, of course, nobody knows, but it turns out—that your son is the first man to land on the Moon? How would you feel?" To which Viola Armstrong answered, "Well, I guess I'd just say, God bless him and I wish him the best of all good luck.") As for the call from Slayton to Neil, "I was happy to get that call," Armstrong related.

Slayton's call really could not have surprised Neil much. As early as midsummer 1962, newspapers had been reporting that Armstrong was going to be named the "first civilian astronaut." NASA officials later denied this, conceding that Armstrong was "definitely on the list" of 32 men out of an applicant pool of 253 who had survived the preliminary screening, but indicating that no final selections had been made. Many close observers of NASA did not believe the denial. All through the selection process, Neil's identification with the ways of NACA/NASA had been an advantage for him. Armstrong was relatively confident that NASA would choose him as one of its next astronauts, but he could not be sure: "A number of us had combat experience. My education level was, I thought, competitive. My experience was rather broad, and having flown the rocket airplanes and things like that, and being involved in a variety of test-flight programs. Nevertheless, the areas that I didn't know how well I compared were physical, emotional, psychological, and perhaps how I was perceived by other people. I didn't know how I would grade in those categories. And any one of those could certainly evict you from the program."

During the four-month stretch from early June 1962, when he turned in his astronaut application, to the day in September when Slayton called him, Armstrong was too busy to worry much about whether he was going to become an astronaut. He spent the second week of June at the Lovelace Clinic in Albuquerque, New Mexico, ostensibly taking his annual NASA test pilot's physical, but with the results of certain tests being relayed—

unbeknownst to him—back to the Manned Spacecraft Center for evaluation as part of the astronaut selection process. Back at Edwards, Neil made a series of flights; picked up his Chanute Award in Los Angeles; and flew a test program for the Saturn rocket then under development.

On July 5, 1962, Armstrong left for an AGARD meeting in France, where he presented a paper coauthored by Ed Holleman on "Flight Simulation Pertinent to Piloted Space Vehicles." Returning from the Paris conference, Armstrong spent all of his time preparing for his final flight in the X-15, which resulted in the highest Mach number Armstrong ever attained in the X-15 program—Mach 5.74, or 3,989 mph. Just at the time Neil was pushing the black rocket plane into the range of maximum speed, smoke started to infiltrate the cockpit, but he managed to land safely.

There was barely enough time for Neil to write up his pilot comments before he had to leave for Brooks Air Force Base in San Antonio. At Brooks he underwent an exhausting week of medical and psychological tests that went a long way toward finalizing the selection of the new astronauts. In Armstrong's opinion, "there were some painful experiences. My sense at the time was that some of these things must have been specially designed to be medical research rather than diagnostic techniques."

A few of the exams were especially diabolical. "There was one," Neil recalled, "where they syringed ice water into your ear for a long period of time until you sort of got uncaged, and another where you had your foot in ice water for quite a while. There were a lot of strange tests like that."

One psychological test that Armstrong remembered was an isolation test: "They put you in a black room where all sensory signals were removed. There was no sound, no light, and no smell. They told you to come out after two hours." Neil applied engineering principles: "I tried to compute a way to figure out how long two hours were. So I used the song, 'Fifteen Men in a Boardinghouse Bed.' I didn't have a watch or anything, but I sang that song until I thought about two hours were up. Then I knocked on the door and shouted, 'Let me out of this place!'" On August 13, Armstrong traveled to Houston's Ellington Air Force Base for a final round of medical and psychiatric tests. There, Armstrong first came before NASA's astronaut selection panel, a group that included Deke Slayton, Warren North, Walt Williams, and Dick Day. Occasionally John Glenn or Wally Schirra drifted in and out of the room. Armstrong related, "I didn't find it at all difficult or pressuring or anything. I found it a natural conversation about the kinds of things I was interested in at the time."

All thirty-two of the finalists (thirteen navy, ten air force, three Marines, and six civilians) gathered one evening for dinner with a small party of leading officials from the Manned Spacecraft Center. Armstrong remembered, "I didn't know too many of the people. I knew Schirra from the XF4H-1 evaluation. I knew some others a little," including Gus Grissom, who had flown at Edwards. Armstrong knew John Glenn and Al Shepard only from the occasional flight-test event. Scott Carpenter was the only Mercury astronaut that Neil had never met.

Only the four Mercury astronauts who had already made their spaceflights—Al Shepard, Gus Grissom, John Glenn, and Scott Carpenter—had already flown farther out of the atmosphere than Armstrong had in the X-15; and Neil was the only one in the entire group who had ever flown a rocket plane or won the Octave Chanute Award. Back at Edwards, Armstrong quietly concentrated on his regular duties. He flew nearly every workday during the three-week stretch prior to Deke Slayton's call.

Neil arrived at Houston's Hobby Airport late on Saturday, September 15, 1962. Neil remembered, "It was completely quiet. Nobody was to know that we were coming in, or that it was going to be announced." As instructed by NASA, Neil checked into the Rice Hotel under the code name "Max Peck," as did all eight of the other selectees, all code-named "Peck." The next morning at Ellington, NASA's new class of astronauts first assembled under Slayton's direction. Walt Williams, the head of flight operations, ran the men through their job description. Bob Gilruth, the director of the Manned Spacecraft Center who had headed the Space Task Group from its inception, told them that with eleven manned Gemini flights on the schedule, at least four Block I Apollos (to be launched on the Saturn I), and a still undetermined number of Block II Apollos, including the one that would make the first lunar landing, "There'll be plenty of missions for all of you." Slayton warned them about some of the new pressures and temptations they would be facing. He told them to be careful about accepting gifts and freebies, especially from companies competing for NASA contracts. Shorty Powers, NASA's public affairs officer, and the "Voice of Project Mercury," ended by briefing the astronauts on the upcoming press conference. He then organized the nine men for the first of what became an interminable sequence of photo shoots.

The University of Houston's 1,800-seat Cullen Auditorium was filled to capacity for the announcement. Reporters and camera crews from all three of the major television networks, from the major radio broadcasting systems, from the wire services, from dozens of national and international newspapers and magazines crammed into the theater, waiting to learn the identities of America's new astronauts. Back on April 2, 1959, the newborn NASA had been taken by surprise by the public sensation surrounding the announcement of the original seven astronauts. This time the more seasoned agency was much better prepared for the media blitz. So, too, were the astronauts themselves.

The "New Nine"—Neil Armstrong, Air Force Major Frank Borman, Navy Lieutenant Charles Conrad Jr., Navy Lieutenant Commander James A. Lovell Jr., Air Force Captain James A. McDivitt, Elliot M. See Jr., Air Force Captain Thomas P. Stafford, Air Force Captain Edward H. White II, and Navy Lieutenant Commander John W. Young—were a truly remarkable group of men. In the opinion of key individuals responsible for the early U.S. manned space program, it was unquestionably the best all-around group of astronauts ever assembled. The educational level of the second group was dramatically higher than that of the Mercury Seven, and with exactly the emphasis on rigorous engineering that NASA's astronaut selection panel had sought. Many had undergraduate engineering degrees, and some had master's degrees. Armstrong had everything but his thesis completed toward a master's in aerospace engineering at the University of Southern California.

The group's experience as pilots and record in the world of flight testing was equally impressive; most had over two thousand hours in the air, and some had set records. Neil had amassed two thousand four hundred hours of flying time, about nine hundred of it in jets. He was the only one of the nine who had done any flying in rocket-powered aircraft.

The group's average age was thirty-two and a half; they weighed an average of 161.5 pounds per man; and their average height was five feet ten. At five-eleven and 165 pounds, Armstrong was slightly above average size for his group. All of the men were married, none of them had ever been divorced, and all of them had children. Armstrong's recollection was "that the questions at the press conference were typical, fairly unsophisticated questions—with answers to match." This two-pronged comment—the second part self-denigrating—illuminates much about what later became the misunderstood character of Armstrong's attitude toward the press.

NASA expected Armstrong to be at Cape Canaveral along with all the other new astronauts for Schirra's Mercury launch, but that was not scheduled until October 3. Leaving most of the preparations for the family's move to Texas in Janet's hands, Neil went right back to his job at Edwards. He flew every working day through the end of the month.

Armstrong's last flight as an FRC employee occurred on September 28, 1962, in an F5D. Following that weekend at home, Neil went by commercial air from Los Angeles, not to Houston, but to Orlando and thence on by car the short distance to Cape Canaveral, where he and the rest of the New Nine watched Schirra's *Sigma 7* Mercury flight go off without a hitch on October 3.

The next day Neil was back at Edwards, as his civil service orders called for his permanent change of station from the Flight Research Center to the Manned Spacecraft Center to be made not until between the eleventh and the thirteenth of the month. In two days' time, he and L.A.-based Elliot See made the 1,600-mile (pre–Interstate Highway System) drive to Texas in See's car. Neil rented a furnished apartment very close to Hobby Airport, then set off with the other new astronauts to inspect the manned space program's contractor facilities nationwide.

Returning to Juniper Hills through Los Angeles on November 3, Armstrong traded in both of his used cars and bought a used station wagon. The family's furniture and clothing had already been shipped to storage in Houston. Neil took Rick with him in the car, while Janet flew to Houston two days later. For the next few months, the Armstrongs lived in their furnished apartment until completion of their home in the new El Lago subdivision, a few minutes east of the Manned Spacecraft Center.

PART FOUR

ASTRONAUT

They say "no man is an island"; well, Neil is kind of an island. . . . Sometimes what he was thinking and his inner thoughts were more interesting to him than somebody else's thoughts were to him, so why should he leave his island, go wading out into the shallows to shake hands with somebody, when he's perfectly happy back in his little grass hut or wherever.

—MICHAEL COLLINS, GEMINI X AND APOLLO 11 ASTRONAUT

CHAPTER 12

Training Days

By the time NASA named Armstrong one of its nine new astronauts in September 1962, the idea of a manned lunar landing seemed probable. Triggering that transformation was a turbulent confluence of dramatic geopolitical events in the spring of 1961 that undermined respect for the fledgling presidency of John F. Kennedy and provoked Kennedy into making his astonishing commitment to a manned Moon landing.

On April 12, 1961, not yet fully three months into JFK's term, the Soviet Union stunned the world by achieving another space first. Just as it had back in 1957 with Sputnik, the U.S.S.R. beat the U.S. to the punch when cosmonaut Yuri Gagarin became the first human space traveler. Three days later, a plot to invade Cuba and overthrow the Communist regime of Fidel Castro failed miserably at the Bay of Pigs.

The resulting international criticism made JFK realize that only dramatic action would restore respect for America. Kennedy turned to the manned space program. The president saw in NASA and its astronauts a means to a political end. "Now it is time to take longer strides—time for a great new American enterprise—time for this nation to take a clearly leading role in space achievement, which in many ways may hold the key to our future on earth." With these historic words, expressed before a joint session of Congress on May 25, 1961, the president threw down the gauntlet: "I believe that this nation should commit itself to achieving the goal, before this decade is out, of landing a man on the Moon and returning him safely to Earth."

Very quickly after their selection, the New Nine (minus Elliot See, who could not attend) got a close-up look at everything NASA was doing to push Apollo ahead. They attended the launch of the third manned orbital Mercury flight made by Wally Schirra on October 3, 1962. Most of the group had never seen a rocket launch before. Nine hours and six orbits later, Schirra's *Sigma 7* splashed down in the Pacific Ocean near the USS carrier *Kearsarge.*

Three weeks later, the new group of astronauts headed out for the first in a series of contractor tours to the Pratt & Whitney Engine Facility at West Palm Beach, Florida, where the fuel cell for the Apollo spacecraft was being developed; to Baltimore, where the Martin Company was assembling Titan II rockets for the Gemini program; as well as to Martin's plant in Denver, where the ICBM version of the Titan II was being built. They then made their way to Aerojet-General Corporation in Sacramento, maker of the Apollo service module propulsion engine; to NASA's Ames Research Center south of San Francisco; and finally to the Lockheed Aircraft Corporation in Los Angeles. The builder of the Apollo launch escape system rocket, Lockheed was preparing to submit a bid for the Apollo lunar excursion module, a contract that eventually went to Grumman. The trips were grueling. The astronauts flew commercially, four on one flight, five on another. "They laid out lots of food and plenty of booze," remembered

Tom Stafford, "but the drinking never got out of hand." Most of the buildings at the Manned Spacecraft Center were still under construction, so for several months all of the astronauts worked out of rented offices in downtown Houston. Every Monday they had a pilots' meeting, chaired by Slayton, in which they got their weekly schedule.

The New Nine spent much of the time on the road. To become familiar with the Apollo launch vehicle—what would become the Saturn V Moon rocket—they visited NASA's Marshall Space Flight Center in Huntsville, Alabama. They met rocketeer Dr. Wernher von Braun for the first time. Just a few months earlier, von Braun had shocked his own people at NASA Marshall by shifting his support from earth-orbit rendezvous (EOR) to the more controversial lunar-orbit rendezvous (LOR) as the best way to land on the Moon. Then the astronauts spent a couple of days at McDonnell Aircraft Corporation in St. Louis. They saw how the Mercury spacecraft were built, and how McDonnell planned to design and build the new Gemini spacecraft. The New Nine received Apollo technical briefings from the Space and Information Systems Division of North American Aviation, Inc., at Downey, California, the prime contractor for the Apollo command and service modules. At Douglas Aircraft Company's facility in Huntington Beach, they saw how the S-IVB upper stage was shaping up for the Saturn IB and V.

The years 1963 and 1964 were about intensive basic training. As Armstrong remarked, "There wasn't anybody that had done this and could tell us how to do it, because nobody had the experience." Specialists in all the various areas pertaining to spaceflight "could tell us what they did know," and those who became systems experts could explain "the details of how the inertial guidance system or the computer or certain kind of engine valves and so on would operate, and how we might handle malfunctions.

"The early part of astronaut training was similar to navy flight training," Armstrong explained. "NASA felt that its new astronauts with little experience with the sophistications of orbital mechanics or the differences between aircraft and spacecraft needed a quick primer."

"With some of those subjects I felt fairly familiar," Armstrong stated. "Orbital mechanics, for example, I had already studied. Overall I didn't find the academic burden to be overly difficult."

Along with the academic curriculum, Armstrong and his classmates went through a number of other formal training programs. In Operations Familiarization, they toured all pertinent launch facilities and studied the rigorous prelaunch procedures at Cape Canaveral and the new Mission Control Center in Houston. In Environmental Training, they were exposed to acceleration, weightlessness, vibration and noise, simulated lunar gravity, and the experience of wearing a pressure suit. Contingency Training involved not only desert and jungle survival schools, but also learning how to use ejection seats and parachutes. Training in Spacecraft and Launch Vehicle Design and Development was accomplished through engineering briefings and mockup reviews.

To keep their piloting abilities and judgment in the cockpit as sharp as possible, the astronauts also went through an Aircraft Flight Training program. This they did by making regular flights in T-33, F-102, and T-38 aircraft assigned to MSC, based at Ellington AFB. They also rode out parabolic trajectories in the "Zero-g Airplane" (aka the "Vomit Comet"), a modified KC-135 aircraft that simulated zero-gravity conditions for roughly thirty seconds at a time. Neil had gone "over the top" in zooms in the F-104A Starfighter, but those flights into weightless conditions provoked nothing like the queasiness brought on by the abrupt changes in gravity due to a parabolic drop. Four days in the Zero Gravity Indoctrination Program conducted at Wright-Patterson AFB during the last week of April 1963 introduced Armstrong to floating free, tumbling and spinning, soaring across the cabin by pushing off the walls and bulkheads, eating and drinking in near zero-g, and learning to use tools.

In late September 1963, the New Nine attended the Water Safety and Survival School at the U.S. Naval School of Pre-flight in Pensacola. For the four naval aviators in the group—Armstrong, Lovell, Conrad, and Young—much of this training, including another confrontation with the Dilbert Dunker, was old hat. New for all of the astronauts was learning how to stay afloat and then get hooked up out of the water into a sling for a helicopter rescue while wearing a bulky pressure suit. (Gus Grissom's Mercury flight on July 21, 1961, had demonstrated just how dangerous such a water rescue could be.) None of the new astronauts had anything close to the experience riding a centrifuge that Armstrong did; many of them had never even seen one. NASA aerodynamics expert and aerospace vehicle designer Max Faget challenged the Mercury astronauts, "If you can get up to twenty g's, you will be my hero for life." As early as 1959, Armstrong had survived forces as high as fifteen g's.

The New Nine got its initiation to the miserable instrument on a four-day visit to Johnsville in late July 1963. During the stay, Neil made eight "dynamic runs" on the centrifuge, his time in the contraption totaling five hours. The astronauts also made several parachute jumps onto land and into water from airplanes at Ellington AFB that took them up and had them jump out from a height of three hundred feet. Furthermore, all of the astronauts had to adhere to the NASA directive to add helicopter flight to preparations for flight simulations of a lunar landing. By the end of the two-week period in November, Neil had flown in various types of helicopters, ending in three hours of solo time. Neil and Jim Lovell were driving back to Houston from their helicopter training in Pensacola on November 22, 1963, the day that President Kennedy was assassinated. Armstrong did not attend Kennedy's funeral; John Glenn was the official representative for the astronauts.

Armstrong's experience exceeded all of the other astronauts' when it came to the critical area of flight simulators. Deke Slayton assigned Neil to simulators when he handed out specialized technical assignments in early 1963.

In projects Gemini and Apollo, astronauts and spacecraft were to be committed to major, complex, and untried maneuvers that, of necessity, had to be carried through to completion, and usually on the first attempt. Simulation was vital to their success. Very little simulation was necessary for Project Mercury, which had as its specific objective the placement of a man in orbit and his return. Project Gemini, on the other hand, which came to life in 1962 as a bridge between Mercury and Apollo, entailed orbital rendezvous and docking. Both were more dangerous and complex maneuvers than sending a capsule into orbit. Being able to chase down another object in space and then linking up with it to take on fuel or other vital components was an absolute requirement. For that reason, learning how to rendezvous and dock, above all else, was Gemini's primary purpose. Without this ability, the other major objectives of Gemini—notably, long-duration flights and EVA were meaningless for Apollo.

No astronaut played a more vital role in the development of flight simulators for Gemini and Apollo than did Armstrong. Often Armstrong found that a simulator did not behave like the spacecraft actually would in flight: "One of the things that I particularly did with all the simulators was to find out if the designers of the simulator had mechanized the equations of motion properly. So I would always be flying the simulator into areas that most people would not ever go, to make sure that when you got to a discontinuity in an equation, there would not be a mathematical error that would cause the simulator to misbehave. I found a surprising number of times that they were not mechanized properly. That responsibility was natural for me because I had done the same work at Edwards."

As they had at Edwards, Armstrong's perspectives as a pilot added vital insights into simulator development. "The guys who were mechanizing the equations—sometimes contractors, sometimes NASA employees—oftentimes did not have the perspective of a pilot," Neil explained. "They couldn't visualize if you were pulling up to a vertical position and then rolling ninety degrees and then pitching forward back toward the ground, what that would mean to the pilot—what the pilot would actually see. Oftentimes they would mechanize the equations without any consideration of what was proper." Armstrong made significant contributions to the Gemini launch-abort trainer, a fixed-base simulator built in the astronauts' group training building at the Manned Spacecraft Center.

In setting up the system of specialization, Slayton understood that far too much was happening too quickly in the program for the astronauts individually to pick up on more than a small fraction of the technical whole. Deke's idea was for the astronauts to share knowledge and experience freely between their various assignments. Another responsibility the astronauts shared was NASA publicity and making appearances before professional audiences, press, and the adoring public. NASA public affairs officers early on accepted the astronauts' own idea of a rotating publicity schedule. Usually lasting a week at a time, the period of public appearances came to be known within the astronaut corps as "the week in the barrel."

Armstrong's first week in the barrel started on July 6, and included stops in Virginia; Washington, D.C.; New York City's World's Fair; and Iowa, where he made five presentations to scientific societies in one day. Exhausted by the incessant glad-handing, he flew back to Houston the next morning. This was an aspect of being an astronaut that he could have lived without.

Armstrong found the transition from research test pilot to astronaut—except for the public celebrity—relatively easy and comfortable. As time passed in astronaut training, Armstrong's peers respected his abilities as a pilot, engineer, and astronaut, admired his intelligence, and they wondered at his unique personality traits.

"My first impression of Neil was that he was quiet," stated Frank Borman. "Because he was so quiet and so thoughtful, when he said something, it was worth listening to. Most of us were, 'We're operational, let's-get-it-done people.' Of course, Neil was operationally oriented, too, but he would be more interested in trying to understand exactly what the inner mechanisms of the system were. Most of us came out of the same mold. But Neil was different." "Neil was a very reserved individual," Mike Collins recalled. "I think he was more thoughtful than the average test pilot. If the world can be divided into thinkers and doers—test pilots tend to be doers and not thinkers—Neil would be in the world of test pilots way over on the thinker side."

"Neil wasn't an expansive guy," Bill Anders offered. "He was totally professional—not overly warm but not cold. I don't remember him and I sitting around having a casual conversation about 'What are your kids doing?' Not that Neil would not have a drink or two with you. But he was a straight arrow in all the ways that counted. In my view, the character of the real person, Neil Armstrong, comes out generally higher than most of his colleagues."

"Neil is as friendly as you can get," said John Glenn. "He was laid-back, friendly, a nice guy, small-town just like where I came from. I don't think either of us put on any airs with one another." Glenn and Armstrong got paired up in early June 1963 for jungle survival training, organized by the USAF Tropical Survival School, at Albrook AFB in the Panama Canal Zone.

What Glenn and everyone else who ever spent any quality time with Armstrong enjoyed, and were surprised by, was Neil's sly sense of humor. John Glenn remembered, "I always got a kick out of Neil's theory on exercise." Armstrong joked with his friends that exercise wasted a person's precious allotment of heartbeats. Dave Scott, Neil's crewmate on Gemini VIII, recalls Armstrong coming into the astronauts' exercise room at MSC when Scott was sweating away pumping iron, getting onto a stationary bicycle, and setting its wheel at its lowest possible tension, and grinning at Dave, saying, "That a boy, Dave! Way to go!"

Dave Scott said, "He was very easy to work with. He was a very smart guy. He could make an analysis of a problem very quickly. The guy was really cool under pressure."

In Buzz Aldrin's words, "Neil was not the boisterous Pete Conrad; and he was not the authoritarian Frank Borman. You mostly had to wait for Neil to make a decision, and often you wouldn't have a clue as to what was going on in his head in the meantime. You just couldn't see through him. But even that opaque quality helped make him a great commander."

CHAPTER 13

In Line for Command

The first members of the New Nine to be assigned to a flight crew were Tom Stafford and Frank Borman. In February 1964, Slayton paired Stafford up with Mercury veteran Al Shepard, the first American in space, as the prime crew for the first manned Gemini mission, designated Gemini III. Assigned as the backup crew for Gemini III were Gus Grissom and Frank Borman. Although as anxious as the next guy for a flight assignment, Armstrong experienced no disappointment. "I had no expectation of getting it. I was so pleased to be associated with the program, because it was going. It was happening. It was exciting. The goals, I thought, were important to not just the United States, but to society in general. I would have been happy doing anything they told me to do." The crew assignments for Gemini III, in fact, had to be changed before preparation for the flight even got going. Because Al Shepard suffered from a chronic inner-ear problem that caused episodic vertigo, Slayton moved Grissom from backup to prime commander, and Gus picked John Young as his new mate. None too happy about the change, Tom Stafford became the backup for Gemini III, under the command of Mercury veteran Wally Schirra. Frank Borman was removed entirely from Gemini III and was held for a later, unspecified Gemini flight.

"I have my own ideas of how Deke assigned the crews," Armstrong remarked, "and it's not easy to explain. I don't think it was a matter of simply switching crews back and forth and alternating. Deke's principle concern was getting a qualified capable commander on each flight. He had the secondary objective of putting people in the other slots so that they would be getting the proper training, preparation, or experience to slide them into a more important slot in their next assignment."

Within the flight crews, "we tried to divide the responsibilities such that each person was about equally loaded. We tried for each person to be able to know how to do everything if he had to, but we divided the responsibilities such that each would go into their area in substantially more depth. The job of the commander differed principally because he had the responsibility for the decisions, just as the commander of a ship or commander of an airliner. He was always responsible for his craft.

"I think the key thought above all else was having commanders coming up that would be right for that job and with the right experience to enable them to have a degree of confidence. Deke always said, and I think he was completely right, that he had to take the position that the guys had all come through the process, were all qualified to fly, should be able to fly, and should be able to accept any task they were given.

"Having said that, Deke did say, and wrote in his own autobiography, that all that being true, he still wanted to get the best people into the best slots that were best suited for them. As an additional, less important technical reason. Deke felt an obligation to his Mercury colleagues. He always put Gus, Al, and Wally, particularly, as his first-line guys—and properly so. They were the first class of astronauts; they had been under the highest scrutiny; they should get their first pick."

Slayton's standard practice was to solicit the commander's input about potential crew members. "One rule we had," Armstrong noted, "was a guy could not be on two flights at the same time. The training preparation period was fairly extensive, so Deke would have a crew and backup crew completely committed for quite a long time period and they couldn't be touched for any other jobs. By the time you got three flights with crews assigned to them, you were using up twelve to about eighteen people, out of not so many. So he tried to think things through ahead of time. He thought that every flight was important, but particularly the early flights of a particular program—the early flights of Gemini, the early flights of Apollo. It was very important that we not stub our toe, because a failure early in a flight program jeopardized the entire program."

Neil was the only member of the New Nine who had a formal administrative responsibility within the Astronaut Office. In the office, Joseph S. Algranti ran aircraft operations, Warren North ran flight crew operations, and Slayton served as coordinator for astronaut activities. Helping out Deke was Al Shepard, who after his dizzy spells grounded him, became chief of the astronauts. Under Shepard in the organization was Gus Grissom, in charge of the Gemini group, and Gordon Cooper, in charge of the Apollo group. Deke gave Armstrong the responsibility for a third group called operations and training. Like Grissom and Cooper, Armstrong had a small number of fellow astronauts working under him. "Deke gave me the assignment of coming up with something that would help him understand how many crews would be needed at any given point in time," Armstrong recalled. "So I took a very simple approach. I took the launch dates as we projected them for Gemini and Apollo. There were several different kinds of Apollo missions at that point in time. So I used that kind of a schedule with just the launch dates and said, 'Okay, if that's right, then how many flight crews do we need?' I started at launch time and went back however many months that you would need the crew to be preparing. And none of these crew members were named—they were just individual A, B, C, D, and so on. I put in all these flights on a time line with block diagrams showing how many people had to be available. At the bottom I toted under each month how many astronauts were in flight status and how many available for flight."

Armstrong's schematic allowed Slayton to determine when additional astronauts needed to be brought into the program, culminating in Houston's announcement in June 1963 that NASA was looking for a new class of ten to fifteen additional astronauts. This third round of astronaut selection set the age requirement at thirty-four rather than thirty-five. Applicants no longer needed to be test pilots, as they would be in service of the broader scientific and engineering requirements of the Apollo lunar landing mission. Eight of the fourteen new astronauts selected in October 1963 turned out to be test pilots, five from the air force, Donn Eisele, Charles Bassett, Michael Collins, Theodore Freeman, and David Scott; two from the navy, Alan Bean and Richard Gordon Jr.; and one from the marine corps, Clifton Williams. The other six were all pilots with wide-ranging academic backgrounds and flying experiences: Edwin "Buzz" Aldrin Jr.; air force fighter pilot William Anders; navy aviators Eugene Cernan and

Roger Chaffee; and two civilians, ex–marine corps pilot Walter Cunningham and former air force pilot Russell Schweickart.

It was in the company of these outstanding astronauts that Armstrong would actually experience spaceflight: with Dave Scott on Gemini VIII, and with Buzz Aldrin and Mike Collins on Apollo 11.

On February 8, 1965, Armstrong received his own first assignment to a flight crew when Slayton named him as backup commander to Gordon Cooper on Gemini V. Although the mission's primary objective was demonstrating preparedness for a rendezvous in space, the astronauts intended to stay in space for eight full days. This was twice as long as the Gemini IV flight then being planned for Jim McDivitt and Ed White.

Serving with Armstrong in the backup role was Elliot See. See supported Pete Conrad, who would sit in the right seat next to Cooper on the prime crew. Armstrong said, "Because everything was based on beating the Russians and getting there by the end of the decade, the schedule was overwhelmingly important." Neil was "really pleased to be assigned to a flight, and quite satisfied to be in the position of backing up Gordon Cooper." After assignment to Gemini V, general training continued for Armstrong but now represented only about a third of his work time. A second third "had to do with planning, figuring out techniques and methods that would allow us to achieve the best trajectories and the sequence of events." The final third of his time involved testing: "thousands of hours in the labs and in the spacecraft and running systems tests, all kinds of stuff, seeing whether it would work and getting to know the systems well. A lot of the testing took place at two o'clock in the morning. We were spelling each other off. The four of us spent enormous amounts of time together, working out the details. I would not say that we never cracked a joke or talked about something off the project, but we were always ninety-eight percent focused on the job we had to do."

Getting ready for the backup role in Gemini V did not preclude Armstrong from serving in a supporting role in Gemini III, a flight made by Gus Grissom and John Young in the spacecraft *Molly Brown*. For Gemini III, the first manned mission in the Gemini program, Neil reported for a week of work at the worldwide satellite network tracking station in Kauai, Hawaii. Designated as a "primary" station, Kauai, the farthest north of the major Hawaiian Islands, transmitted verbal commands to the orbiting Gemini spacecraft. "Secondary" stations, such as the Caribbean Sea tracking station on Grand Bahama Island (GBI), handled radar and telemetry information only.

In the view of some NASA folks, such assignments were partly a way for Slayton to give his astronauts a little rest and relaxation. For Gemini III, Neil traveled to Hawaii a week before the launch to help perform tracking and communications simulations. Gemini III's objective was to demonstrate the ability of a spacecraft to change orbits by firing its maneuvering thrusters, a fundamental requirement in the rendezvous maneuver, which in turn was essential to the Moon landing. Specifically, Gemini III was to demonstrate the ability to move around effectively in space by making three carefully executed "burns," or timed firings of its rocket engines. The only real problems in the flight came at the end. The spacecraft landed about fifty miles short of its target, and the jerking deployment of the spacecraft's parachute threw the astronauts into their instrument panel, shattering Grissom's faceplate. During the twenty-one-week stretch between the launches of Gemini III and Gemini V, Neil spent twenty-six days at the McDonnell plant in St. Louis where the Gemini V spacecraft was being tested and prepared for flight. Another twenty-plus days were spent in Florida at the Kennedy Space Center. Mixed in between were trips to California, North Carolina, Virginia, Massachusetts, Colorado, and Texas.

Armstrong and the other members of the GT-5 crew traveled, during this roughly five-month period, well over sixty thousand miles. Some of the flying was done commercially, but the astronauts did a significant amount of it themselves. This helped them keep up their proficiency as pilots.

The Original Seven astronauts learned much of their celestial navigation at the Morehead Planetarium, on the campus of the University of North Carolina at Chapel Hill. For the Mercury program, a brilliant planetarium director by the name of Tony Jenzano designed and constructed versions of the Link flight trainer that mirrored the view from inside the space capsules. From two barber's chairs within a "spacecraft" constructed from plywood, cloth, foam rubber, and paper, astronauts controlled the movement of a star-field projection that simulated spacecraft pitch and roll. The chairs tilted slightly to simulate the action of rocket thrusters that produced left and right yaw.

Armstrong paid numerous visits to the Morehead Planetarium. His last visit came on February 21, 1969, five months before the launch of Apollo 11. No Mercury, Gemini, or Apollo astronaut spent more time studying the stars at Morehead than Armstrong. Neil said that the time spent at Morehead helped the astronauts to recognize the stars and constellations, paramount in the Gemini program for navigational computations and astronomy-related experiments. Apollo flights, with their improved computer capabilities, required crew members to have "a good visual representation" to perform sextant sightings and navigational computations involving all thirty-six stars being used as the basis for NASA's celestial navigational system.

Flying cross-country together in T-38s in preparation for their March 1966 Gemini VIII flight, Armstrong and Dave Scott regularly tested each other's knowledge of the stars. "We would be flying at a high altitude of 40,000 feet and we would turn the lights completely down in the cockpit," Neil remembered. "You got a wonderful view of the sky and it was a great opportunity to practice." On Apollo 9 in March 1969, Scott would perform a lot of excellent star work that kept the onboard guidance and navigation computer properly aligned.

The Gemini V team of Cooper, Conrad, Armstrong, and See, as well as the later Gemini VIII crew of Armstrong, Scott, and their backups (Pete Conrad and Dick Gordon), grew into closely knit units. Gemini V launched on August 21, 1965. The Titan II rocket shot the spacecraft aloft from Launchpad LC-19 just a few seconds before 9:00 A.M. EST, following a two-day hold due to weather conditions at the Cape and because of problems loading the cryogenic fuel. Gordon Cooper remembered: "Ours was the first spacecraft to go into space with a fuel cell: an on-site photochemical generator that produced its own energy. Previous spacecraft had relied on batteries, which would be too cumbersome and heavy given the amount of electronics the more advanced spacecraft were now carrying. In Gemini V, for example, we were taking into space the first onboard radar, and first computer, both of which drew substantial electric power. Proving we could fly with a fuel cell was paramount."

Following backup crew member procedure, Armstrong and See were at the Cape for the launch, then returned to the Manned Spacecraft Center. On their third orbit, Conrad noticed that the oxygen pressure in the fuel cell had dropped from 800 to 70 pounds per square inch, just when "we'd released a rendezvous pod, had it on radar, and were just getting ready to intercept it—an experiment designed to provide crucial information about never-before-attempted space rendezvous." The pressure in the fuel cell eventually recovered, but by then the chance to demonstrate a rendezvous had passed. The flight of Gemini V ended only an hour and five minutes short of eight days, splashing down ninety miles away from its rescue ship because someone on the ground had sent up incorrect navigation coordinates to the onboard computer. The mission amassed impressive data on the physiological effects of weightlessness (it took two days for the cardiovascular systems of Cooper and Conrad to recover), but the disappointing rendezvous outcome underscored that aspect for the next Gemini flights.

Three weeks after the splashdown of Gemini V, on September 20, 1965, NASA formally named the crew for Gemini VIII. Armstrong would serve as the command pilot, as he had in the backup crew for Gemini V. Rather than Elliot See, who had been with Neil on the Gemini V backup crew, Slayton paired Neil with Dave Scott, the first member of the third class of astronauts to get a flight assignment. Assigned as backup to Neil and Dave were Pete Conrad, the pilot of the just completed Gemini V, and Dick Gordon, also new to the Gemini program.

With his assignment in September 1965 to command Gemini VIII, the first phase of Armstrong's career as an astronaut came to an end. For the next six months, until the launch of Gemini VIII on March 16, 1966, Armstrong and Scott trained almost without interruption for their first spaceflight, the most complex ever tried to that point in the American space program—and one that almost cost them their lives.

CHAPTER 14

Gemini VIII

Cape Kennedy, Florida. 9:41 A.M. EST, Wednesday, March 16, 1966. This is Gemini Launch Control. We are T minus 114 minutes for Gemini VIII on Pad 19 and nineteen minutes away from the Atlas/Agena liftoff on Pad 14. Prime pilots for the mission, Astronauts Neil Armstrong and David Scott, were over the hatch and into the Gemini VIII spacecraft at thirty-eight minutes past the hour. They are now hooking up. . . .

Three and a half years into his career as astronaut, Neil Armstrong, thirty-five years old, finally entered a spacecraft, atop a fully fueled Titan II rocket, ready to make his first space shot. Gemini VIII, the fourteenth flight in the U.S. manned space program, was definitely worth waiting for. A rendezvous in space had been made only once before, just four months earlier, and had never been managed by the Russians. It had happened in December 1965 when astronauts Wally Schirra and Tom Stafford in Gemini VI coasted up from their orbit to stop only a few yards away from Gemini VII, with Frank Borman and Jim Lovell aboard. Now Gemini VIII was to perform not just a rendezvous but the first actual docking in space, by joining up with the specially designed, unmanned Gemini Agena Target Vehicle (GATV).

The Gemini VIII mission also called for thirty-three-year-old, Texas-born pilot Dave Scott to perform a far more complicated EVA than Ed White had accomplished in America's first space walk during Gemini IV in June 1965. Also promising to occupy the crew during their scheduled seventy-hour, fifty-five-orbit flight were onboard experiments involving zodiacal light photography, frog egg growth, synoptic terrain photography, nuclear emulsion, and atmospheric cloud spectrophotography. "In ancient Greek mythology, Gemini meant the twins, Castor and Pollux," Armstrong explained. Armstrong and Scott designed the patch for Gemini VIII "having a ray of light emanating from Castor and Pollux going through a prism and reflecting the full spectrum of spaceflight." Gemini VIII's fundamental objective was preparing for the Moon landing. When NASA in the summer of 1962 decided that the lunar-orbit-rendezvous (LOR) method was the only way to get to the Moon by the end of the decade, it became absolutely essential to learn how to rendezvous and dock with another spacecraft.

It was up to Armstrong, as commander, to pull off those critical maneuvers for the first time.

As originally developed by Lockheed for the U.S. Air Force, the Agena was a second-stage rocket. As such, it proved so reliable that NASA mission planners as early as 1961 contemplated using it as a target vehicle in a rendezvous experiment, an idea that blossomed into Project Gemini. The repurposed Agena needed a three-way data communications system, a radar transponder and other tracking aids, an attitude stabilization system, and a docking collar. Most complicated of all, the GATV needed a restartable engine capable of no less than five start-and-stop cycles in space, enabling the docked pair of spacecraft to be maneuvered in any direction. Just eleven days before the launch of Armstrong's mission, a modified Agena was certified for launch.

A petty prelaunch problem inside their spacecraft almost cost Neil and Dave their chance to go after the Agena, now streaking into space: "Just after Dave and I slid through the hatches and into our couches, one of the guys in the flight preparation crew found some epoxy in the catcher mechanism on Dave's harness. It was very hard for us to do anything about it, so restricted we were in our seats, but Pete Conrad, our backup commander, and pad leader Guenter Wendt, after a little sweating, got the catch unglued."

Janet stayed home with her two small boys in Houston and watched the liftoff nervously on television. Neil had gotten motel reservations for his parents, who were taken with other VIPs to the Cape Kennedy viewing stands by a NASA bus. June and her husband, and Dean and his wife, were also present. Neil himself experienced a "counterbalance" to the buildup of anxiety and anticipation in that "most times in airplanes when you're going to go fly, you go fly. But in spacecraft a lot of times you go to the launchpad and just sit for a couple of hours and then get out of the spacecraft and go back to your quarters. It happened so often that it was always a surprise when you really launched. You didn't really expect it" until you felt the rocket's anchoring bolts shear off for breakaway.

"When the Atlas/Agena went on time," Armstrong recalled, "that was a great sign. Then we went precisely on time with our Titan as well, which was a good sign, too, because it meant that our rendezvous schedule was going to be just like we'd practiced for."

"The Titan II was a pretty smooth ride," Neil remembered, "a lot smoother than the first phase of the Saturn V would be in Apollo. The launch was very definite; you knew you were on your way when the rocket lit off. The g levels got to be pretty high in the first stage of the Titan—something like seven g. First, all you see is blue sky and then as you get into the pitch-over program—you're upside down and you're pitching so that your feet are going up towards the sky and you see the horizon coming down through the top of the window. It's quite a spectacular sight because you're going over the Caribbean and you see all those blues and greens and occasionally an island here and there. It would be nice to enjoy the view, but you're too worried about the engine keeping running."

Intermittent voice communications with Gemini VIII happened a few minutes at a time as the spacecraft circled the globe on its easterly path. Relaying between the cockpit and Houston was a worldwide tracking network with stations on Ascension, a British island in the South Atlantic; at Tananarive in the Malagasy Republic, on the island of Madagascar off the east coast of Africa; at Carnarvon, in western Australia; at Kauai, the northern-most Hawaiian Island; and at Guaymas, in Mexico on the Gulf of California. Not until the astronauts were over Hawaii did they try to do much sightseeing. Armstrong was able to make out Molokai, Maui, and the big island of Hawaii. Both men started looking for the shoreline of Texas, hoping to see Houston and to pinpoint the location of their homes. But the job at hand was to chase down the Agena, presently some 1,230 miles away from Gemini VIII and moving in a separate, higher orbit.

The first task Armstrong needed to perform was aligning the spacecraft's inertial platform, a fixed base that measured angles—and thus directions—in the void of space where all directions are relative. The inertial platform consisted of three gyroscopes mounted at right angles to one other. As the spacecraft moved relative to the gyroscopes, the inertial measuring unit fed pitch, roll, and yaw angles to the onboard computer tracking the Agena via radar. Three accelerometers mounted in tandem with the gyroscopes measured the spacecraft's reaction to thruster firings.

A five-second burst of Gemini VIII's forward thrusters would slow the spacecraft into a position where its orbital inclination—the angle between the plane of its orbit and that of the equator—matched up precisely with the Agena's. This critical moment came at one hour and thirty-four minutes elapsed time into the mission.

"A fundamental requirement of rendezvous," Armstrong explained, was "to get your orbit into the same plane as the target's orbit, because if you're misaligned by even a few degrees, your spacecraft won't have enough fuel to get to its rendezvous target. So the plan is to start off within just a few tenths of a degree of your target's orbit. That is established by making your launch precisely on time, to put you in the same plane under the revolving Earth as is your target vehicle." But no matter how precisely the two launches are timed, the angles of inclination in the orbits of the two spacecraft would be slightly askew. In the case of Gemini VIII, a .05-degree difference between its inclination and that of the Agena's needed to be burned off.

Even under ideal circumstances, chasing down a target in space required unusually keen piloting. Without extensive simulator time, it is doubtful that any astronaut could ever have been truly ready to perform a space rendezvous. A guidance computer was needed to compute the location of the two spacecraft, to define the best transfer arc into the GATV's orbit, and, during the final phases of rendezvous, to solve precise mathematical problems based on radar lock-on with the Agena. Built for NASA by Federal Systems Division of IBM in Owego, New York, the Gemini guidance computer was among the world's first computers to use digital, solid-state electronics for the purpose of assisting with the real-time guidance, navigation, and control of a flying machine. "This was a teeny-tiny computer," Armstrong related. Measuring nineteen inches long and weighing fifty pounds, the computer fit inside the front wall of the spacecraft. Within this compact unit, tiny doughnut-shaped magnets comprising the computer's core memory stored 159,744 bits of binary information; less than 20,000 bytes. Adding only slightly to this capacity was a tape drive by which the astronauts could put alternate programs into the computer. Gemini VIII was the first space mission to benefit from the alternate-tape system. Even with what was then the most current computer technology, it was incumbent on the mission planners to reduce the complexities of rendezvous.

Mathematical models, simulations, and early Gemini flight experiences determined that the optimal altitude difference between the two spacecraft was fifteen miles, and that the ideal transfer angle—the angular distance the Gemini spacecraft needed to traverse during its rise to the higher orbit of the Agena—was 130 degrees. As Armstrong explained, "What our mission planners worked out was an approach path that allowed us to arrive at the Agena when it appeared to be a great big star fixed in the middle of the background and things weren't all going every which way. This technique gave us the advantage of having a much easier approach to our target vehicle, because we didn't have the background moving on us. With our target frozen against the star background, we could know we were on the right path. It automatically told us something important if the target started moving; it told us that we had a velocity component that we needed to take out." As for the best lighting conditions, it was found that the Sun should be behind the Gemini spacecraft during its braking phase to rendezvous. From these stipulations, the mission planners worked backward to design launch times, ascent trajectories, and orbital parameters that set up the optimum conditions for Gemini's terminal phase of rendezvous leading to docking.

From the time of the spacecraft's first burn at one hour and thirty-four minutes into the mission to the point in time that the spacecraft began terminal phase, it took approximately two hours and fifteen minutes. Armstrong and Scott then decided to eat a meal. Inside the meal packet labeled Day 1/Meal B was a freeze-dried chicken and gravy casserole. But a call from CapCom Jim Lovell in Houston, relayed through the tracking station at Antigua in the British West Indies, told the crew to get ready for their next burn—a phasing adjustment, or slight in-plane repositioning, requiring another platform alignment. Armstrong and Scott used patches of Velcro to stick their packaged food on the ceiling of their spacecraft until the burn was complete. Retrieving their food half an hour later, the astronauts found the casseroles still dry in spots. Armstrong next tried a package of brownies, only to have crumbs float all over the cabin.

The next maneuver, a plane-change burn, came over

the Pacific Ocean just before completing a second orbit, at 2:45:50 elapsed time. Punching the aft thrusters, Armstrong produced a horizontal velocity change of 26.24 feet per second, which brought Gemini VIII's nose down, perhaps imprecisely:

2:46:27 Armstrong: *I think we overdid it a bit.*

Not until the spacecraft was over Mexico was Neil's gut feeling confirmed. Lovell told him to add two feet per second to his speed by making another very short burn. The men got the spacecraft in plane and into an orbit in which they were below the Agena and catching up to it. When they got into the GATV orbit, they made computations of range and range rate with the computer, with charts and also with the ground. After making a number of adjustments, they "could hopefully arrive at the target with the target having zero relative motion against the stars and with us approaching at a reasonable rate that had us using a minimum amount of fuel in decelerating for final approach."

The terminal phase could not begin until Gemini VIII had a solid radar lock-on with the Agena. Commander Armstrong kept range and range rates constantly in mind so as not to overshoot the target by closing in on it too fast. At 3:08:48 elapsed time, Armstrong reported, "We're getting intermittent lock-on with the radar." Thirty-five minutes later, with the spacecraft over Africa, Neil reported a solid radar lock. Next Armstrong needed to perform another burn. The transfer arc in which Gemini VIII had been moving for the past couple of hours in order to catch up with the Agena had been elliptical, the pathway that was dictated by the gravitational field of *one* body.

Armstrong nosed down his spacecraft and applied the aft thrusters. The burn resulted in a velocity change of 59 feet per second, which circularized Gemini VIII's orbit and put it more precisely in plane with the Agena.

It took a while before the crew could see its target. Armstrong explained, "At some point we knew we would see the target. But we had to be pretty close. According to the mission plan, what we wanted to do is be in the dark throughout 130 degrees of the transfer arc—or at least 125 degrees or so. Then at roughly ten miles out, the target would go into daylight. At that point it lit up like a Christmas tree. We could see it against that dark sky just like a gigantic beacon. When that happened, the star background became less important because we would be on a good trajectory, so we could make the final adjustments visually." Soon thereafter, Scott radioed the crew's sighting of an object seventy-six miles distant that was gleaming in the sunlight. They assumed it was the Agena. With the target located ten degrees above Gemini VIII, Armstrong needed to align the inertial platform once again, in preparation for one of his last translation maneuvers. In it Neil would pitch up the spacecraft's nose some thirty degrees and cant the vehicle roughly seventeen degrees to the left. When that maneuver was completed successfully, he had time to take another look at the Agena.

A few minutes later, the Agena vanished from view as it entered twilight, soon to reappear for the astronauts when the acquisition lights on the target vehicle, by command from Gemini VIII, blinked on. "Once we completed our transfer arc," explained Armstrong, "we had to make final adjustments that would get us exactly into the same position and to the same speed as the Agena, so that we would be flying in formation. From that point, we did what was called 'station keeping.' This meant we stayed about one hundred fifty feet apart. We flew around the target but never got very far away from it. We had to stay in the same orbit as the Agena, because if we went astray by even just a little bit, the errors propagated. So we had to fly essentially in formation."

High over the tracking ship, which was positioned near the Caribbean island of Antigua, the crew of Gemini VIII prepared to apply the brakes to their spacecraft so that it would not close too quickly on the Agena and fly right by it. Delicately, Armstrong handled the braking by intermittently firing his aft thrusters in very short bursts, while Dave Scott called out Gemini VIII's range and rate. Two minutes and twenty-one seconds later came the glare of the Agena's lights. Edging ahead at the glacial pace of five feet per second, Gemini VIII bore down on the Agena. Armstrong's excitement was evident:

5:53:08	Armstrong:	*I can't believe it!*
5:53:10	Scott:	*Yes, I can't either. Outstanding job, Coach!*
5:53:13	Armstrong:	*Way to go, partner!*
5:53:16	Scott:	*You did it, boy! You did a good job!*
5:53:17	Armstrong:	*It takes two to tango.*

Two minutes later, CapCom Lovell, who had kept quiet so as not to bother Armstrong and Scott during the critical braking phase, broke in and asked the crew for an update on the rendezvous.

5:56:23 Armstrong: *Flight Houston. This is Gemini VIII. We're station keeping on the Agena at about 150 feet.*

With relative velocity between the two vehicles canceled out, rendezvous—only the second ever made in the brief history of the Space Age—had been achieved.

Station keeping posed no particular problem for Armstrong. "It was very easy to fly close. We flew around the vehicle and took pictures of the Agena from different perspectives, in different lighting." Armstrong always used the term "we" when it came to flying any aircraft or spacecraft, but the requirements of the Gemini mission were too serious for him to share the piloting responsibilities with Scott, at least not yet. Neil planned to let Scott fly the spacecraft sometime later in the mission, after they undocked or after Scott did his EVA. Armstrong and Scott kept their rendezvous station across most of that "day," knowing that the plan was to proceed with the docking before they moved into the next "night," when docking conditions would be far from optimal. In the orbit they were in, daylight lasted for about forty-five minutes.

The rendezvous began just west of Hawaii. Gemini VIII's location at daylight put it in the vicinity of USS *Rose Knot Victor*, tracking the spacecraft from off the northeastern coast of South America, at the very moment Armstrong was easing the spacecraft toward the docking at the barely perceptible closing rate of three inches per second.

6:33:40 CapCom: *Okay, Gemini VIII. It looks good here from the ground. We're*

		showing CONE RIGID. Everything looks fine for the docking.
6:33:52	Armstrong:	*Flight, we are docked! Yes, it's really a smoothie.*

Celebration broke loose in Mission Control for a few mad seconds. CapCom congratulated the men and reported that the Agena was stable, with no noticeable oscillations.

During the first minutes of the docking, both the crew and the flight controllers focused on the performance of the Agena, given how riddled with problems the GATV had been. Houston had difficulty verifying that the Agena was receiving and storing the commands uplinking for an upcoming yaw maneuver. Flight also wondered why the Agena's velocity meter did not seem to be operating. These two mysteries suggested a malfunction in the Agena's attitude control system. CapCom told Armstrong that if it went wild, to turn it off and take control with the spacecraft. Six minutes after this warning, the Tananarive tracking station lost the spacecraft's signal as it moved into a dead zone. For the next twenty-one minutes, there would be no communications with Gemini VIII, now coupled in flight with the Agena as one integrated spacecraft.

Then came the next chilling words from Gemini VIII:

7:17:15	Scott:	*We have serious problems here. We're . . . we're tumbling end over end up here. We're disengaged from the Agena.*

Armstrong recalled the sequence of events leading to the in-flight emergency, the first potentially fatal one ever experienced in the U.S. space program: "We had gone into night just shortly after the docking was made. You didn't see a lot on the night side. You saw stars up above, and down below you might see lights from a city or lightning areas embedded in thunderstorms, but you didn't see much. Dave noticed from the ball indicator, and called to my attention, that we were not in level flight like we were supposed to be, but rather in a thirty-degree bank angle."

As their spacecraft had moved into nighttime conditions, the astronauts had turned up the lights in their cockpit as far as they could go, making it almost impossible to detect any changes in their horizon line unless they were looking directly at instruments: "I made some efforts to reduce the bank angle, mainly by triggering short bursts from the Orbit Attitude and Maneuvering System [OAMS]. Then the banking started to go again, so I asked Dave to shut off the controls to the Agena. Dave had all the controls for the Agena on his side of the spacecraft."

To no avail, Scott commanded the target vehicle to turn off its attitude control system; he jiggled the target vehicle switches and cycled them on and off again; he energized and deenergized the entire Agena control panel. Armstrong relates, "I really believed that we wouldn't have any trouble with the docking, based on the simulations we did," but no one had conjured a simulation in which a *coupled* Gemini-Agena experienced such deviant motions. "If we had been able to practice in such a situation," Armstrong felt, "I'm sure we would have figured it out much more quickly.

"We had a couple of flights in the Gemini program under our belt by this point," Armstrong noted. "So it was natural to suspect that if there were a problem or mistake, it would come from the Agena, which had had quite a few problems in its development."

Reinforcing the bias against the Agena was the warning that Jim Lovell had issued, just moments prior to docking, that at any sign of trouble Armstrong and Scott were to get off the Agena and take control of their own spacecraft. Neil simply said to his crewmate, "We're going to disengage and undock," and Dave immediately agreed.

"Go," Armstrong said to Scott. "We disengaged successfully," Neil explained years later, "but I was a bit concerned because I didn't want to have a re-impact immediately afterward with the Agena. So I pulled away sharply, hoping I could increase the distance before one of us rotated back into the other one. That worked fine. We then immediately tried to get control of our own spacecraft, which we found we couldn't do. Immediately it was obvious that the problem was not the Agena's. It was ours."

The real villain was one of Gemini VIII's OAMS thrusters—specifically, thruster number eight, a small rocket with twenty-three pounds of thrust used to roll the aircraft. Apparently sometime while Armstrong had been using the OAMS to maneuver the Gemini-Agena combination, a short circuit stuck the thruster open.

"I didn't know at the time," related Armstrong, that "you only hear the thruster when it fired; you didn't hear it when it was running steadily."

Gemini VIII was spinning dangerously out of control. According to Armstrong, "The rate of rotation kept increasing until it reached the point where the motions began to couple. In other words, the problem became not just a precariously high rate of roll but also the coupling of pitch and yaw," in engineering terms, the same sort of control dilemma as the inertial roll coupling that had so plagued the design of early supersonic aircraft.

"Our spacecraft turned into a tumbling gyro, the fastest motion of which was our roll rate. Our roll rate indicators only went up to twenty degrees per second, and all the roll rate indicators had shot up against the peg, so we were clearly beyond twenty degrees per second in all axes—although sometimes they mysteriously came swinging back all the way across." When the revolutions surpassed over 360 degrees per second, "I became very concerned that we might lose our ability to discriminate accurately," Armstrong recalled. "I could tell when I looked up above me to the controls for the rocket engine that things were getting blurry. I thought I could, by holding my head at a certain angle, keep the controls in focus, but I knew we were going to have to do something quickly to make sure that we could work on the problem without losing our vision or our consciousness."

Armstrong found his options narrowed to one, "to stabilize the spacecraft in order to regain control. The only way I could do that was to engage the spacecraft's other control system." This was the reentry control system (RCS), which was up in the nose of the spacecraft. As the RCS had two individual rings that were coupled, "its propellant tanks were not normally pressurized until shortly before their normal use. There was a single pushbutton switch that energized pyrotechnic valves that allowed high-pressure gas to pressurize the UMDH/N204 propellant tanks. Once the tanks were pressurized, each redundant ring (A and B) could be operated individually using electrical switches. Once we blew the squib valves, we used both rings to regain control. Then we shut off one of the rings to save its propellant for the entry phase. Mission rules dictated that once the squibs were blown, we were obliged to land at the next available landing site.

"We turned off the other control systems, the ones in the back end, and stabilized the spacecraft with only the front-end system," Neil related. "It didn't take an awful lot of reentry control fuel to do that, but it took enough."

With the spacecraft now stabilized thanks to his firing of the RCS, Armstrong energized the thrusters, one by one. When he hit the switch for thruster number eight, Gemini VIII immediately started to roll again. "We found the culprit," Armstrong noted, "but we didn't have a lot of fuel in the back-end system left available to us at that point.

"Murphy's law says bad things always happen at the worst possible times," Armstrong would later say. "In this case, we were in orbits that didn't go over any tracking stations. We were out of radio contact almost all of the time, and for the short stretches when we were in contact, the ships at sea had limited ability to communicate back with Mission Control or to transmit data to Houston. By the time we went over a tracking station or two and were able to convey the nature of our problem so Mission Control knew what was going on, there wasn't any way they could help much at that point." Finally having put a stop to the maverick spinning, Armstrong took his first chance to explain what had happened, and Scott told Houston they hadn't seen the Agena since they had undocked.

Neil recalled his decision to start up the reentry control system: "I knew what the mission rules were. Once we energized the RCS and the integrity gets broken in both RCS rings, we had to land—and land at the next convenient opportunity. I had to go back to the foundation instincts, which were 'save your craft, save the crew, get back home, and be disappointed that you had to leave some of your goals behind.'" Houston told them they would terminate the flight and bring it down in the west Pacific. A destroyer was about six hours away and heading toward that splashdown area. Neil and Dave knew that rescues on open ocean did not always happen quickly. Even on land and with modern communications, it was hard to find something as small as a space capsule. Stories had circulated in NASA that it sometimes took the Russians forty-eight hours to find their astronauts after their parachuting down into places like Kazakhstan or Siberia.

The two astronauts had much to do to prepare for emergency reentry and splashdown: "Dave and I understood that we probably had several hours to get ready. From the ground we were given the retrofire location time, which was over Africa and on the night-side of Earth, so we prepared for retrofire activity. We were flying over the tracking station in Kano, Nigeria, when Houston started giving us the countdown for the retrofire time. We lost communication with the ground midway through that count, so they didn't really know if the retro had come off or not. But the retrofire was stable and our readings of the retro change in velocity—that is, the amount of slowing down that we had done—was proper for the target that we wished to hit. Our guidance system seemed to be working properly, so we steered a course for Okinawa."

As Gemini VIII came into daylight, "We appeared to be dropping at a prodigious rate," Neil recalled. "We could almost see those big mountains [Himalayas] coming up at us." The spacecraft's main chute deployed on time, orienting them with their view up rather than down, so "there was a mirror that we used, a small flight pocket mirror, and by looking into that I could look down over the side and see that we were, thankfully, over water. Being an old navy guy, I much preferred coming down in the water to coming down in Red China," Armstrong remembered with a smile.

While they were coming down under the chute, Neil was the first to hear the sound of propeller airplanes in the vicinity. "We assumed it was friendly."

The splashdown itself turned out to be, in Neil's words, "not too bad." A C-54 rescue plane arrived quickly and dropped navy frogmen into the rough waters to attach a big flotation collar around the spacecraft. Nothing remained but to wait for the destroyer *Leonard Mason.* The wait turned into a nauseous ordeal.

"The Gemini was a terrible boat," Neil explained, "a good spacecraft, but not a good boat." Much to their regret, neither Armstrong nor Scott took their tablets of meclizine, used to avoid motion sickness. "So both of us really got seasick." Fortunately, they did not have much in their systems to regurgitate.

After more than two hours, the frogmen, themselves queasy from inhaling the stench from Gemini VIII's burnt heat shield, opened the spacecraft's hatches, and the astronauts climbed out. Neil reluctantly accepted handshakes from the crew. "I was very depressed at this point. We had not completed all the things we wanted to do. We'd lost Dave's chance to do all those EVA marvelous jobs. We'd spent a lot of taxpayers' money, and they hadn't gotten their money's worth out of it. I was sad, and I knew Dave was, too." It took about fourteen hours for the ship to get them to Okinawa.

After a good night's sleep in Okinawa, the astronauts were flown to Hawaii. They arrived back at Kennedy

Space Center on March 19, three days after launching from the Cape. Not until March 25 did Armstrong and Scott return to their homes in Houston. The next day NASA convened the crew's first post-flight press conference. Even several days of talking over technical matters with his associates did not alleviate Neil's depression.

International media paid a great deal of attention to the unprecedented spaceflight ordeal of Gemini VIII. All the networks in the United States broke into their regular evening programming with emergency news bulletins. (ABC's interruption of an episode of its immensely popular prime-time *Batman* series was rewarded with more than one thousand phone calls from complaining viewers.) The next morning's *New York Daily News* carried the banner headline "A Nightmare in Space!" Even staid *Life*, with its exclusive contract for the astronauts' personal stories, elevated the events into melodrama. Initially, the magazine positioned its coverage as "Our Wild Ride in Space—By Neil and Dave," but Armstrong put a stop to it. He called Hank Suydam, a *Life* writer assigned to Houston, who then wired his boss, Edward Thompson, *Life*'s editor-in-chief:

> I JUST HAD A PHONE CALL FROM NEIL ARMSTRONG WHO WAS VERY UPSET AT THE ADVANCE BILLING IN THIS WEEK'S MAGAZINE WHICH READ "OUR WILD RIDE IN SPACE." HE ASKS THAT THE HEADLINES YOU USE WITH THEIR ACTUAL PIECE NOT CONCENTRATE SOLELY ON THE EMERGENCY AND NOT BE PHRASED IN WHAT HE CONSIDERS AN OVERLY JAZZY WAY. I TOLD HIM WE APPRECIATE HIS POINT. I EXPLAINED, HOWEVER, THAT WE DO HAVE TO USE HEADS TO CRYSTALLIZE THE ESSENCE OF VARIOUS PHASES OF THE STORY. I GAVE HIM A

GENERAL ASSURANCE THAT WE WOULDN'T REPEAT THE ONE IN THE BILLING AND WOULD PROBABLY UTILIZE, FOR THE MOST PART, QUOTES FROM THEIR OWN PIECE.

The editor at *Life* obliged, but only partway. He toned down the piece, took the astronauts' byline off it, and changed its title to "High Tension Over the Astronauts." *Life* continued to run articles on Gemini VIII in its next two issues. For the second, a version of the title that had upset Neil reappeared as "Wild Spin in a Sky Gone Berserk." The third article, entitled "A Case of 'Constructive Alarm,'" gave the astronauts their bylines, with their words so heavily edited that Armstrong again complained. In particular, Neil was upset at the cut of his final quote: "I think we'd put this almost identically, so I'll speak for both of us. We were disappointed that we couldn't complete the mission, but the part we did have, and what we did experience, we wouldn't trade for anything."

Worse than the hype in the media was the sniping that came from some fellow astronauts. According to astronaut Gene Cernan, "It didn't take long for some of the guys around the Astronaut Office to criticize Neil's performance. 'He's a civilian pilot, you know, and maybe he has lost some of the edge. Why didn't he do this, or why not do that? He wouldn't have gone in the spin if he would have stayed docked with the Agena.' Screwing up was not acceptable in our hypercompetitive fraternity and, if you did, it might cost you big-time. Who knew if the criticism might reach Deke's ears and change future crew selections in favor of the person doing the bitching? Nobody got a free ride when criticism was remotely possible. Nobody."

"Everybody second-guessed everybody," recalled

astronaut Alan Bean, who was backing up John Young and Mike Collins on Gemini X with crewmate Clifton C. Williams at the time. "Don't forget, you're dealing with really competitive people. You almost had to find something wrong in the other guy's performance. It was part of the way it was."

Buzz Aldrin, who was then preparing as backup pilot for Gemini IX, in retrospect has agreed that it is only with twenty-twenty hindsight that anyone can criticize anything that Neil did during the emergency. On the other hand, Aldrin has offered the following conjecture: "I think there may have been a slim chance that they could have avoided activating one ring of their reentry system."

"I didn't hear any of the criticism," stated Frank Borman, who along with Wally Schirra accompanied the Gemini VIII crew back to Hawaii after greeting them in Okinawa following their rescue. "I wouldn't have participated in that crap if there was. I think Neil and Dave did a good job. I don't think anybody realizes how close that came to utter disaster. In retrospect, that was probably as dangerous as Apollo 13. Not as time consuming, but if they had run out of reaction control fuel in stopping their spinning, they would have been dead." Schirra felt the same way about it: "The decisions that Neil and Dave made were all good decisions."

Gene Kranz was just taking over as flight director from John Hodge during a shift change in Mission Control when Scott's urgent report came in over the radio. In retrospect, according to Kranz, "It would have been tough for the controller in a very dynamic situation to track that the solid-on one was the problem. But he might have done it." Rather than blaming the crew for any measure of failure, Kranz placed the blame on himself and on the other flight directors and planners in Houston: "I was damn impressed with Neil, as was virtually everyone that had anything to do with the program." In the debriefing he gave to his flight controllers after the Gemini VIII mission, Kranz asserted: "The crew reacted as they were trained, and they reacted wrong because we trained them wrong. We failed to realize that when two spacecraft are docked they must be considered as one spacecraft, one integrated power system, one integrated control system, and a single structure. We were lucky, too, damned lucky, and we must never forget this mission's lesson." In retrospect, treating docked spacecraft as a single system was, in Kranz's judgment, one of the most important lessons to come from the entire Gemini program: "It had a profound effect on our future success as flight controllers." It was a lesson that proved invaluable when the second potentially fatal in-flight emergency happened, in 1970 during Apollo 13.

In total agreement with Kranz about where the fault lay, Chris Kraft asserted, "We tricked the astronauts on that one. I think Neil and Dave did absolutely what I would have had them do. As for criticizing them afterward for doing that? I guess maybe a few astronauts might have said, 'I'm better than that.' But they're only fooling themselves."

No one was ever a tougher, more honest critic of his technical piloting performance than Armstrong was himself: "I always felt as though if I had been a little smarter I would have been able to figure out the right diagnosis and been able to come up with something more quickly than I did. But I didn't. I did what I thought I had to do and recognized the consequence of that. You do the best you can." Following his return to Houston, he found that, just a day or two before their launch, there had been a problem with the environmental control system in the spacecraft. This resulted in technicians pulling out the system to replace one or two parts. Curiously, the wiring for the damaged control system was part of the same cable that operated what turned out to be the faulty rocket. "So my guess," said Neil, "was that, sometime during that process, the technicians did something that put a nick in that cable, which allowed it to short. To my knowledge, they were never able to isolate that problem. Of course, the back end of the spacecraft—the adapter—did not come back to Earth with you. So if it really had been something in the back section of Gemini VIII, we never had a chance to examine it."

Much more vigorously than Neil, Dave Scott has defended the wisdom of what he and his commander did in space: "There was never any doubt in my mind that we had done everything right. Otherwise we would never have survived." If Gemini VIII had turned tragic, as Neil would later speculate, "It's conceivable that what happened to us would forever have remained a mystery." Scott concurred: "They wouldn't have known what happened because they wouldn't have gotten any downlink. They wouldn't have known it was the Gemini because they never would have gotten any data, because it would have been turning too fast." Such a mysterious tragedy "would have caused a big glitch in the program. It would have taken us a long time to figure out what happened, if we ever would have." Without knowing, it would have been very difficult to proceed into the Apollo program. Then, if the Apollo fire, too, had occurred, just ten months later, killing three more astronauts, national support for the manned space program would likely have vanished, along with prospects for a Moon landing. As Dave Scott said, "If we had not recovered from the spin, it could have been a showstopper."

Turning out as well as it did, the broader political repercussions of the Gemini VIII flight were minor. "In the flight, both of them came across as being pretty much what we thought of them before," Mike Collins explained. "There was certainly nothing in the aftermath that affected their crew assignments, absolutely not. And there would have been if they'd screwed up big-time." Astronaut Bill Anders, whose first mission was to be Apollo 8's historic circumlunar flight in December 1968, agreed: "Not only was Neil quick-thinking, he certainly wasn't shy about doing things that well could have worked against him." According to Chris Kraft, the way Armstrong handled himself during the emergency gave NASA "even greater confidence in Neil's abilities."

Two weeks after the flight, the Gemini VIII Mission Evaluation Team "positively ruled out" pilot error as a factor in the emergency. In revealing the team's findings, Bob Gilruth commented, "In fact, the crew demonstrated remarkable piloting skill in overcoming this very serious problem and bringing the spacecraft to a safe landing." There was no question that Armstrong would be given another assignment as a mission commander. NASA presented both men with its Distinguished Service Medal. From the air force Dave also received a Distinguished Flying Cross. Major Scott also received a promotion to lieutenant colonel, whereas Neil received a $678 raise that brought his salary to $21,653, making him, thanks to his twelve years in the civil service system, the highest paid astronaut. On March 21, 1966, just two days after he arrived back from Gemini VIII,

NASA named him the backup commander and William Anders the backup pilot for Gemini XI, a rendezvous and docking flight made by Pete Conrad and Dick Gordon six months later.

It would be his last crew assignment prior to Apollo.

CHAPTER 15

The Astronaut's Wife

To the seven thousand folks back in Wapakoneta, their native son was a "space hero." On April 13, 1966, three weeks after his townsfolk had nervously sat around their TV sets awaiting news of their boy's splashdown from Gemini, the little Ohio burg played host to 15,000 attendees of a gala homecoming in Neil's honor.

Armstrong was in no mood to celebrate, but Wapakoneta made the request, NASA gave its seal of approval, and the event was on. For his old friends and neighbors, the astronaut put on his best face. Though it was a raw early spring day, Neil and Janet smiled and waved their way through the town in an open convertible from the airport to the fairgrounds. Following a brief press conference, the parade drove through Wapakoneta's flag-bedecked downtown business district to Blume High School, where Neil had graduated. Neil thrilled everyone by saying, "You are my people, and I am proud of you." Neil termed the homecoming "magnificent" and repeatedly told the crowds that the reception was "more than I deserve." In attendance was Governor James Rhodes, who announced that the state would join Auglaize County in building an airport named for Neil. Neil's parents beamed with pride, relieved that their son had made it home after the near-disastrous mission.

If not for NASA's unwritten rule that wives best not be at the Cape for launches, Janet Armstrong could have been in Florida that awful night. Instead, she was at home in El Lago, caring for her young boys (Mark Stephen Armstrong had been born on April 8, 1963) while hosting her sister and a few other guests. NASA equated keeping the wives away from the launch with "protecting" them. If a disaster occurred at the launchpad, no one wanted a wife to be exposed to a television audience of millions.

For the astronauts, the rationale for keeping the wives at home was different. Deke Slayton did not want the wives at the Cape. In the nervous days leading up to launch, a wife's presence could only divert her husband's attention. No astronaut wanted to risk Deke's ire. Some wives suspected their husbands were having extramarital affairs; a few wives might have known it with certainty. Members of the press who covered the NASA beat knew about a few of the indiscretions, but such things were not reported in 1960s America. Infidelity was not something that Janet spent much time worrying about. Staying at home alone was also nothing new to her. "When the men are preparing for a flight," Janet explained in an interview with *Life* magazine's Dodie Hamblin in March 1969, "they are really home hardly at all. They come on weekends and even then they have work to do. We're lucky if they have a chance to come in and sit down and say hello before they go off again a day later. Having them for eight hours is a privilege during times like this."

As for the dangers in Neil's line of work: "Certainly I realize that there are risks involved in his profession. I suppose we spend years trying to prepare ourselves for a possible tragedy, because the presence of danger is there. But I have a tremendous amount of confidence in the space program. I know that Neil has confidence, and so have I."

Yet the pressures of Neil's first space shot in March 1966 had been different, more extreme. For Gemini VIII, television cameras had not been allowed inside her home, but they were positioned to start filming whenever she went outside. In her living room sat a *Life* magazine photographer. Janet realized that she was constantly on display, as were all of the wives of astronauts during a space mission. When Neil and Dave got into trouble, initially Janet had set out for Mission Control; joining her was the NASA Public Affairs officer who had been assigned to her family and to stay with her during the mission. As soon as the trouble with Gemini VIII was known, NASA had turned off the squawk boxes the agency supplied to astronaut families, leaving her as well as leaving Lurton Scott, at her home in nearby Nassau Bay, in the dark as to what was happening. The PAO drove an insistent Janet to the Manned Spacecraft Center, but she was denied entrance. Janet was understandably furious that an astronaut's wife would not be allowed into a secure place to follow what was going on inside Mission Control.

"Don't you ever do that to me again!" Janet would tell Deke Slayton. "If there is a problem, I want to be in Mission Control, and if you don't let me in, I will blast this to the world!" As for turning off the squawk boxes, Janet understood: "NASA did not know who was in our homes listening to the squawk boxes. There might have been information that would be leaked to the public that NASA did not want leaked in a critical situation, which is why they had a policy for terminating communications in our homes during a crisis. This was totally understandable for security reasons." What was not understandable to Janet was why an astronaut's wife would not be allowed into a secure place to follow what was going on inside Mission Control. "Okay, the men there would have felt bad if something awful happened to our husbands and it might have been difficult for them to see us there, but my comment to Deke was, 'Well, what about the wives?' "

Life's version of Janet's experience that night infuriated her almost as much as NASA's treatment of her did. In its initial story about the Gemini VIII flight, the magazine ran a melodramatic picture of Janet down on her knees, "listening but not watching" as she leaned over a living room TV set. According to the caption, the picture was taken just as "the word came that the astronauts had been picked up and were back in good shape." The caption quoted Janet accurately as saying, "I simply knew they were going to make it. But also I am a fatalist." The truth ended there. "The picture published in *Life* with me kneeling at the TV was because the squawk box was there." (The shot was taken in her home before the squawk box was turned off.) "I was on my knees there with my eyes closed trying to concentrate on what was being said, but it came out that I was in a praying position and blah, blah, blah. Well, that's not true."

Given the tragic deaths of Elliot See and Charlie Bassett just days before Gemini VIII, NASA should have shown far greater consideration for the astronauts' wives. See and Armstrong, the two civilians chosen by NASA for the New Nine back in 1962, had become quite close working together as the backup crew for Gemini V. In that role, they had spent a lot of time together, as had Janet with Elliot's wife Marilyn. Not since Chet

Cheshire back in Korea had Neil grown so close personally to another man: "Elliot was a hard worker, diligent. He really worked hard on Gemini V. He had good ideas and would express them. He may not have had the same personality as most of the astronauts, but being of a little bit different personality is not necessarily bad. I heard from others that they thought his piloting—particularly his instrument skills—were not as good as they should have been. I flew with him a good bit, and I don't recall anything that was of substantial concern to me."

The deaths of Elliot and fellow Gemini XI crew member Charlie Bassett had occurred on February 28, 1966, as they were coming in for a landing at St. Louis's Lambert Field in a T-38 airplane. The two men had flown up from Houston, accompanied by Tom Stafford and Gene Cernan in another T-38, in order for the four of them to get in some practice time on McDonnell's rendezvous simulator. Approaching the field in bad weather, both planes overshot the runway. Stafford climbed out of the fog, circled, and landed safely. Hoping to keep the field in sight, See banked to the left to stay below the clouds. His T-38 slipped too low. The aircraft smashed into Building 101, the same building in which McDonnell technicians were working on the Gemini IX spacecraft. Elliot and Charlie died instantly; no one else died.

On March 2, 1966, two weeks to the day before the launch of Gemini VIII, Neil and Janet joined a large group of mourners at two separate memorial services held for their deceased comrades. The following day, with every one of the astronauts present, the two astronauts were buried in Arlington National Cemetery outside Washington, DC. In 1964, astronaut Theodore C. Freeman had been the first U.S. astronaut to lose his life when his T-38 trainer crashed after flying into a flock of geese. First on the scene with the horrible news was a reporter from one of the Houston papers. Hearing her husband had been killed, Faith Freeman became inconsolable. The press was hardly more civil to Marilyn See and Jeannie Bassett. Spared the gory details by NASA and supported emotionally virtually 24/7 by a circle of other astronauts' wives, Jeannie Bassett learned that her husband had been decapitated in his crash when it was reported in *Time* magazine.

Six months before the Freeman tragedy, on April 24 at 3 A.M., Janet had awakened to the smell of smoke. She woke Neil, who jumped up to investigate. Seconds later came his shouts that the house was on fire. Unable to connect to an operator or the emergency number, Janet ran into the yard and called out to their next-door neighbors and friends, Ed and Pat White.

The Whites and the Armstrongs had arrived together in Houston in the fall of 1962 as members of the New Nine. Several other astronauts also lived in the area, as did a number of NASA managers. The Bormans, Youngs, Freemans, and Staffords built homes in the El Lago subdivision, just down the block and around the corner from the Whites and Armstrongs. The Sees, Carpenters, Glenns, Grissoms, and Schirras also lived nearby. Together, the two neighborhoods amounted to an astronaut colony.

Neil and Janet grew quite close to their next-door neighbors, the Whites. Separating the two backyards was a six-foot-tall wooden fence. Ed and Pat heard Janet through their bedroom window. Fortunately, the Armstrongs had open windows, too. As Janet explained, "The reason the children were not asphyxiated was that our air conditioning wasn't working and it was a warm night and I had closed the doors and opened the windows."

Janet vividly recalls the image of the former hurdler Ed White clearing her six-foot fence. Ed flew to the rescue with a water hose. Neil brought twelve-month-old Mark out, and meanwhile Pat managed to phone in the alarm. The living room wall was glowing red, and window glass was cracking. Ed passed the hose to Janet, took Mark from Neil, and handed the child over the fence to Pat so he could get another hose. The heat was now so intense that Janet had to hose down the concrete just to be able to stand on it in her bare feet. Parked in the garage, the fiberglass body of Neil's new Corvette began to melt.

Neil made a second trip into the fire. "The first time I just held my breath the whole time; the second time I had to get down lower and put a wet towel over my face. I was still trying to hold my breath. I couldn't completely. When you take a whiff of that thick smoke, it's terrible." He would later say to Janet that the twenty-five feet he traversed to save Rick was "the longest journey" he ever made in his life, because he feared what he might find when he got there. But six-year-old Rick was fine. Neil took the wet towel from his own face, put it over his elder son's, and scrambled out into the backyard with the boy in his arms. Then both men picked up the hoses and continued their firefighting. Neighbors found their dog "Super" alive and well.

The volunteer firefighters began arriving some eight minutes after Pat White's call and took the rest of the night to drench the flames. The Armstrongs lived with the Whites for a few days before moving everything worth rescuing into a nearby rental home. The Armstrongs lost many valuable possessions, including family photographs, particularly pictures of Karen. They stayed in a rental home while a new house was built on the same lot, this time by a fire specialist. After the fire, the inspectors found the cause, but not without Neil's input. The builder had not sealed the wall paneling, so that moisture had warped the boards. When they fixed the warped boards, they had unknowingly knocked a nail into a wire. This created a short, with a small current flowing for some months. The temperature built up gradually until it ignited. It wasn't until Christmas 1964 that their new home was ready. In the fire Neil lost most of his prized boyhood collection of airplane models as well as all of his handwritten notebooks filled with drawings of aircraft and aircraft design specifications, plus cartons full of issues of old aviation magazines.

Janet harbored no illusions: "We could have easily all been consumed by the smoke. It was real sickening." Even Neil characterized the danger in stark terms: "It could have been catastrophic. Had we started to become asphyxiated before we woke up, then we probably would not have made it."

But more tragedy was still to come. Nine months later, on January 27, 1967, the vigilant neighbor Ed White would die in the Apollo 1 launchpad fire along with crewmates Gus Grissom and Roger Chaffee.

"People are always asking me what it is like to be married to an astronaut," Janet told *Life* magazine during interviews conducted from 1966 through 1969. "What it's like for *me* to be the wife of *Neil Armstrong* is the more appropriate question. I'm married to Neil Armstrong, and being an astronaut happens to be part of his job. To me, to the children, to our families and close friends, he will always be Neil Armstrong, a husband and father of two boys, who has to cope with the problems of urban living, home ownership, family problems, just like everybody else does."

Janet did not coddle Neil, but she did keep his clothes clean and made meals for the family. "It never, never shows in Neil that he's had a very distressed day. He does not bring his worries home. I don't like to ask him questions about his work," Janet related, "because he lives with it too much already. But I love it when someone else asks him about his work, and I can sit and listen to it all. The only way we wives can participate, really participate, in what the men are doing is to know as much as we can about it in advance and then follow it closely on radio and television and through the communications with the ground."

Janet worked hard to keep her boys grounded, as did Neil: "You don't want your children to go around with their thumbs under their armpits and saying, 'I'm an astronaut's son.' For this reason we try to make everything we do very common and everyday. We feel that is very important for them not to be favored by their classmates. We want them to grow up and have a regular life—a normal life. Kids are kids, and you want them to be kids, and yet this program has demanded an awful lot of our children. When you put your children in public, they really have to be very sophisticated children."

Janet's mantra became, "Living in the present is most important. We take our lives day by day. As for planning and organizing for the future, it is very difficult. I have a husband whose schedule is changing day by day, sometimes minute by minute, and I never know whether he's coming or going, particularly during flight time when he's on a crew."

The pressure on all the astronauts' wives was extraordinary. Each bore a heavy burden, trying as they did to appear before the public as Mrs. Astronaut and the All-American Mother. They knew what NASA and even the White House expected of them. For an astronaut's wife, deciding what to wear was about much more than just a woman's sense of style or even her vanity. It was about maintaining the wholesome and sanctified image of the entire U.S. space program, and of America itself.

"Our lives were dedicated to a cause, to try to reach the goal of putting a man on the Moon by the end of 1969. It was an all-out effort on everyone's agenda. It wasn't just our astronaut families that had put our lives on hold; thousands of families were in the same mode." NASA might have been wise to have established a formal counseling program for its astronauts' families, given that thirteen of the twenty-one marriages of astronauts who went to the Moon ended in divorce or separation.

Janet never participated very actively in any wives' clubs, being more of a loner, as Neil was. In the coming years, Janet's struggle for identity would only intensify, because she was no longer just *any* astronaut's wife; she bore the extra burden of being the wife of the first man on the Moon.

CHAPTER 16

For All America

Even before he completed his debriefings on the Gemini VIII flight in late March 1966, Armstrong was named backup commander for Gemini XI. So quickly and thoroughly did he get into his training for the new role that he was not even able to stay overnight with his family in Wapakoneta the day of his hometown's gala.

The Gemini IX and Gemini X missions took place within a span of seven weeks in June and July 1966, repeating the pattern of a relatively easy rendezvous followed by a problematic docking. The Agena intended for Gemini IX never made it into space. It spiraled deep into the Atlantic Ocean after its Atlas booster failed shortly after launch. For the Gemini X flight of July 18 to 21, Armstrong served as a CapCom in Houston. This time the docking worked, as Commander John Young nestled his machine to a solid hookup with a brand-new Agena. This was the first time that a manned spacecraft had fully embraced a target vehicle since Armstrong's Gemini VIII flight, and the first time it ever *stayed* embraced. Later in the mission, pilot Mike Collins performed a remarkable EVA lasting an hour and a half, an extremely welcome result.

For Armstrong, training as the backup commander was more about teaching than learning since XI was his third run-through after Gemini V and VIII. What most interested and concerned Armstrong about Gemini XI were those untested aspects of the mission, particularly regarding pilot maneuvers. Rendezvous with the Agena was supposed to occur on the spacecraft's first revolution about the planet, with a two-second launch window. This maneuver simulated the type of rendezvous that might be used by a lunar module with a command module after the LM returned from the surface of the Moon. A quick rendezvous was needed due to limitations on the ascent stage due to the LM's fuel supply. Some of the mission planners called it a "brute force" technique, since the spacecraft would be approaching the target vehicle at very high speed, whereas all earlier rendezvous flights had closed in on the target rather leisurely, waiting until the start of the fourth orbit before beginning to station keep. The other major novelty in Gemini XI was the experimental tethering of the Gemini spacecraft with the Agena via a nearly one-hundred-foot Dacron cord. One goal of the tether experiment, according to Armstrong, was to "find out if you could keep two vehicles in formation without any fuel input or control action." Another goal was to see whether tethering enhanced the stability of both spacecraft, thereby lessening the risk of their bumping into each other.

In the summer leading up to the launch, Armstrong and Anders helped Pete Conrad and Dick Gordon, the prime crew, work on the techniques required to carry out all the aspects of the Gemini XI mission. Much of that time the four men spent together in a beach house on the Cape. According to Neil, "We'd go out on the beach and work out trajectory procedures and rendezvous procedures by drawing diagrams on the sand and walking around our drawings and essentially acting out the procedure and working out the difficult parts that we didn't quite understand. This was a very relaxing but useful endeavor. Sometimes we'd have our cook from the astronaut quarters put together a picnic lunch and we'd take it out there with us, spend a few hours with no telephone to bother us, and we'd really concentrate on something."

Gemini XI launched on September 12, 1966. The rendezvous technique worked well. The spacecraft shattered the 475-mile-high world altitude record set just two months before by John Young and Collins in Gemini X when it rose to an orbital apogee of some 850 miles. The tether exercise caused several nervous moments. Dick Gordon's connecting the tether from the Gemini to the Agena during his first EVA turned into a major athletic contest. Nearly blind from sweat, Gordon sat perched upon the nose of the spacecraft trying to connect the tether to the target vehicle to which it was docked. Conrad ordered Dick to come back in after only thirty of the planned 107 minutes, so tired did Gordon appear. Even releasing the one-hundred-foot tether from its stowage container proved to be a chore, as the Dacron line got hung up on a patch of Velcro. Once linked, the line rotated oddly, occasionally causing such oscillations that Conrad needed to steady the vehicle with his controls. After being hogtied to the Agena for three hours, Conrad and Gordon happily put an end to the puzzling experiment by jettisoning the docking bar. On Gemini XII, Buzz Aldrin and Jim Lovell successfully completed the tether experiment and proved that the differential gravity between two orbiting vehicles tethered at slightly different altitudes permitted station keeping without the use of fuel.

Neil followed the Gemini XI mission from the CapCom station at Mission Control in Houston. With the flight's successful conclusion on September 15, and following his participation in several of the debriefings, Neil's responsibilities in the Gemini program came to an end.

There was one last Gemini flight, Gemini XII, from November 11 to 15, 1966. Jim Lovell, the commander, and Buzz Aldrin, the pilot, carried out an impressive rendezvous and docking flight involving fifty-nine revolutions of the planet. The most notable achievement of the flight was Aldrin's very successful five-hour EVA.

Most space program analysts concur that Gemini was a vital bridge between Mercury and Apollo. Indeed, all of the specified goals of Gemini had been achieved, and then some: demonstration of the ability to rendezvous and dock with a target vehicle; demonstration of the value of a manned spacecraft for scientific and technological experiments; performance of work by astronauts in space; use of a powered, fueled satellite to provide primary and secondary propulsion for a docked spacecraft; long-duration spaceflights without extraordinary ill effects on the astronauts; and precision landing of a spacecraft. Major records set during the Gemini program included the longest manned spaceflight (330 hours and 35 minutes), the highest altitude (851 miles), and the longest total EVA time for one astronaut (5 hours and 28 minutes, compiled by Aldrin on his three separate EVAs during Gemini XII). By the time Lovell and Aldrin reentered the atmosphere, bringing Gemini XII and the entire program to a close, time spent in space by a piloted U.S. spacecraft stood at 1,993 hours.

It chagrined Neil Armstrong to know that his abbreviated Gemini VIII flight accounted for only some ten hours of them.

Such disappointment was trivial to the tragic personal losses that Neil and Janet continued to suffer. On June 8, 1966, Neil's boss and best friend from his days back at Edwards, Joe Walker, was killed in a freak midair collision over the Mojave. It happened when Walker's F-104N Starfighter inexplicably flew too close to a plane with which he was flying in formation—the XB-70A Valkyrie, a $500 million experimental bomber that North American Aviation had designed for Mach 3–plus speeds—and became caught in the mammoth plane's extraordinarily powerful wingtip vortex. Walker died instantly. One of the Valkyrie pilots, air force major Carl S. Cross, died in the wreckage of the bomber. The other XB-70A pilot, Al White, a test pilot for North American, survived via the plane's ejection capsule, but not without some serious injuries. Magnifying the tragedy was that the deaths came during what amounted to a publicity shoot for General Electric.

In Houston, Armstrong got a phone call from Edwards shortly after the accident happened. It had only been three months since the fatal crash that killed his good friend Elliot See in the company of Charlie Bassett. Squeezed between those two fatal airplane accidents, Neil had survived his own near-disaster in Gemini VIII. Neil and Janet were among the seven hundred persons who attended Walker's emotionally charged funeral. "All my adult life had been interrupted by the loss of friends," Neil remarked.

• • •

In early October 1966, Neil took off on a twenty-four-day goodwill tour of Latin America. Touring with Armstrong was Dick Gordon, just off his Gemini XI flight, and Dr. George Low, the former deputy director for manned spaceflight at NASA Headquarters, who a few months earlier had become head of Apollo Applications at MSC. Joining them were their wives as well as other NASA personnel and members of other agencies such as the State Department. The entourage traveled fifteen thousand miles through eleven countries and made appearances in fourteen major cities. Everywhere the astronauts went, throngs of humanity lined the streets. Crowds rushed them. Throughout Latin America, they found the people "spontaneous, friendly, and extremely warm."

This trip was Neil's first brush with the iconic status that would later change his life so dramatically. In Colombia, the second country visited, "the reception was overwhelming," George Low wrote in his journal. In Quito, the capital of Ecuador, the people "were not satisfied to stay on the sidewalks" and gave the motorcade "just barely enough room for the cars to pass through." In São Paolo, Brazil, the entourage saw people hanging out of nearly every window. In Santiago, Chile, little old ladies clapped their hands overhead and yelled "Viva!" More than two thousand five hundred guests showed up at a formal dinner reception in Rio de Janeiro, each one of whom expected to shake hands with the astronauts. At the University of Brasília, one thousand five hundred people crowded into a five-hundred-seat auditorium to hear the astronauts speak. Over the course of the three-and-a-half-week journey, untold millions got a look at the visiting American astronauts. "Whenever possible," Low wrote, "Neil and Dick were out of their cars shaking hands, signing autographs, and fostering a very personal relationship." All over South America, the tour made front-page headlines and national television, as when the president of Venezuela, Raul Leona, and his children welcomed the Americans at La Casona, the presidential palace on the outskirts of Caracas. Venezuela, Colombia, Ecuador, Peru, and Bolivia mounted heavy security precautions. In La Paz, Bolivia, armed troops stood every quarter mile from the airport to the center of town. In Brazil, Paraguay, and Uruguay, there was almost no military security, but a police escort performed crowd control. In a few places, like Buenos Aires, crowds overwhelmed the visiting Americans. In several instances, the men from the State Department, USIA, and NASA were forced to handle security.

Other than being mobbed regularly by autograph seekers, there were surprisingly few incidents aside from several Vietnam War protests. As soon as Neil heard about the trip, he had enrolled in a Spanish conversation class. He'd also spent many evenings with a set of encyclopedias to learn all he could about the differences in the eleven countries they'd be touring. He and the other astronauts gave slide show presentations and answered questions about everything from technical queries to whether flying in space changed their view of God.

Armstrong's successful presentations were not what most impressed George Low about Neil. "Neil had a knack of making short little speeches in response to toasts and when getting medals, in response to questions of any kind," Low recalled. "He never failed to choose the right words." In his travel journal Low concluded, "All I can say is that I am impressed. Neil made a tremendous hit with the people."

Given the important role that George Low would play in future discussions about Apollo crew assignments, and later about which astronaut should be the first to step onto the Moon's surface, his extremely positive evaluation of Armstrong was an influential factor in Neil's subsequent fortunes as an astronaut.

Inside the State Department, USIA, and NASA, politically minded officials felt that the goodwill tour through Latin America had struck a blow for "the American way."

PART FIVE

APOLLO COMMANDER

As the day clock was ticking for takeoff, would you every night, or most nights, just go out quietly and look at the Moon? I mean did it become something like "my goodness"?

"No, I never did that."

—NEIL ARMSTRONG IN RESPONSE TO QUESTION POSED BY HISTORIAN DOUGLAS BRINKLEY DURING AN INTERVIEW IN HOUSTON, TEXAS, SEPTEMBER 19, 2001

CHAPTER 17

Out of the Ashes

By New Year's Day 1967, many believed that President Kennedy's "by the end of the decade" deadline for landing a man on the Moon might be achieved a couple of years ahead of schedule. The Gemini program had finished on a roll. Most of the Apollo spacecraft hardware was well on its way to being built. The powerful Saturn rocket that was to boost the Apollo spacecraft on its way to the Moon was getting closer to being operational. Although a number of astronauts had died in airplane crashes, nothing about the space program itself had been at fault. Everything about the program appeared to be proceeding. Beating the Russians to the Moon seemed a safe bet. Then, on January 27, 1967, a devastating accident occurred at Kennedy Space Center in Florida.

Between 6:31 and 6:32 P.M. EST, a fire flashed through the Apollo Block I command module as it sat atop its uprated Saturn IB rocket on Pad 34. In the cockpit were Apollo 1 astronauts Gus Grissom, Roger Chaffee, and Ed White. The crew was going through a dress rehearsal for a launch that was not scheduled to occur for three more weeks, when a stray spark erupted into an inferno. Seconds later, all three men were dead.

When the clock struck midnight, it was the Armstrongs' eleventh wedding anniversary, and that of the day Karen died.

An electrical wire on the floor of the spacecraft's lower equipment bay had become frayed, probably due to the procession of technicians in and out of the spacecraft in the days before the test. A spark from the frayed wire jumped into some combustible material, likely foam padding or Velcro patches. In the 100 percent oxygen atmosphere, even a momentary flicker became a firebomb. The three astronauts died from asphyxiation in a matter of seconds.

For a brief, horrible moment, the astronauts realized what was happening to them. Roger Chaffee yelled first through his radio, "Fire in the spacecraft!" followed by White's "Fire in the cockpit!" and then again by Chaffee's "We're on fire! Get us out of here!"

Fifteen seconds after Chaffee's first words, the Apollo 1 command module blew apart.

When the Apollo fire occurred, Armstrong was at the White House as part of a delegation of astronauts that included Gordon Cooper, Dick Gordon, Jim Lovell, and Scott Carpenter to witness the signing of an international agreement known by the complicated title Treaty on Principles Governing the Activities of States in the Exploration and Use of Outer Space. The astronauts called it the "non-staking-a-claim treaty" because it precluded land claims on the Moon, Mars, or any other heavenly body. The treaty—signed simultaneously in Washington, London, and Moscow and still in effect today—outlawed the militarization of space. It also assured the safe return of any astronauts making an unexpected landing in another country.

Following the signing was a reception in the Green Room of the White House, hosted by President Johnson and wife Lady Bird and attended by many dignitaries from the world over. The astronauts "worked the crowd," as per NASA directive. Afterward, Armstrong and the other astronauts went to their hotel.

When they entered their rooms at about 7:15 P.M., the astronauts saw the red message light on their telephones. The front desk relayed to Neil the urgent need to call the Manned Spacecraft Center. Dialing the number, he reached the Apollo program office. The man on the phone in Houston shouted to Neil, "The details are sketchy, but there was a fire on Pad 34 tonight. A bad fire. It is probable the crew did not survive." The NASA employee then told Neil not to leave the hotel, in order to avoid the media.

The astronauts headed into the hallway to find out what the others had heard. The considerate hotel owner arranged for them to occupy a large suite near their rooms.

Before congregating in the suite, each astronaut tried to call home. Neil could not reach Janet. Astronaut Alan Bean had telephoned Janet shortly after the accident and told her to get over to the Whites'. Pat White was not home when Janet arrived; she was picking up her daughter Bonnie from ballet class. Janet was waiting near the Whites' carport when mother, daughter, and son Eddie drove into the driveway. Janet "did not know anything when I went over there. I only knew there was a problem. When Pat and her children arrived, all I could say was 'There's been a problem. I don't know what it is,' and I didn't."

NASA sent over astronaut Bill Anders to tell Pat the horrible news. Janet recalled that a number of other friends arrived and they stayed until three A.M., comforting the distraught family members. Back in their suite, Neil and his fellow astronauts downed a bottle of scotch. Late into the night, they talked about what must have happened to cause such a disaster.

None of the astronauts liked the Block I spacecraft that North American Aviation had built for NASA, the early version of the Apollo command module that was to be test-flown in Earth orbit prior to any lunar mission. Certainly not Gus Grissom, who after one checkout run at the manufacturer's plant in Downey, California, had left a lemon on top of the CM simulator. As the long night in the Georgetown Inn wore on, the topic of conversation, as Jim Lovell remembers, moved "from concern for the future of the program, to predictions about whether it would now be possible to get to the Moon before the end of the decade, to resentment of NASA for pushing the program so hard just to make that artificial deadline, to rage at NASA for building that piece of crap spacecraft in the first place and refusing to listen to the astronauts when they told the agency bosses they were going to have to spend the money to rebuild it right." "I don't blame people for anything," Neil thought. "These types of things happen in the world we are living in, and you should expect them to happen. You just try your best to avoid them. And if they do happen, you hope you have the right kind of procedures, equipment, knowledge, and skill to survive them. I've never been a blamer."

As for the deaths of Grissom, Chaffee, and especially his good friend and neighbor, Ed White, "I suppose you're much more likely to accept a loss of a friend in flight, but it really hurt to lose them in a ground test." According to Neil, "that was an indictment of ourselves. It happened because we didn't do the right thing somehow. That's doubly, doubly traumatic. When certain things happen in flight, there is just nothing you can do to handle them. You are doing what you want to be doing, so injuries and even deaths are easier to accept than in a ground test, where there should be escapes available for any accident that occurs. As to why all the brainpower at NASA and the aerospace industry missed the danger of ground testing the spacecraft in a 100 percent oxygen-rich environment, well, it was *some* bad oversight. We'd been getting away with it for a period of time, as we had tested that way all through the Gemini program, and I guess we just became too complacent."

Four days after the fire, two separate funerals were held in honor of the three fallen astronauts. Naturally, Neil and Janet attended both of the funerals. The first, for Grissom and Chaffee, took place with utmost gravity and military dignity at Arlington National Cemetery. The funeral for Ed White took place later the same day inside the Old Cadet Chapel at West Point, as both Ed and his father had graduated from the U.S. Military Academy. Neil served as one of the pallbearers, along with four other members of what had been the New Nine: Borman, Conrad, Lovell, and Stafford. Buzz Aldrin also served as a pallbearer.

As tough as the deaths were on their fellow astronauts, it was much harder, of course, on the widows of the fire. Pat White attended both services, the only one of the widows to do so. She came to remember hardly any of it. What made it worse: NASA had tried to make her bury Ed in Arlington with Gus and Roger, instead of at West Point where the family knew he would want to be. For many months afterward she barely functioned. As next-door neighbor and best friend Janet remembered, Pat was "absolutely devoted" to Ed. She cooked him gourmet meals, handled all his outside correspondence. She was "the perfect wife and loved every minute of it." In late 1968, after Pat failed to show up for an exercise class and could not be reached by phone, Janet and Jan Evans (wife of Ronald Evans, a Group V astronaut who became Command Module Pilot for Apollo 17), knowing of Pat's chronic depression, feared the worst. They broke into her house and found Pat clutching a bottle of pills, which had to be wrestled from her hand. Other members of the "Astronaut Wives Club" rallied around her and her two children, and Janet remained her confidante until Pat's death in 1983, an apparent suicide following a bout with cancer.

The day after the fire, Dr. Robert Seamans, NASA's deputy administrator, announced the formation of an accident investigation board; the only astronaut to serve on the panel was Frank Borman. The Johnson administration allowed NASA to keep the investigation entirely in-house. The review board quickly found what had caused the accident. By April 5, the panel submitted its formal report stating that an arc from faulty electrical wiring in an equipment bay inside the command module had started the fire. In the 100 percent oxygen atmosphere, the crew had died of asphyxia caused by inhalation of toxic gases. The board report concluded with a list of eleven recommendations for hardware and operational changes.

It would take NASA two years to fix all the problems with Apollo. A special Apollo Configuration Control Board, chaired by George Low, eventually oversaw 1,341 design changes for the spacecraft. Never again would a grounded spacecraft risk the highly explosive 100 percent oxygen-rich atmosphere. On the launchpad, the cabin would hold an atmosphere of 60 percent oxygen and 40 percent nitrogen, while the astronauts breathed 100 percent oxygen through their separate suit loops. The cabin nitrogen would be bled off as the spacecraft ascended. Apollo was granted months to not only fix the spacecraft, but also rethink its previous decisions, and change many things for the better.

The Monday morning after the review board's report, Deke Slayton called together a group of his astronauts at the Manned Spacecraft Center. Slayton had invited only eighteen astronauts to attend, though the corps now numbered nearly fifty. Only one of the original Mercury astronauts, Wally Schirra, sat at the conference table. The rest came from the second and third groups of astronauts. Five of the men had not yet flown in space: Bill Anders, Walt Cunningham, Donn Eisele, and Clifton Williams. (Williams would be killed a few months later when his T-38 airplane crashed.) The other thirteen were veterans of at least one Gemini flight: John Young, Jim McDivitt, Pete Conrad, Schirra, Tom Stafford, Frank Borman, Jim Lovell, Gene Cernan, Armstrong, Dave Scott, Mike Collins, Dick Gordon, and Buzz Aldrin.

Slayton told them straight out: "The guys who are going to fly the first lunar missions are the guys in this room."

Every astronaut around the table knew that he had qualified as a finalist in the competition for the first lunar landing. The most likely candidates were the seven men from the New Nine who had already served as a commander for a Gemini flight: McDivitt, Borman, Stafford, Young, Conrad, Lovell, or Armstrong. Wally Schirra had upset Slayton by complaining about the nature of the Apollo 2 mission to which he had originally been assigned back in 1966. Deke responded by moving Schirra's crew out of the prime role for Apollo 2 to the backup crew for Apollo 1.

At the meeting, Slayton laid out the course of the entire Apollo program. The first manned Apollo mission, the one delayed by the fatal fire, would take place in approximately a year and a half, after a series of major equipment tests. NASA was now calling this first manned mission Apollo 7. In honor of Grissom, White, and Chaffee, there would be no other Apollo 1. There was also to be no Apollo 2 or 3. Slayton told his astronauts that the upcoming Apollo flights would proceed from type A through type J. The A mission, to be performed by the unmanned flights of Apollo 4 and Apollo 6, would test the three-stage Saturn V launch rocket as well as the reentry capabilities of the command module. The B mission, involving Apollo 5, would be an unmanned test of the lunar module. The C mission—which Apollo 7, the first manned flight, was to satisfy—would test the Apollo command and service modules (CSM), the Apollo crew accommodations, and the Apollo navigation systems in Earth orbit. D would test the combined operations of the CSM and the lunar module (LM), also in Earth orbit. E would also test the combined operations but do it in deep space. F amounted to a full-dress rehearsal for the lunar landing, while G would be the landing itself. Following the first landing came H, with a more complete instrument package aboard the LM for improved lunar surface exploration, followed by I, originally conceived as lunar-orbit-only flights with remote-sensing packages inside the CSM and no lander. NASA had made no plans beyond the J mission, which repeated H but with a lander capable of staying on the lunar surface for a longer period of time.

Slayton then named the first three Apollo crews. To the surprise of some, Deke called on Schirra to command Apollo 7, along with his crew of Eisele and Cunningham. Backing up Wally's crew would be Tom Stafford, John Young, and Gene Cernan. After

Schirra's crew had been moved out of Apollo 2 to serve as backup for Apollo 1 back in 1966, Stafford's crew had become the backup for Jim McDivitt's Apollo 2 crew. Now, however, McDivitt was to become commander for Apollo 8, the proposed first test of the lunar module. Serving on McDivitt's Apollo 8 crew were Dave Scott and Rusty Schweickart, with Pete Conrad, Dick Gordon, and C. C. Williams serving as backups. (Al Bean would replace Williams after Williams's death in December 1967.) Comprising the crew for Apollo 9, which was to be a manned test of the CSM and LM in high Earth orbit, was Frank Borman, Mike Collins, and Bill Anders. Armstrong, Jim Lovell, and Buzz Aldrin would serve as the Apollo 9 backup crew.

This took the Apollo crew assignments up through the D mission. For Armstrong, it was clear that his command of an Apollo mission could come no sooner than Apollo 11, as an astronaut had never moved from a backup crew to the very next prime crew. Considering that missions E and F would have to be accomplished before NASA moved on to the actual Moon landing, or the G mission, it appeared that the historic first step onto the Moon would not happen at least until Apollo 12. If Neil ended up getting the command for Apollo 11, he would be flying the dress rehearsal for the landing, not the landing itself.

CHAPTER 18

Wingless on Luna

Armstrong had begun to study the problem of how to land a flying machine on the Moon some seven and a half years before he became the commander of Apollo 11. "We knew that the lunar gravity was substantially different [roughly one-sixth that of Earth's]," Armstrong recalls of the engineering work begun at Edwards following President Kennedy's commitment in May 1961. "We knew that all our aerodynamic knowledge was not applicable in a vacuum. We knew that the flying characteristics of such a vehicle were going to be substantially different from anything we were accustomed to."

The astronauts had to attack the unique stability and control problems of a machine flying in the absence of an atmosphere, through an entirely different gravity field. "That was a natural thing for us," perhaps especially for Neil, "because in-flight simulation was our thing at Edwards. We used in-flight simulations to try to duplicate other vehicles, or duplicate trajectories."

The assistant director of research at the Flight Research Center, Hubert Drake, got the small group organized. Back in the early 1950s, Drake had played a similar catalytic role in conceptualizing ways to attain speeds of Mach 3 and altitudes over 100,000 feet in a research airplane, an initiative that led to the hypersonic X-15 program. Also attacking the problem of a lunar landing research vehicle were research engineers and frequent collaborators Gene Matranga, Donald Bellman, and Armstrong, the only test pilot involved.

The first idea that the Drake group considered was some form of helicopter, because of the helicopter's abilities to hover and to take off and land vertically. Unfortunately, helicopters could not replicate the consequences of lunar gravity. Another idea was to suspend a small lunar landing research vehicle beneath a giant gantry and "fly" the vehicle tethered. An even safer option was to go the route of an electronic, fixed-based simulator. Ultimately, NASA used all three methods—helicopters, a tethered Lunar Landing Research Facility (LLRF) at NASA Langley in Virginia, and different electronic fixed-base simulators—to study the problems of lunar landing and to train Apollo astronauts. Ultimately, Drake's group opted for VTOL (vertical takeoff and landing) technology. In VTOL, an aircraft equipped with translatable engines flew with some helicopter-like traits.

The Drake group mounted a jet engine in a gimbal placed underneath the test vehicle so that the thrust produced by the jet always pointed upward. The jet would lift the test vehicle to the desired altitude, whereupon the pilot would throttle back the engine to support five-sixths of the vehicle's weight, simulating the Moon's one-sixth gravity. The vehicle's rate of descent and horizontal movement would be handled by firing two throttle-able hydrogen peroxide lift rockets. An array of smaller hydrogen peroxide thrusters would give the pilot attitude control in pitch, yaw, and roll. If the primary jet engine failed, auxiliary thrust rockets could take over the lift function, temporarily stabilizing the machine. What was so radical about the concept was that aerodynamics—the science on which all flying on Earth was done—played absolutely no part. In this sense, the lunar landing test vehicle that Armstrong helped to conceptualize in 1961 was the first flying machine ever designed for operation in the realm of another heavenly body, yet one that could also fly right here on Earth.

Given the complexity of the project, Armstrong said they decided that "what we should do first was build a little one-man device that just investigated the qualities and requirements of flying in a lunar environment. With that, a database would grow from which we could build the bigger vehicle carrying the mockup of the real spacecraft." Through the summer and fall of 1961, the Drake team devised such a craft. According to Neil, "It looked like a big Campbell Soup can sitting on top of legs, with a gimbaled engine underneath it."

Unknown to the Drake group, another team of engineers at Bell Aerosystems in Buffalo, New York, was also exploring the design of a free-flight lunar landing simulator. The descendant of the company that had built the X-1 and other early X-series aircraft, Bell was the only American aircraft manufacturer with any significant experience in the design and construction of VTOL aircraft using jet lift for takeoff and landing. Drake heard about the Bell initiative from a NASA official, and Bellman and Matranga traveled to Buffalo, where they rode the company's Model 47 helicopters on simulated lunar descents. What the FRC engineers saw confirmed their suspicion that helicopters just could not fly the descent trajectories and sink rates required for a lunar lander.

NASA contracted with Bell to draw up blueprints for a small, relatively inexpensive lunar landing test vehicle whose design would be independent of the actual Apollo configuration, since the configuration had not yet been decided upon. Bell's job was to lay out a machine with which NASA could investigate the inherent problems of lunar descent from altitudes up to two thousand feet with vertical velocities of up to two hundred feet per second.

Not until July 1962 did NASA settle on how to go to the Moon. Many qualified engineers and scientists envisioned getting there and back in one brute rocket ship roughly the size of the Empire State Building. It would fly to the Moon, back down rear-end first to a landing, and blast off for home, in a mission mode called Direct Ascent. With twelve million pounds of thrust, the proposed Nova rocket was by far the most powerful booster ever built. A second major option for the lunar landing—and one that many spaceflight experts, including Dr. Wernher von Braun, came to favor—was Earth Orbit Rendezvous, or EOR. According to this plan, a number of the smaller Saturn-class boosters being designed by the von Braun team at Marshall Space Flight Center in Alabama would launch components of the lunar-bound spacecraft into Earth orbit, where those parts would be assembled and fueled for a trip to the Moon and back. EOR required far less complicated booster rockets, ones nearly ready to fly. To the surprise of many, NASA selected neither Direct Ascent nor Earth Orbit Rendezvous. On July 11, 1962, officials announced that a concept known as Lunar Orbit Rendezvous, or LOR, would be used. LOR was the only mission mode under consideration that called for a customized lunar excursion module to make the landing.

The LOR decision was made over the strenuous objections of President Kennedy's science adviser, Dr. Jerome Wiesner. Like other skeptics, Wiesner felt that if rendezvous had to be part of the lunar mission, it should be attempted only in Earth orbit. If rendezvous failed there, the threatened astronauts could be brought home simply by allowing the orbit of their spacecraft to decay. In the end, NASA's mission planners determined that LOR was no more dangerous than the other two schemes, likely even less dangerous, and that it enjoyed several critical advantages. It required less fuel, only half the payload, and somewhat less new technology. It did not require the monstrous Nova, and it called for only one launch from Earth, whereas the once-favored EOR required at least two. Trying to bring down a behemoth like the upper stage of a Nova onto the cratered lunar surface would be next to impossible. A Moon landing via EOR looked only marginally easier. After months of study, there was no choice but to go with LOR.

The greatest technological advantage of LOR was that it turned the lander into a "module." Only the small, lightweight lunar module (LM), not the entire Apollo spacecraft, would have to land on the Moon. Also, because the lander was to be discarded after use and would not return to Earth, NASA could customize the LM's design solely for maneuvering flight in the lunar environment and for a controlled lunar landing. In fact, with LOR, NASA could tailor all of the modules of the Apollo spacecraft independently—the command module (CM), service module (SM), and LM. The LM would be a two-stage vehicle that would descend to the surface using a throttle-able rocket engine. But the lower module, holding the landing legs, descent engine, and associated fuel tanks, would remain on the lunar surface and act as the launch platform for the upper or ascent stage, with its separate fixed-thrust engine, associated tankage, attitude control rockets, and cockpit.

Most important, LOR was the only mission mode by which the Moon landing could be achieved by Kennedy's deadline. For NASA, that was the clincher. Armstrong remembered the phrase, "LOR saves two years and two billion dollars." Overnight, a landing module became one of the most critical systems in the program. The big Saturn V rocket could propel astronauts inside their snug command module into lunar orbit, but Apollo was all about landing.

Immediately, serious work on the LM began. In November 1962, the Grumman Corporation of Long Island, New York, won the contract. The path to a finished LM involved many alterations. A long string of test failures kept the Grumman team busy fixing and refining its extraordinary machine for nearly seven years. Not until March 1969 was the first LM ready to test-fly. It took place in Earth orbit, as the primary task of Apollo 9.

With the LOR decision in hand, the requirements for the Flight Research Center's lunar landing research vehicle became much more explicit. Strictly by chance, the characteristics, size, and inertias of the original LLRV design were very much like what Grumman needed to build into the LM. Bell Aerosystems began fabricating two LLRVs of the same design in February 1963. On April 15, 1964, the machines arrived at Edwards disassembled and in boxes, because FRC technicians wanted to install their own research instruments. Ten feet tall and weighing 3,700 pounds, the LLRV had four aluminum truss legs that spread thirteen feet. The pilot sat out in the open air, behind a Plexiglas shield, in a rocket ejection seat built by Weber Aircraft. Weber's seat was so effective that it operated successfully at "zero-zero," the lowest point in an ejection envelope, and could do so safely even if the LLRV was moving downward at thirty feet per second. No ejection seat ever performed better, which was a good thing given that it would have to be used more than once in the LLRV program.

The first pilot to fly the LLRV was Neil's former boss Joe Walker. Walker made the inaugural flight on October 30, 1964. This flight consisted of three brief takeoffs and landings totaling just under a minute of flight time.

Between 1964 and the end of the LLRV test program in late 1966, some two hundred research flights were carried out at Edwards. Pilots could operate the vehicle in one of two modes. They could fly it as a "conventional" VTOL with the jet engine locked in position and providing all the lift: the "Earth mode." Or they could fly it in the "lunar mode" in which the engine could be adjusted in flight to reduce the apparent weight of the LLRV to its lunar equivalent. In the lunar mode, lift was provided by a pair of controllable five-hundred-pound-thrust rockets that were fixed to the fuselage outside the gimbal ring. The pilot could modulate the angle and thrust of the engine to compensate for aerodynamic drag in all axes. Generally, the pilots preferred flying the Earth mode. On the other hand, the sensitive throttle for the rocket engine made altitude control much better in lunar simulation.

The LLRV closely duplicated what it was like to fly over the Moon, though its highest altitude reached just under eight hundred feet, and its longest flight lasted less than nine and a half minutes. Amazingly, no serious accidents occurred during the LLRV program.

Armstrong had left Edwards for Houston in September 1962, so he was unable to stay as informed about the LLRV program as he would have liked. However, he was chosen by Houston to be the engineering pilot focal point, ensuring that the LLRV met the astronauts' needs. NASA did not want any of the astronauts flying the risky machine. Ground simulators offered considerable help. As Neil explained, "Traversing large pitch or roll angles required more time or larger control power. It was expected that control characteristics ideal on Earth might be not at all acceptable on the Moon." The astronauts found that good control could be obtained with "on-off" rockets that had been mechanized for rate command—that is, for the vehicle's angular rate (or rate of change) proportional to control deflection—but they were still, in Neil's words, having "some difficulty in making precise landings and eliminating residual velocities at touchdown, probably due to a pilot's natural reluctance to make large attitude changes at low altitudes."

The Lunar Landing Research Facility at Langley was an imposing 250-foot-high, 400-foot-long structure that had become operational in June 1965 at a cost of nearly $4 million. "It worked surprisingly well," said Armstrong. "The flying volume—180 feet high, 360 feet long, and 42 feet wide—was . . . adequate to give pilots a substantive introduction to lunar flight characteristics." To make the simulated landings more authentic, its designers filled the base of the huge eight-legged, red-and-white structure with dirt, and modeled it to resemble the Moon's surface. Often testing at night, they erected floodlights at the proper angles to simulate lunar light, and installed a black screen to mimic the airless lunar "sky." Technicians sprayed the fake craters black so the astronauts could experience the shadows that they would see during the Moon landing. Though "the engineers at Langley did some wonderful work trying to create a flexible [cable and pulley] system that allowed it to feel like a real flying spacecraft," control for pitch and roll could be, in Neil's words, "excessively sluggish." "The LLRF was a clever device," in Armstrong's judgment. "You could do things in it that you would not want to try in a free-flying vehicle, because you could be saved from yourself."

In 1964, the Astronaut Office looked around to see what VTOL machines might be available as possible lunar landing simulators. Deke Slayton asked Armstrong to look into the potential of the Bell X-14A, the small and versatile aircraft that engineers at NASA's Ames Research Center were using to simulate lunar descent trajectories. In February 1964, he made ten evaluation flights, but concluded that another class of training vehicle was required.

"Having no flying machines to simulate lunar control characteristics was frustrating," Armstrong recalled. The only effective alternative was to try the Flight Research Center's LLRV, however risky some considered it. Heading the LLRV program at Houston was Dick Day, the simulations expert from the Flight Research Center who back in 1962 had helped Neil to become an astronaut.

The decision to turn the LLRV into a trainer, or LLTV, came early in 1966, just prior to Armstrong's Gemini VIII flight. By this time, Grumman had come a long way toward finalizing the design of the LM. Although the LLRV predated the LM by five years, it was not that different in size and control-rocket geometry from Grumman's vehicle. Relatively quickly and inexpensively, NASA got Bell to produce an advanced version of the LLRV that even more closely matched the characteristics of the LM, whose first test flight, designated Apollo 5, was scheduled for January 1968.

The decision to build LLTVs brought Neil back into lunar landing studies. In the summer of 1966, as he was preparing for his backup role in Gemini XI, Houston ordered three LLTVs at roughly $2.5 million each. At the same time, the Manned Spacecraft Center requested that the Flight Research Center prepare its two LLRVs for shipping to Houston as soon as the FRC engineers were done with them. Neil participated in discussions with Bell on what was needed in the LLTV design. He was on the scene when LLRV number one arrived in Houston from Edwards on December 12, 1966. When FRC test pilot Jack Kleuver came to Houston to verify that the machine was working, Armstrong observed. When the first familiarization flights were made at Ellington AFB, Neil watched the operation and studied their ground rules. He spent January 5 to 7, 1967, participating in the LLTV Design Engineering Inspection at Bell. A few days later, he helped to review the final results of the LLRV program. While in California, Neil flew some LM trajectories in a Bell H-13 helicopter. He also witnessed an LLRV flight. Immediately after attending the funerals for the Apollo 1 crew in late January, Armstrong and Buzz Aldrin flew a T-38 to Langley Field in order to make simulated lunar landings on the LLRF. It was Neil's first time on Langley's gadget, and it would not be his last. On February 7, 1967, he and Buzz flew a T-38 to Los Angeles to be custom-fitted for an LLRV ejection seat at Weber Aircraft. Later in the month, he went again to Los Angeles, this time to North American (with Bill Anders), to review the design for the tunnel through which the astronauts would move back and forth between the Apollo command and service module (CSM) and the LM. In March 1967, he reviewed the LM landing radar program at Ryan Aircraft in L.A. and San Diego. During these months, he also got in a good bit of helicopter time in order to prepare for training in the LLTV. Helping transform the research vehicle into a training vehicle was a challenge for which Armstrong as an engineer, test pilot, and astronaut was extremely well suited. Back in 1961, he had contributed to the machine's original concept. Bell built the LLTV essentially on the same structure as the LLRV, but now the main goal was to replicate as closely as possible the trajectory and control systems of the LM. Certain flying characteristics of the LM could not be replicated, however. Most notably, it was impractical, if not impossible, to design the LLTV so that it provided the rate of descent that the LM had.

Another goal was to make the LLTV as much like the LM in terms of critical design features. For example, Bell built the new LLTVs with an enclosed cockpit that enjoyed LM-like visibility. To match the LM configuration, it also moved the control panel from the center of the cockpit to the right side, and set up the same array of visual displays. The LLTV was given a three-axis side-arm control stick comparable to what Grumman was placing into the LM, and a rate-command/attitude-hold control system closely approximating the handling characteristics anticipated for the LM. The LLTV also incorporated a compensation system that sensed any aerodynamically induced forces and moments, and provided automatic correction through the engine and attitude rocket system. In this way, the motions of the LLTV more closely approximated flight in a vacuum. Improvements were made in the electronics system to take advantage of the same miniaturized, lightweight components that were being used in the LM. Other improvements included an improved ejection seat, more peroxide for the rockets to increase their duration, a slightly upgraded jet engine, and a modified attitude to be more like the LM.

Two older modified LLRVs, dubbed A1 and A2, were also used for astronaut training. The three new machines, the first of which arrived from the Bell factory in December 1967, became LLTV B1, B2, and B3. Before the astronauts were allowed to fly any of them, they received a couple of months' flight instruction. The astronauts that Slayton designated as potential LM crewmen, including Armstrong, then went to helicopter school for three weeks, to Langley's LLRF for a week, and finally to fifteen hours in a ground simulator before they got their first chance to fly an LLTV, always at nearby Ellington. As Neil had already gone to navy helicopter school in 1963 and had built up quite a bit of "helo" time over the next four years, he only had to brush up on his helicopter skills. While flying helicopters was not optimal training for lunar flight, it was nonetheless valuable to understand trajectories and flight paths.

As an experienced engineering test pilot Armstrong did an outstanding job thinking through, often counterintuitively, what it took to fly in the unusual lunar environment, and not letting his helicopter training dominate his piloting decisions. Eventually, all prime and backup commanders of Apollo lunar landing missions practiced on the LLTV. As the program went on, there was not enough LLTV time available and the backup commanders were cut short. The astronauts who flew the LLTV besides Armstrong were Borman, Anders, Conrad, Scott, Lovell, Young, Shepard, Cernan, Gordon, and Haise.

Neil's initial LLTV flight came on March 27, 1967, when the machine first came to Ellington Field; he made two flights in LLTV A1 that day. Due to technical problems, the machine was not flown again after a few more flights in March. (None of the three new LLTVs were ready for flight testing until the summer of 1968.) When the LLTV came back on line, Armstrong was the first to get checked out in LLTV A1. Between March

27 and April 25, 1968, Neil made ten flights in the converted LLRV. The LLTV was a dangerous machine to fly. "Without wings," as Buzz Aldrin noted, "it could not glide to a safe landing if the main engine or the thrusters failed. And to train on it properly, an astronaut had to fly at altitudes up to five hundred feet. At that height, a glitch could be fatal." Armstrong found out just how unforgiving the machine could be on May 6, just fourteen months before the Apollo 11 landing.

"I wouldn't call it routine, because nothing with an LLTV was routine, but I was making typical landing trajectories during the flight that afternoon, and as I approached the final phase of one of them, in the final one hundred feet of descent going into landing, I noted that my control was degrading. Quickly, control was nonexistent. The vehicle began to turn. We had no secondary control system that we could energize—no emergency system with which we could recover control. So it became obvious as the aircraft reached thirty degrees of banking that I wasn't going to be able to stop it. I had a very limited time left to escape the vehicle, so I ejected, using the rocket-powered seat. The ejection was somewhere over fifty feet of altitude, pretty low, but the rocket propelled me up fairly high. The vehicle crashed first, and I drifted in the parachute away from the flames and dropped successfully in the middle of a patch of weeds out in the center of Ellington Air Force Base."

During the explosive ejection, the first he had experienced since abandoning his crippled Panther jet over Korea seventeen years earlier, Neil accidentally bit hard into his tongue. That was his only injury, except for a bad case of chiggers from the weeds, but it was a close call.

Those who observed the accident or who subsequently heard about it felt that Armstrong was very lucky to be alive. The cause of the accident turned out to be a poorly designed thruster system that allowed Armstrong's propellant to leak out. Loss of helium pressure in the propellant tanks caused the attitude rockets to shut down, producing loss of control. The fact that NASA was flying the vehicle in such windy conditions was a major contributing factor. At Edwards, the FRC engineers had put a fifteen-knot limit on wind speed for LLRV flying, but the Houston staff felt they had to raise it to thirty knots in order to be able to use the machine on a regular basis.

After his accident Armstrong behaved, typically, as if absolutely nothing out of the ordinary had just happened. Upon returning from a late lunch, astronaut Al Bean saw Neil at work at his desk in the office the two men shared. A little later, Bean went out in the hallway and walked over to a group of colleagues who were talking; he thought he heard them say that somebody had just crashed the LLTV. According to Bean, "I'm saying, 'What happened?' and they said, 'Well, the wind was high and Neil ran out of fuel and bailed out at the last minute and the ejection seat worked and he lived through it.' I said, 'When did this happen?' They said, 'It just happened an hour ago.' 'An hour ago!' I said, 'That's bullshit! I just came out of my office and Neil's there at his desk. He's in his flight suit, but he's in there shuffling some papers.' And they said, 'No, it was Neil.' I said, 'Wait a minute!' So I go back in the office. Neil looked up and I said, 'I just heard the funniest story!' He said, 'What?' I said, 'I heard that you bailed out of the LLTV an hour ago.' He thought a second and said, 'Yeah, I did.' I said, 'What happened?' He said, 'I lost control and had to bail out of the darn thing.' "

Bean continued his story: "Offhand, I can't think of another person, let alone another astronaut, who would have just gone back to his office after ejecting a fraction of a second before getting killed. He never got up at an all-pilots meeting and told us anything about it. That was an incident that colored my opinion about Neil ever since. He was so different than other people."

Once more, as had been the case in his Gemini VIII flight, Armstrong rightfully came out of the experience with an enhanced reputation for being able to handle an emergency situation. Houston grounded the LLTV pending the findings not only of the MSC's accident investigation team but also of a special review board. By mid-October 1968, the two reports were out, urging LLTV design and management improvements, yet clearing the program to continue.

Four minutes into a planned six-minute flight on December 8, 1968, MSC chief test pilot Joe Algranti was forced to "punch out" from LLTV 1 when large lateral-control oscillation developed as he descended from a maximum altitude of 550 feet. Ejecting at two hundred feet, Algranti, who had flown the LLTV more than thirty times, landed by parachute uninjured, while the $1.8 million vehicle crashed and burned several hundred feet away. Once again, Houston convened an accident investigation board, headed this time by astronaut Wally Schirra.

MSC director Bob Gilruth and MSC's Director of Flight Operations Chris Kraft both felt that it was only a matter of time before an astronaut would be killed in the instrument. "Gilruth and I were ready to eliminate it completely," Kraft noted, "but the astronauts were adamant. They wanted the training it offered."

LLTV flying resumed in April 1969. When nothing went wrong in the first few flights involving only MSC test pilots, routine training flights for the astronauts began again. Even after the lunar landings began, either

Kraft or Gilruth "grilled every returning astronaut, hoping to find some way to get the LLTV grounded forever." They lost every time, because the astronauts wanted it.

For three straight days in mid-June 1969, less than a month before the launch of Apollo 11, Armstrong flew one of the new LLTVs while Kraft and other NASA managers held their breath. Over the course of those three days, he took the LLTV up for lunar descents eight separate times. In all, he made nineteen flights in the converted LLRVs and eight flights in the new LLTVs. No other astronaut before or after Armstrong flew the vehicle so much.

CHAPTER 19

Amiable Strangers

The trio of Frank Borman, Jim Lovell, and Bill Anders were farther away from home than any human beings had ever been. Gradually slowing from a top speed of approximately twenty-five thousand miles per hour, the crew of Apollo 8 had just passed the point where the gravity of the planet and its natural satellite balanced out. Now Apollo 8 would be "falling" toward the Moon.

It was midafternoon, December 23, 1968, just following the crew's live television transmission of a grainy yet very recognizable view of the Earth from over two hundred thousand miles away, or more than four-fifths of the way to the Moon. In Houston's Mission Control, a new shift of flight controllers, the so-called Maroon Team under flight director Milton L. Windler, was preparing for the spacecraft to reach the critical point where the astronauts could insert their spacecraft into humankind's first lunar orbit. If the burn failed, Apollo 8 would by default swing around the Moon on a slingshot path back toward Earth.

Armstrong stood in the back of Mission Control, quietly pondering the upcoming lunar orbit insertion. As the backup commander for Apollo 8, Armstrong had spent every moment of the last two and a half days deeply involved with the details of the circumlunar flight. At the Cape on the morning of the launch on December 21, Neil had awakened at 3:00 A.M. so he could eat breakfast with the prime crew. When Borman, Lovell, and Anders were painstakingly suiting up, Neil hustled over to Launchpad 39A. It was customary for one or two members of the backup crew to monitor the prelaunch sequence from inside the cockpit and to set and check all the switches.

Launch came just a few minutes later than scheduled, at 7:51 A.M. The first manned flight of the Saturn V "Moon Rocket" was something to behold. Neil watched the slow, fiery ascent of the gigantic booster from the big window inside Launch Control Center in the company of Buzz Aldrin and Fred W. Haise Jr., his fellow backup crew members.

Into the early afternoon, Armstrong monitored the progress of the flight, through its two Earth orbits, through its translunar injection, and well on its way toward the Moon. Then, along with Aldrin and Haise, Neil boarded a NASA Gulfstream and headed back to Houston, arriving there at about 7:00 P.M.. In the plane with them were their wives, Janet, Joan Aldrin, and Mary Haise. The women had watched the launch from the VIP viewing stands, providing moral support for the astronauts' wives who were also in attendance.

After a quick trip home to El Lago to shower and change clothes, Neil drove over to Mission Control. Though he stayed there late, the next morning he returned early. Spotting Armstrong in the big room full of consoles, Deke Slayton approached with a pressing topic: Neil's next assignment.

Of course, no one was yet sure what the mission for Apollo 11 would be. For Apollo 11 to become the first lunar landing mission, not only would Apollo 8 need to complete its bold around-the-Moon flight successfully, but Apollo 9 and Apollo 10 would also have to come off without a hitch. If anything went wrong, the G mission, the first landing, could easily fall back to Apollo 12 or even to Apollo 13. If the deadline got too tight, NASA might even move the landing up to Apollo 10. In the wake of Apollo 8's audacity, even something as daring as that lived in the realm of possibility. Still, Armstrong's assignment to Apollo 11 looked fortuitous. He left the brief meeting with Slayton knowing there was a chance that he was going to be commanding the first Moon landing attempt.

It took an extraordinary turn of events for the Apollo missions to line up the way they did. In the original schedule that Slayton in April 1967 had first outlined to the Apollo astronauts, there was to be no such circumlunar flight. Following the first manned Apollo flight, or C mission—handled in October 1968 by Schirra, Eisele, and Cunningham in Apollo 7—the D mission was supposed to test the combined operations of the CSM and LM in Earth orbit. But Grumman's LM was not ready to fly. Wanting to keep the momentum of the program going, a few risk-takers in NASA, notably George Low, proposed a radical stopgap. Since the LM was not yet ready, why not expedite the flight sequence by flying the CSM around the Moon?

The idea was so bold that initially NASA leadership in Washington strongly resisted it. Yet as soon as Apollo 7 in October 1968 proved to be an unqualified success, that is exactly what NASA decided to do. The fact that the Soviet Union in September 1968 had just sent Zond 5 on a lunar flyby, and was preparing Zond 6 for the same sort of flight in November, helped cure the indecision. Zond was large enough to carry a cosmonaut, and the idea in the American mind ever since Sputnik was that the Soviets would do everything they could to keep upstaging America in space. Inside NASA, the intrepid move of a circumlunar flight met with significantly less than universal acceptance. Armstrong supported the radical redirection of Apollo 8, as did virtually all of the astronauts, but not until he was convinced that the problems with the Saturn V were fixed.

On December 23, 1968, as things were settling down following the end of the TV transmission from Apollo 8, Armstrong and Slayton retired to one of the back rooms at the Mission Control Center for what would prove to be a historic conversation.

"Deke laid out his thinking about Apollo 11, and asked how I felt about having Mike Collins and Buzz Aldrin as my crew. We talked about it a little bit, and I didn't have any problem with that. And Deke said that Buzz wasn't necessarily so easy to work with, and I said, 'Well, I've been working with him the last few months, and everything seems to be going all right.' But I knew what Deke was saying. Then he said he wanted to make Jim Lovell available for Apollo 11 even though it would be a little bit out of sequence, but that's what he'd do, if that is what I thought I needed. I would have been happy to get Lovell. Jim was a very reliable guy, very steady. I had a lot of confidence in him. It would have been highly unusual for the crew assignment to have worked out this way, but Deke offered the possibility that it would be Jim Lovell and Mike Collins as my crew."

Armstrong took until the next day to give Deke his answer. By then, the crew of Apollo 8, with Jim Lovell piloting the command module, was in orbit around the Moon. Lovell would never learn that if Armstrong's answer had been different, he would have become a member of the Apollo 11 crew. "Jim had already been commander of Gemini XII," Neil stated, "and I thought he deserved his own command. I thought it would be not right of me to pull Lovell out of line for a command, so he ended up with Apollo 13. To this day, he doesn't know anything about that. I have never related these conversations that I had with Slayton to anyone. As far as I know, Buzz doesn't know about them, either." If Armstrong had taken Lovell for Apollo 11, Aldrin would have been pushed back to a later crew, probably to the ill-fated Apollo 13.

Neil answered Slayton in the way he did because he had had no trouble working with Buzz, and Lovell deserved his own command.

One might wonder how Fred Haise felt about the development—after all, Haise was part of Neil's backup crew on Apollo 8 as the LM pilot. "Deke didn't think that Fred was quite ready for a prime crew," Armstrong recalled. "We had talked a little bit about the mission, and Deke said it could be a lunar module landing attempt—could be—though I thought that was a bit remote at that point." If Haise had been put on Neil's Apollo 11 crew, then Aldrin would have remained in the job of CM pilot. The critical factor was that Slayton wanted to get Collins, who had been out due to neck surgery, back in the sequence. Prior to their assignment as backup for Apollo 8, Armstrong had not had much interaction with either Aldrin or Haise. They had never served on a crew together before. Neil had worked on Gemini XI at the same time that Lovell and Aldrin were working on Gemini XII, so Neil and Buzz were at the Cape together a lot and saw each other frequently. Fred Haise he had seen much less. As part of his collateral responsibilities, Haise worked a lot on the lunar modules, and it was clear to Neil that Fred knew a lot about that area. Haise became the LM pilot for the Apollo 11 backup crew. Providing the rest of the backup crew was Apollo 8's Jim Lovell, the backup commander, and Bill Anders, the backup command module pilot. Frank Borman had retired.

Armstrong was content that Mike Collins and Buzz Aldrin would serve with him on Apollo 11.

That Christmas Eve was extraordinary, not just for Armstrong but for everyone who was glued to their television sets watching the live transmission from Apollo 8 in lunar orbit. People would remember the day for the rest of their lives.

During their live broadcast that evening, Borman, Lovell, and Anders took turns reading the first ten verses of the Book of Genesis while television viewers watched wondrous pictures of the Moon's surface passing surreally underneath, in an almost godlike view. The astronauts next pointed their lightweight TV camera back at the home planet to show the awesome and delicate beauty of a waxing Earth "rising" gloriously above the lunar surface. The astronauts concluded their lunar vigil with the hopeful message "Merry Christmas and God bless all of you, all of you on the good Earth." A few hours later, on Christmas morning, Apollo 8 ignited its service propulsion system, the main rocket engine of the command and service module, and accelerated out of lunar orbit. Delighted to be homeward bound, Lovell remarked as the spacecraft came around the back side of the Moon, "Please be informed, there is a Santa Claus."

Apollo 8 splashed down safely on the morning of December 27, six days and three hours, two Earth orbits, and ten lunar orbits after launch. It had been a truly historic flight. Not only were Borman, Lovell, and Anders the first humans to break the bonds of Earth's gravity, their journey proved that astronauts could travel the nearly quarter of a million miles separating the home planet from its nearest neighbor. Their mission proved that course-correction maneuvers could be done out of line of sight and out of communications with Earth, that a craft could be tracked from an immense distance, and that it could successfully orbit the Moon and return.

The year 1968 had been extraordinarily traumatic for America, not that 1967 had been much better. The paroxysms started in January 1968 when North Korea seized the USS *Pueblo*, claiming the American ship had violated its territorial waters while spying. The Vietcong launched the Tet Offensive, and in March, the world learned of the Mȳ Lai massacre. The same month, Lyndon Johnson announced that he would not seek reelection. In a span of nine weeks, Martin Luther King Jr. and Senator Robert F. Kennedy were assassinated. At the Democratic National Convention in Chicago that August, Mayor Richard Daley's police department clashed with angry crowds of demonstrators. Internationally, Israel and Jordan clashed in a border dispute. Violence ignited between Protestants and Catholics in Northern Ireland, initiating "The Troubles"; in Paris, protesters nearly brought down the government; and Soviet tanks rolled into Czechoslovakia. Amid such upheaval, spending billions on trips to the Moon, in the view of many people, seemed a misguided priority. A popular saying began: "If we can go to the Moon, why can't we" and finished with any number of objectives: end injustice, eradicate poverty, cure cancer, abolish war, and clean up the environment.

Even amid the clamor, criticism, and declining support for the space program, the astronauts remained the object of public adulation.

• • •

On January 4, 1969, Slayton called Buzz Aldrin and Mike Collins into his office where Armstrong was already on hand, and told them they were being named the prime crew for Apollo 11. Deke said it was "conceivable and may work out that this will be the first lunar landing attempt." Deke then added that he wanted them to conduct their preparation for the flight on the assumption that the landing was going to happen, so that, if it did, they would be totally ready to carry it out.

NASA announced the Apollo 11 crew to the public five days later, on January 9. The announcement came following ceremonies in which the Apollo 8 crewmen were awarded medals from President Johnson at the White House and received standing ovations at a joint meeting of Congress attended by the president's cabinet, the Supreme Court justices, and the diplomatic corps. Armstrong, Collins, and Aldrin were not present in Washington when, as many newspapers put it, the "Moon Team Is Named." They did appear at an Apollo 11 press briefing in Houston the next day. Mike Collins later called himself, Armstrong, and Aldrin "amiable strangers," a description that made Apollo 11 unique in terms of the way in which the crew related to one another.

Collins was the trio's most lighthearted member. His father, General James Collins, had fought with General Pershing in the Philippines and won the Silver Star in WWI. His brother, James Jr., was a field artillery battalion commander in WWII, and later became a brigadier general. An army brat, Mike moved regularly with his family from state to state. Yet despite moving often, Mike was always popular with other kids and with his teachers. He had a gift for leadership, for getting along with others, for thinking clearly, and for expressing himself well. He graduated from West Point in 1952, in the same class as Ed White. Frank Borman was two years ahead of him, Buzz Aldrin one year behind. He went into the air force and by 1956, First Lieutenant Collins was a part of a fighter squadron flying F-86s in France. In 1957, he married a young woman from Boston named Patricia. While still based in Europe, Collins applied for the air force test pilot school at Edwards. Not until 1961 did the school admit him. The school had changed its name to the USAF Aerospace Research Pilot School (ARPS) and begun to build a program designed to train U.S. military test pilots for spaceflight. Collins became part of ARPS Class III. With him in Class III were Charlie Bassett, the astronaut killed with Elliot See in 1966, and Joe Engle, who eventually became the only person to fly into space in two different winged vehicles, the X-15 and the Space Shuttle. In all, twenty-six ARPS graduates earned astronaut's wings by flying in the Gemini, Apollo, or Space Shuttle programs. When NASA named its third group of astronauts in 1963, Collins was one of them, specializing in pressure suits and extravehicular activity. For the December 1965 flight of Gemini VII, he was Jim Lovell's backup. Mike's first spaceflight came in July 1966 on Gemini X, an exciting mission that achieved a successful docking with the Agena target vehicle during which Collins performed a spacewalk that retrieved a micrometeorite package that Dave Scott on Gemini VIII had been unable to bring back. Mike's first Apollo assignment was backup to Walt Cunningham on the second Apollo flight. In the shakeup after the launchpad fire, Collins was to be the command module pilot for what became Apollo 8. Due to surgery for a bone spur in his spinal column, however, he was replaced by Jim Lovell. Thanks to a rapid recovery, he got paired with Armstrong and Aldrin for Apollo 11.

In under eight years, Collins went from a fighter pilot who had a hard time getting into test pilot school, to the command module pilot for the first lunar landing. Armstrong naturally liked Collins, who was good-humored and liked to joke, yet was also thoughtful, articulate, and learned. Long after the Apollo 11 mission, Mike would comment: "A closer relationship, while certainly not necessary for the success or happy completion of a spaceflight, would seem more 'normal' to me. Even as a self-acknowledged loner, I feel a bit freakish about our tendency as a crew to transfer only essential information, rather than thoughts or feelings."

Born January 30, 1930, in Glen Ridge, New Jersey, Edwin Eugene Aldrin Jr. was the third child and only son of a pilot in the army air corps during World War I. A highly educated man, before going into the military, Gene Aldrin earned a doctor of science degree from MIT. Resigning from the air corps in 1928, Aldrin Sr. became a stockbroker who got out just before the crash.

Gene Aldrin met his wife Marion, daughter of an army chaplain, while serving in the Philippines. Buzz's autobiographical 1973 book *Return to Earth*, best known for its candid account of his battle with alcoholism and depression in the years immediately following the Apollo 11 flight, represents his growing up as a classic case of a boy desperately seeking the approval of a strong father. Settling his family in Montclair, Gene Aldrin became an executive with Standard Oil of New Jersey and was rarely home. Leaving Standard Oil in 1938, he became an independent aviation consultant. Among his professional associations were Charles Lindbergh, Howard Hughes, and Jimmy Doolittle.

With so much talk about aviation in the Aldrin home, Buzz naturally took an interest in flying. When the United States entered World War II, Buzz was eleven. Aldrin Sr. returned to duty as a full colonel in the South Pacific, and later, in Europe, he studied antisubmarine warfare. When the war ended, his father was serving as chief of the All Weather Flying Center at Wright Field in Ohio. After high school Buzz went to West Point, and graduated third in his class; his father wanted to know who finished first and second. Buzz fought in Korea with the 51st Fighter Wing; his outfit, flying the F-86 interceptor, arrived in Seoul the day after Christmas 1951. That same day, as cold winds hit nearly one hundred miles per hour in the Sea of Japan, Neil Armstrong, on the carrier *Essex*, left Yokosuka for his third round of combat flying over North Korea. By the time the final cease-fire was negotiated in 1953, Aldrin had flown a total of sixty-six missions, including three encounters with Soviet MiGs.

Back in the States, Aldrin reported to Nevada's Nellis Air Force Base as a gunnery instructor. He married and the following year, 1955, he applied and received a three-month assignment to Squadron Officer School in Alabama. Finishing the officers' school, the Aldrins went to Colorado Springs, where Buzz served as aide to General Don Z. Zimmerman, dean of faculty at the new U.S. Air Force Academy. In August 1956, Buzz joined the 36th Fighter-Day Wing, stationed in Bitburg, Germany, and lived there for three years. Buzz flew the F-100, the most sophisticated fighter in the air force, and practiced nuclear strikes against targets behind the Iron Curtain. By now the Aldrins had three children. One of the friends Buzz made at Bitburg was Ed White. In

1958, when White completed his time at Bitburg, Ed enrolled at the University of Michigan for graduate work in aeronautics. By now Aldrin had set his sights on the air force's experimental test pilot school at Edwards, but like White, he wanted more education first. He asked the air force to send him to MIT, and in three years' time, he finished a doctor of science degree. In the spring of 1962, as he was starting work on his doctorate, Aldrin had applied for the second class of astronauts, to which Ed White and Neil Armstrong were selected. By the time NASA announced that it would be selecting a third class of astronauts, Buzz had orders to work on Defense Department experiments that would be flying aboard the Gemini spacecraft, but he got to Houston another way. Passing the battery of psychological and physical tests, Major Edwin E. Aldrin Jr. became one of the fourteen astronauts announced to the public on October 17, 1963, the same class as Mike Collins. Assigned early in his training to mission planning, Aldrin became a member of the MSC panel on rendezvous and reentry.

As an astronaut, Aldrin exhibited a curious combination of ambitiousness and naïveté, of maneuvering and total directness. Unsure how Deke Slayton selected crews, Buzz decided to ask him. However, this tactic backfired, and Aldrin was assigned to back-up behind Cernan for Gemini X. Under the prevailing custom, he would skip two flights and be on the prime crew of Gemini XIII. But with the program to end with Gemini XII, there would be no Gemini XIII.

The deaths in February 1966 of Elliot See and Charlie Bassett, the original crew for Gemini IX, altered the order. Jim Lovell and Buzz moved up from backup Gemini X to backup Gemini IX. With all the assignment shifting, that inked Lovell and Aldrin into the slots as the prime crew for Gemini XII, the program finale.

The backyard of the Aldrin home in Nassau Bay connected to the backyard of Charlie and Jeannie Bassett. The two families and their children had become good friends. One day Jeannie Bassett took Buzz aside and reassured him that "Charlie felt you should have been in it all along. I know he'd be pleased."

Buzz went on to make one of the most successful flights in the Gemini program, featuring his remarkable five-hour EVA.

Armstrong still did not know Aldrin very well when the two men, along with Jim Lovell, were assigned as backup for Apollo 9, the prime crew for which was originally Frank Borman, Mike Collins, and Bill Anders. Eventually, when NASA decided to make Apollo 8 the circumlunar mission, Apollo 9 became Apollo 8, and Apollo 8, scheduled as the first flight to go with the LM, became Apollo 9. Mike Collins's back surgery then shifted the crew assignments. Lovell replaced Collins on Apollo 8, and Fred Haise took Lovell's place on the backup crew with Armstrong and Aldrin. In Aldrin's view, it was automatic that Buzz would stay with Armstrong when their turn came for a prime crew, probably as Apollo 11. Yet Slayton gave Armstrong the option of replacing Aldrin with Lovell.

Slayton may have originally teamed Aldrin with Armstrong on the backup crew for Apollo 9 (which became Apollo 8) because he felt that the other commanders would not work as well with Buzz as Neil would. Deke recognized that Aldrin's personality grated on several of the other astronauts. "I'm not sure I recognized at that point in time what might be considered eccentricities," Armstrong related. "Buzz and I had both flown in Korea, and his flying skills, I was sure, were good. His intelligence was high. He was a creative thinker, and he was willing to make suggestions. It seemed to me that he was a fine person to work with. I really didn't have any qualms with him at that point." As the chief technician responsible for the final sealing of the astronauts into their capsule, Guenter Wendt saw all the Apollo crews in action. The Apollo 11 crew "just didn't seem to gel as a team. Usually when a mission crew was named, they stuck together like glue. You saw one, you saw all three, together. But these three! If we broke for lunch, they always drove away separately. There did not seem to be much camaraderie between the three men. I've always said that they were the first crew who weren't really a crew."

CHAPTER 20

First Out

The first question asked by a reporter at the Houston press conference introducing the crew of Apollo 11 on January 9, 1969, got right to the issue: "Which of you gentlemen will be the first man to step out onto the lunar surface?" Deke Slayton replied for the astronauts, "I don't think we've really decided that question yet. We've done a large amount of simulating, and I think which one steps out will be dependent upon some further simulations that this particular crew runs."

Thus emerged a critical issue in the life of Neil Armstrong, one that has provoked questioning, speculation, and controversy from 1969 to the present. How exactly did NASA decide which of the two astronauts inside the LM would be the first to step out onto the Moon?

Through the first months of 1969, there is no doubt that Aldrin believed he would be the first to step out onto the lunar surface. As Buzz has explained, "Throughout the short history of the space program the commander of the flight remained in the spacecraft while his partner did the moving around. I had presumed that I would leave the LM and step onto the Moon ahead of Neil." Leading metropolitan papers ran a story by a space-beat correspondent whose headline read "Aldrin to Be First Man on the Moon." During the Apollo 9 mission a few weeks later, Dr. George E. Mueller, NASA's associate administrator for manned space flight, told a number of people, including some reporters, that Aldrin would be the first out on Apollo 11.

Buzz felt confident about the situation until he heard rumors of a different "outcome" in the days following Apollo 9's splashdown. Aldrin heard through the MSC grapevine that it had been decided that Armstrong would go out first rather than he. Initially, the news only puzzled him. When he heard that NASA wanted Neil to do it because he was a civilian as opposed to someone who was still serving in the military, however, Buzz became angry.

For a few days, a chagrined Aldrin mulled over the situation, consulting only with his wife. Feeling that "the subject was potentially too explosive for even the subtlest maneuvering," Buzz decided to take the direct approach. He went to Neil.

If Aldrin expected a definitive resolution from Armstrong, he was sadly mistaken. "Neil, who can be enigmatic if he wishes, was just that," Buzz recalled. "Clearly, the matter was weighing on him as well, but I thought by now we knew and liked each other enough to discuss the matter candidly." In his 1973 autobiography *Return to Earth*, Aldrin wrote that Neil "equivocated a minute or so, then with a coolness I had not known he possessed he said that the decision was quite historical and he didn't want to rule out the possibility of going first." Aldrin has since claimed that the description of this incident in his book was exaggerated by his 1973 coauthor. "I understood that it was typical not to get anything decisive on this from Neil, particularly when it was really not his decision to make. His observation about the historical significance of stepping out first, which he did make, was perfectly valid, and I understood it as such. It was also clear to me that Neil did not want to discuss the matter further. There was absolutely no indication from him of, 'Yes, I think you're right. I think I'll push someone to make a decision.' There was no indication at all that that was going to happen."

Aldrin tried in vain to curb his mounting frustration, "all the time struggling not be angry with Neil." Pushing him hard, as always, was Buzz's own father. In a phone conversation with his dad, Buzz had suggested that the decision might very well go in favor of Neil's egressing the lunar module first. The senior Aldrin became "instantly angry," telling his son that he "intended to do something about it." According to Buzz, "it took a great deal of persuasion, but I finally got him to promise he'd stay out of it." The promise was not kept, however, as Gene Aldrin called on some influential friends who had connections inside NASA and the Pentagon.

As if to get ahead of his father, Buzz approached a few of his fellow astronauts, particularly those like Alan Bean and Gene Cernan who he imagined might be sympathetic because they were in the same position as he, as lunar module pilots for Apollo 10 and Apollo 12. Instead of a constructive reaction to his overtures, however, Aldrin's private conversations led to the general notion that Buzz was lobbying behind the scenes to be first. According to Gene Cernan, Aldrin had "worked himself into a frenzy" over who was going to be the first man to walk on the Moon. "He came flapping into my office at the Manned Spacecraft Center one day like an angry stork, laden with charts and graphs and statistics, arguing what he considered to be obvious—that he, the lunar module pilot, and not Neil Armstrong, should be the first down the ladder on Apollo 11. Since I shared an office with Neil, who was away training that day, I found Aldrin's arguments both offensive and ridiculous. Ever since learning that Apollo 11 would attempt the first Moon landing, Buzz had pursued this peculiar effort to sneak his way into history, and was met at every turn by angry stares and muttered insults from his fellow astronauts. How Neil put up with such nonsense for so long before ordering Buzz to stop making a fool of himself is beyond me."

Apollo 11 crewmate Mike Collins recalled a similar incident. "Once Buzz tentatively approached me about the injustice of the situation," Collins remembered, "but I quickly turned him off. I had enough problems without getting into the middle of that one. Although Buzz never came out and said it in so many words, I think his basic beef was that Neil was going to be the first to set foot on the Moon." Aldrin has insisted that his fellow astronauts misinterpreted his motive. "I didn't really want to be first," claimed Buzz, "but I knew that we had to have a decision."

With highly unflattering talk building inside the MSC about what many believed to be Aldrin's lobbying campaign, Slayton tried to put an end to it. Deke dropped by Buzz's office to say that it would probably be Neil who would be out first. At least Slayton gave Buzz a more palatable reason for the pecking order. "Neil was a member of the second group of astronauts, the group ahead of the group to which I belonged," Aldrin related. "As such, it was only right that he step onto the Moon first, as Columbus and other historical expedition commanders had done. For the decision to have been the other way, to have the commander sitting up there watching the junior guy go out, kicking up the dust, picking up the contingency sample, saying the famous words and all that, the mission would have been so criticized by all sorts of people. It would have been so inappropriate."

According to Aldrin, he was okay with Slayton saying it would be Neil; what had frustrated him all along were the effects of everyone's not knowing: "Whether or not I was going to be the first to step onto the Moon was personally no great issue. From a technical standpoint, the great achievement was making the first lunar landing, and two of us would be doing that." Buzz fully understood that "the larger share of acclaim and attention would go to whichever of us actually made the step," and, according to his own testimony, that was fine with him if it was to be Neil, because he was not after the acclaim. What he did resent, however, was how "the decision was stalled and stalled, until finally it was the subject of gossip, speculation, and awkward encounters" in which friends, family, and reporters kept asking him, "Who is going to be first?" Armstrong possessed the type of stoic personality that easily handled such ambiguity and uncertainty, whereas Aldrin did not.

Buzz felt compelled to make one last appeal for clarity: "I went finally to George Low, director of the Apollo program office, and explained what I had heard. I said I believed I understood their need for careful consideration and added I'd happily go along with whatever was decided. It was no huge problem for me personally, but it would be in the best interest of both morale and training if their decision would be made as soon as possible." Low assured Aldrin that it would.

At an MSC press conference on April 14, George Low indicated that "plans called for Mr. Armstrong to be the first man out after the Moon landing. . . .

A few minutes later, Colonel Aldrin will follow Mr. Armstrong down the ladder." It has been Aldrin's understanding ever since that NASA, in the end, decided on the order of the astronauts' egress based solely on the LM's interior design and the physical positions of the two astronauts inside the LM's cockpit—an engineering rationale that seemed to make total sense and came closer to satisfying Aldrin's sensibilities. According to Buzz, he had talked to Neil about the matter and together they had "speculated" as to how the decision should be made. "Our conclusion," Aldrin explained, "was that the decision as to who would be first would be determined by the allocation of tasks on the lunar surface and by our physical positions inside the LM itself. Unless something changed, as LM pilot I'd be on the right, a pilot's usual position, and Neil would be on the left, next to the hatch opening. It was not practical, and it was an added complication to change positions with Neil after the landing. And that, to the best of my knowledge, is how the matter was finally decided." Aldrin also claimed that, as soon as the announcement was made, he was okay with the situation. Mike Collins remembered otherwise: "Buzz's attitude took a noticeable turn in the direction of gloom and introspection shortly thereafter." Several other NASA officials also recalled Buzz being extremely disappointed, including the head of launch preparations at the Cape, Guenter Wendt: "Buzz had it in mind that he should be the first one to exit the LM and place the historic footprint in the lunar soil. He alienated a lot of people, management and astronauts alike, arguing his case. Neil, the mission's commander, just plugged along, nose to the grindstone, trying to stay focused on the job."

Neil always asserted that he never spoke to Buzz

about the matter in any detail, nor in the weeks leading up to the "first out" decision had he spoken about it with anyone else, not even with Janet. Armstrong's version of the "first out" decision, in fact, differed from Aldrin's narrative in several salient respects. First, Neil was never as concerned about the matter as Aldrin came to believe Neil was. As for the conversation described by Aldrin (or Aldrin's writer), "I can't remember the exact conversation," asserted Armstrong. "I can recall one point in time—and I don't know if this was the same time—when he asked me what I thought about it and my reply was, 'I'm not going to take a position on that. It's for our simulations and other people to decide.' The reality was that it was not something that I thought was really very important. It has always been surprising to me that there was such an intense public interest about stepping onto the lunar surface, let alone who did it first. In my mind the important thing was that we got four aluminum legs safely down on the surface of the Moon while we were still inside the craft. To me, there wasn't a lot of difference between having ten feet of aluminum leg between the bottom of the spacecraft in which we were standing and the surface of the Moon and having one inch of neoprene rubber or plastic on the bottom of our boots touching the lunar surface."

"In the Gemini program," Neil related, "the copilot had always been the guy who did the EVA, and that was principally because the commander was always so loaded with jobs that it was impractical to try to have the commander do all the work necessary to prepare himself to do it. The copilot had much more time available, and it was a much more logical thing for him to do. When we first did simulations on the ground for the Apollo surface activity, that was the way we tried to do it, probably just as a result of the Gemini experience. We tried doing it the same way. Accordingly, Buzz may have felt that, because that's the way it had been done in Gemini, that's the way it should be done, and it would be his responsibility to do it. He may have thought that it was important to him." However, as more and more ground simulations were done, it became clearer and clearer to everyone that it would be easier and safer for the commander to leave the LM first. In the end, the ground simulations showed that the technique of having the right-hand pilot, which was Aldrin, go out first just did not work very well. According to Neil, "I think that most people felt that there was an inherent risk involved in the LM pilot stepping around the commander and going out the hatch first that was avoidable doing it the other way." When the key people in charge of the simulations, notably MSC engineers George Franklin and Raymond Zedekar, concluded that the risk was far less when the commander went out first, the mission planners scratched the Gemini procedures and wrote new ones for Apollo. "And that's the way we did it," Armstrong explained. "On subsequent Apollo flights, where it wouldn't matter symbolically and historically which astronaut came out first, they all did it exactly the same way, with the commander stepping out first."

In fact, the technicalities of hatch design and LM interior layout were not truly the primary matters that those in charge of the Manned Space Program considered when it came to who was going to be first out. Those technicalities certainly did not need to determine the outcome, as Alan Bean, the lunar module pilot for Apollo 12, clearly explained: "Here's what you could have done if you were Buzz or myself. Before you put on your bulky EVA backpack, you're standing in there and you can move wherever you want. Buzz moves from right to left and puts on his stuff over there, and Neil moves from left to right and puts on his stuff where Buzz had been. Then they exchange backpacks; Buzz takes Neil's backpack off the rack and hands it over to him, and Neil hands Buzz's backpack to Buzz. 'I want to get out the door, let's change.' It was a nothing thing. No matter what NASA said at the time, Buzz could have gotten out first easily; it was all about where you put your backpack on." What Bean suggests is that NASA used the technicalities of hatch design and LM interior layout as a way of shutting the door on the entire "first out" controversy—and calming down an upset Aldrin in the process: "My opinion is, they were looking for technical reasons because they didn't want to say directly to Buzz or anyone else that 'we just want Neil to go out first.' Look, NASA knew both these guys. Slayton probably approached Neil at some point early on and said to him, 'Look, we picked you to command this mission and we want the commander to go out first.' Neil wouldn't have told anyone about that sort of conversation, if it happened, but Deke very easily could have come to him and said, 'I don't want us to talk about this to anyone, but I want you out there first. And I don't want us to ever mention this again.' Knowing Deke and knowing Neil, neither of them ever would, or ever did, talk about it, as far as I know."

Beyond all that which Al Bean explains, there was also a political decision in favor of Armstrong. Sometime in the middle of March 1969, in the heady days following the successful completion of Apollo 9, an informal meeting took place between Deke Slayton, the director of flight crew operations; Bob Gilruth, the

MSC director; George Low, the Apollo program manager (who had traveled throughout South America with Armstrong in 1966); and Chris Kraft, the director of flight operations. Kraft related the gist of the meeting: "Right around the time of Apollo 9, George Low and I had the same revelation, the way things looked it was going to be Aldrin out of the LM first on Apollo 11 because he was the lunar module pilot and he was doing all the training with the scientists and with the experiments package that was going to be put on the Moon, and Buzz knew about all that in detail. When we realized that, we called a meeting specifically to discuss the matter. For things like this in that time period, it was usually just the four of us that got together—Gilruth, Slayton, Low, and myself. Look, we just knew damn well that the first guy on the Moon was going to be a Lindbergh. He's going to be the guy for time immemorial that's going to be known as the guy that set foot on the Moon first. And who do we want that to be? The first man on the Moon would be a legend, an American hero beyond Lucky Lindbergh, beyond any soldier or politician or inventor. It should be Neil Armstrong. Neil was Neil. Calm, quiet, and absolute confidence. We all knew that he was the Lindbergh type. He had no ego. He was not of a mind that, 'Hey, I'm going to be the first man on the Moon!' That was never what Neil had in his head. The most Neil had ever said about it might have been that he wanted to be the first test pilot on the Moon or the first flier to land on the Moon. If you would have said to him, 'You are going to be the most famous human being on Earth for the rest of your life,' he would have answered 'Then I don't want to be the first man on the Moon.' On the other hand, Aldrin desperately wanted the honor and wasn't quiet in letting it be known. Neil had said nothing. It wasn't his nature to push himself into any spotlight. Neil Armstrong, reticent, soft-spoken, and heroic, was our only choice.

"It was unanimous. Collectively, we said 'Change it so the lunar module pilot is no longer going to be the first one out.' Bob Gilruth passed our decision to George Mueller and Sam Phillips at NASA Headquarters, and Deke told the crew. In our meeting, we had told Deke to do that. He did not argue with us. He did it, and I'm sure he did it in his most diplomatic way.

"Buzz Aldrin was crushed, but he seemed to take it stoically. Neil Armstrong accepted his role with neither gloating nor surprise. He was the commander, and perhaps it should always have been the commander's assignment to go first onto the Moon. Buzz probably thought that he was a better-trained man for the EVA job and had more capabilities than Neil to do the job on the lunar surface—and frankly, he may have been right. In the end, Neil gave Buzz a lot of the responsibility for surface activities. He expected Buzz to do them well and knew that Buzz could do them better than he. But nothing about performing surface activities had anything to do with the reason why we made the decision about who should be first out."

At no time in talking over the situation from every angle did Slayton, Gilruth, Low, or Kraft express the first word about the LM's interior layout or its hatch design. As Kraft attested, that was "an engineering side to it that we hadn't considered. That was a fortuitous excuse." Slayton, in particular, wanted the decision explained in technical terms. "That was Deke," Kraft explained. "He didn't want to be known as the guy that had made the decision that Buzz was not going to do it and that Neil was."

In fact, none of the four men present at the March

1969 meeting ever felt very comfortable confessing the truth about what was said there. For example, in a memorandum for the record prepared by George Low in September 1972 following a personal meeting in his office with Aldrin, Low wrote, "Aldrin asked me whether the decision as to who would be the first man on the Moon had been made by NASA Houston, NASA Headquarters, or whether it was an externally imposed decision on NASA. I told him that this was a Bob Gilruth decision based on a Deke Slayton recommendation." Obviously, Low's version of events did not precisely match up with the story subsequently told in Chris Kraft's 2001 autobiography *Flight: My Life in Mission Control*.

Clearly, then, it was hard for Aldrin to let it all go: as late as 1972, Buzz was still bothered enough by it to ask George Low how the "first out" decision had actually been made. That can only mean that Buzz was not totally convinced that the technical reasons he had been given back in 1969 were really as determinative as he and the rest of the world had been led to believe.

Buzz never knew about the Gilruth-Slayton-Low-Kraft meeting until Chris Kraft published his autobiography, nor did Armstrong. Even after becoming aware of the behind-the-scenes, nontechnical factors in the decision making, Neil remained convinced that engineering considerations related to the interior layout of the LM played a primary role in determining who should be the first man out: "It just seems to me that the fact that all six Moon landings were done the same way is pretty strong evidence that that was the proper way to do it. Otherwise, they would have changed it. I can't imagine the other commanders, especially someone like Al Shepard [commander of Apollo 14], agreeing to something if it wasn't the right way to do it. Knowing their nature, the other commanders would have done it or certainly attempted to do it differently, if they had thought there was a better way. I would have felt the same about it myself."

CHAPTER 21

Dialectics of a Moon Mission

Buzz Aldrin's concern over who should be the first out did nothing to help the working relationship of the Apollo 11 crew. At the same time his sour feelings never seriously impaired the crew's training for its historic mission because Armstrong's stoic personality did not allow it to. If the commander had been a confrontational sort like Frank Borman or Alan Shepard, the situation with Aldrin might easily have turned highly injurious to the mission.

"Neil would have regarded that kind of infighting as sort of beneath him," explained Mike Collins. "I never heard Neil say a bad thing about Buzz. Their working relationship, as I saw it, was always extremely polite and, from Neil to Buzz, in no way critical. What Neil actually thought about Buzz, God only knows."

The training for the first Moon landing was intense enough to challenge the patience and goodwill of the entire NASA team, not just the Apollo 11 crew. Not only did the astronauts need to be made ready, so too did the entire NASA ground apparatus, including the Mission Control Center, the tracking network, the quarantine facility to house a crew that might bring back lunar "bugs," not to mention the Saturn V rocket, Command

Module No. 107, and Lunar Module No. 5. Armstrong, Collins, and Aldrin trained fourteen-hour days, six days a week for six full months. They often worked another eight hours on Sundays.

Beginning January 15, 1969, until July 15, 1969, the day before the launch, the crew of Apollo 11 logged a total of 3,521 actual training hours. That equates to 126 hours per week, or 42 hours per crew member, in specified training programs and exercises. Roughly another 20 hours per week were taken up by poring over mission plans and procedures, talking to colleagues, traveling to training facilities, and other routine work. Armstrong and Aldrin logged 1,298 and 1,297 training hours, respectively, while Collins recorded some 370 hours less. Half of Collins's hours came inside the CSM simulator, where he worked physically apart from Armstrong and Aldrin. On the other hand, there were very few hours when Neil and Buzz were not working side by side. Nearly a third of their training time was inside the cramped quarters of the LM simulator.

The overriding objective of Apollo 11 was to get the Moon landing done. Training for surface activities represented less than 14 percent of the astronauts' time. That included preparing Armstrong and Aldrin to collect geological samples and to set up all the planned lunar surface experiments, as well as to learn how to handle the Extravehicular Mobile Unit (EMU). This vital piece of equipment, the EMU, was composed of all the protective apparel and paraphernalia worn during lunar surface work. During EMU training, the astronauts "checked out" every part of this assembly.

"We practiced the lunar surface work until we were reasonably confident in our ability to carry out the surface plan," Armstrong stated. "If the descent and the final approach to landing were rated a nine on a ten-point scale of difficulty, I would put the surface work down at a two. Not that there weren't some high risks involved with it, because there were. Certainly, we were completely dependent on the integrity of our pressure suits, and there were significant questions about the thermal environment—whether we would have overheating problems, because it was going to be warm out there on the lunar surface, over 200 degrees Fahrenheit. We did some of our surface simulation work in the altitude chamber, with thermal simulation, and those had worked well. So we reached a confidence level that it was going to work fine. The only real concerns involved the unknowns that we couldn't simulate, because we didn't know what they were. In the end, the ground simulations proved to be pretty good even though the lunar gravity conditions could not be matched."

Ever since the selection of the New Nine in 1962, "All of us were exposed over several years to Geology 101. We had very fine instructors who were very knowledgeable about astrogeology and selenology, the astronomical study of the Moon. We went to Hawaii and Iceland, great places to focus on volcanic rocks. The assumption was that on the Moon we would encounter tectonic formations principally, or remnants of volcanic and tectonic lava flows, that sort of thing. I was very tempted to sneak a piece of limestone up there with us on Apollo 11 and bring it back as a sample. That would have upset a lot of apple carts! But we didn't do it."

Another reality for Apollo 11 was that Armstrong and Aldrin would not have much time to spend on the lunar surface. "That was principally dictated by the fact that we didn't know how long our supply of water for the cooling of our suits would last," Armstrong explained. "Neither that nor our metabolic rate could be duplicated on Earth in a lunar gravity. As it turned out, we were able to stay out a little bit longer than our plan stipulated. After getting back in the craft, we drained the water tanks to see how much water remained. From that we got a useful data point against the time that we had been out." Though Neil enjoyed geology, he found the nature of the discipline a bit puzzling: "The geologists had a wonderful theory they called the 'theory of least astonishment.' According to the theory, when you ran into a particular rock formation, you hypothesized how it might have occurred and created as many scenarios as you could think of as to how it might have gotten there. But the scenario that was the least astonishing was the one you were supposed to accept as the basis for further analysis. I found that fascinating. It was an approach to logic that I had never experienced in engineering." Yet it was precisely Armstrong's systematic engineering approach that Harrison "Jack" Schmitt, the Harvard-trained geologist and later Apollo 17 lunar module pilot, who directed Neil and Buzz's training for lunar-rock collecting, connects with Neil's geologic capabilities, citing Neil's collection of rock samples as "the best that anybody did on the Moon."

All of the Apollo training was important, but no aspect of it was more critical to mission success than the work done in the flight simulators. The two main simulators were the command module simulator, built by North American, and the lunar module simulator, built by the LM designer, Grumman. Collins spent the lion's share of time in the former, Armstrong and Aldrin in the latter. With all of the controls and instruments hooked up to a bank of computers in a back room as well as to consoles in Mission Control, the CSM simulator—nicknamed

"The Great Train Wreck" for its jumbled array of differently shaped boxes and compartments—was dynamic and totally interactive. Looking out the windows as they were making their "flights," the astronauts saw rough displays of the Earth, sky, Moon, and stars. "On balance the simulations were quite good," remembered Neil, though the fidelity between simulation and the real thing was known to be imperfect. They "did a good enough job to give us the level of confidence we needed."

During his Apollo 11 training, Armstrong spent 164 hours in the CSM simulator, which was only about one-third of the time that Collins, the CM pilot, spent in it. Naturally, given his primary responsibility in the mission, Neil spent considerably more time practicing lunar landings, 383 hours in the LM simulator and thirty-four in the LLTV or LLRF, for a total of 417 hours of Moon landing simulation. His grand total of 581 hours in a simulator equates to over seventy-two days—more than ten full weeks—of eight-hour days in a simulator. Aldrin compiled eighteen more hours in the CM simulator and twenty-eight more in the LM than did Neil. Unlike Neil, Buzz did not fly either the LLTV or LLRF during the six-month-long preparation for Apollo 11. "You are trying to build simulators to be exactly like the real thing, but they never are able to get it to the degree of reality that it flies as easily as a real machine. People who had not been involved in simulator development during their career usually just tried to 'win.' They tried to operate perfectly all the time and avoid simulator problems. I did the opposite. I tried actively to encourage simulator problems so I could investigate and learn from them. I'm sure that some of the guys were well aware of my approach," surely more so following what has become a rather notorious incident involving a particularly taxing "run" in the LM simulator.

Mike Collins told the story: "Neil and Buzz had been descending in the LM simulator when some catastrophe had overtaken them, and they had been ordered by Houston to abort. Neil, for some reason, either questioned the advice or was just slow to act on it, but in any event, the computer printout showed that the LM had descended below the altitude of the lunar surface before starting to climb again. In plain English, Neil had crashed the LM and destroyed the machine, himself, and Buzz.

"That night in the crew quarters Buzz was incensed and kept me up far past my bedtime complaining about it. I could not discern whether he was concerned about his actual safety in flight, should Neil repeat this error, or whether he was simply embarrassed to have crashed in front of a roomful of experts in Mission Control. But no matter, Buzz was in fine voice, and as the scotch bottle emptied and his complaints became louder and more specific, Neil suddenly appeared in his pajamas, tousle-haired and coldly indignant, and joined the fray. Politely I excused myself and gratefully crept off to bed, not wishing to intrude in an inter-crew clash of technique or personality.

"Neil and Buzz continued their discussion far into the night, but the next morning at breakfast neither appeared changed, ruffled, nonplussed, or pissed off, so I assumed it was a frank and beneficial discussion, as they say in the State Department. It was the only such outburst in our training cycle."

Aldrin's version of the late-night exchange is a little different. "The three of us often ate dinner quite late in our quarters. Afterward, Mike and I sat around having a drink and talking while Neil had gone off to bed. Mike said something like, 'Well, how did it go? What did you guys do in the simulator today?' and I said, 'Well, we lost control during an abort.' Now just how loud I said that, I can't really say. But what I said, I felt was between the two of us, between Mike and me. I didn't feel I needed to express my feelings about this to Neil, because that was just not Neil's and my relationship. In the normal course of what we were doing, I did not critique him. But Mike asked me a question about the simulation, so I told him what happened. It was a surprise to both of us when Neil came out of his bedroom and said, 'You guys are making too much noise. I'm trying to sleep.'" Neil did not say a word at the time to defend what he had done in the simulation, why he chose not to abort. "That wouldn't have been Neil," explained Aldrin.

What Aldrin was explaining to Collins when Neil came out to quiet them down was that "I thought we were playing a game and we should make an attempt to do everything we could to win the game, and the sooner we did it when we saw that things were going bad for us, the better off we'd be and the more in keeping with what we'd actually do in a real situation like that." The most important thing in every situation, Aldrin said to Collins, was not to crash. "I felt analyzing this and that system and whatever was not playing the game properly as far as the simulator people were concerned. If they threw a failure at us and we were losing control of the LM, would we in real flight actually go on and land? I'm not sure we would. The same way that if something disabled the commander, or disabled the primary guidance, or disabled the landing radar, why, we wouldn't land on the first try, we'd abort and come back. Clearly, there was a difference between Neil and I in how we reacted to the simulation. Neil had his reason for doing what he did. It was between him and the simulator people to decide what he got out of that. As for me, I was there to support what was going on in the training, almost as an observer. Thus my answer to Mike's question about my evaluation of what had happened."

Some versions of the simulation story that have been told over the years suggest that Aldrin also urged Armstrong to abort, but Neil explained, "I don't recall that Buzz asked me to abort—ever—I don't remember that. What I do remember is that the descent trajectory that we were on during the simulation and the information we had available to us had become seriously degraded, and I thought that it was a great time to test the Mission Control center, 'Okay, guys, let's see what you can do with this.' I knew that I could abort at any time—and probably successfully—but then you lose the mission, the rest of the simulation. This was a chance to test the control center. Buzz took that as a black mark against us. He thought it was a mark against his ability to perform, a mark against both of us and against our crew ability. I didn't look at it that way at all. It was a complete difference of opinion, and he expressed his concern to me later that night."

As for precisely what went on during that late-night exchange between him and Aldrin, "I don't really remember the details of that, but I do remember that Buzz expressed his displeasure. He had a different way of looking at the sims. He never liked to crash in a real simulation, while I thought it was a learning experience for all of us. Not just for the crew but also for the control center personnel. We were all in it together."

Interestingly, this story of the simulation-that-Armstrong-would-not-abort is reminiscent of Neil's April 1962 flight in the X-15, the one in which his aircraft ballooned up and ended up dangerously over Pasadena. In both cases, Neil was trying to promote technological learning through dialectical experimentation. "If we couldn't come up with a solution or the ground controllers couldn't come up with a solution, that was an indication to me that, for one, I needed to understand that part of the flight trajectory better." As a matter of fact, as a result of his crashing into the lunar surface during the particular simulation under review, Neil "constructed a plot of altitude versus descent rate with bands on it that I hadn't had before, so that I could tell when I was getting into a questionable area. If I had aborted when everyone wanted me to, I probably would not have bothered to even make that." At the same time, the "botched" simulation caused the flight director and his people to reevaluate how they had analyzed the situation. "I'm sure they improved their approach to understanding it, too, and knowing when they were getting into a dangerous area," Armstrong stated. "So it did serve a valuable purpose. I was a little disappointed that we didn't figure it out soon enough, but you learn through the process. These were the most extensive simulations I had ever encountered—and they needed to be. The Moon landing was a bigger project, a more extensive project, with more people involved, than any of us had ever encountered."

Four months into training, Apollo 10 flew to the Moon. Launched on May 18, 1969, with a crew composed of three veterans of Gemini rendezvous missions—commander Tom Stafford, command module pilot John Young, and lunar module pilot Gene Cernan—the eight-day mission was a very successful full-dress rehearsal of the Moon landing. Apollo 10 achieved a number of space firsts, including the first CSM-LM operations in the cislunar and lunar environment, the first CSM-LM docking in translunar trajectory, the first LM undocking in lunar orbit, the first LM staging in lunar orbit, and the first manned LM-CSM docking in lunar orbit. About the only thing Apollo 10 did not accomplish was the lunar landing itself, though its LM—nicknamed Snoopy—did swoop down to within a mere 50,000 feet of the proposed Apollo 11 landing site before shooting back to orbit and redocking with Charlie Brown, the command module.

Apollo 10 aided preparations for Apollo 11 in a number of ways. First, as Armstrong explained, "There was the matter of lunar module handling qualities, LM responsiveness, and LM engine operations. We wanted to know how similar or different was flying the actual LM from flying the simulator and from flying the LLTV."

There was also the matter of the lunar environment itself, especially the possibly significant gravitational effects that "mascons" might have on flight paths. Mascons were areas beneath the visible lunar surface, generally in the mares, that because the interior rock was of greater density than that of the surrounding area, exerted a slightly higher gravitational force. From the flights of the five unmanned Lunar Orbiter spacecraft of 1966 and 1967, telemetry data indicated that the Moon's gravitational pull was not uniform. Perturbations likely caused by mascons had led to slight dips in the paths of the Lunar Orbiters. Apollo 10 documented the influence of the mass concentrations on Apollo 11's exact trajectory.

"As a result of all the fine photographs from Apollo 10, Buzz and I developed a very high level of confidence in our ability to recognize our flight path and principal landmarks along the way. By the time we launched in July, we knew all the principal landmarks on our descent path by heart and, equally importantly, we knew all the landmarks on our way prior to the point at which we would ignite our descent engines. That was as important as a cross-check, to be able to determine that we were, in fact, geographically over the exact place we wanted to be over—and as close to the scheduled time on the flight plan as possible."

Finally, the success of Apollo 10 meant that Apollo 11 would certainly be the first landing mission. The only uncertainty was the date of the launch. A few weeks after Apollo 10, Deke Slayton asked Armstrong if he was ready. Armstrong answered: "Well, Deke, it would be nice to have another month of training, but I cannot in honesty say that I think we have to have it. I think we can be ready for a July launch window." On June 11, 1969, NASA announced that the Apollo 11 astronauts had received the go-ahead for the landing attempt. Their launch would come on July 16 with the historic landing scheduled for Sunday afternoon, July 20.

The conscientious, highly professional, and vigorous manner in which Armstrong, Collins, and Aldrin pursued every item on their six-month training agenda gave NASA great confidence in the crew. Yet the Apollo 11 mission was replete with unknowns, uncertainties, and unexplored risks—some technological, others human. How would individual astronauts perform during a crisis? In Commander Neil Armstrong, NASA managers took a calculated risk that, in order to achieve the landing, he might push the envelope, his luck, or his abilities a little too far.

NASA established its "mission rules" as a system of preventive checks. The genesis of the concept had come early in the Mercury program from veteran engineers in the NASA Space Task Group. Early on, they decided they had better formally record every one of their important thoughts and observations about the

Mercury capsule, about the rocket that was to launch Mercury, about each flight control system, and every possible flight situation. As Chris Kraft related, "We noted a large number of what-ifs, too, along with what to do about them. Then we printed the whole bunch in a booklet and called it our mission rules." In preparation for Apollo 11, it took many months for teams of mission planners, flight directors, simulation experts, engineers, and astronauts to talk over, debate, review, redraft, and finalize the rules for what was to be the first Moon landing. The first complete set of rules for Apollo 11 was not published until May 16, 1969, two months before launch. Rules were then updated weekly as ongoing simulations revealed where changes were needed. Although the 330-page third (or "C") revision came out five days before the launch, it was not the end to the changes. On the day of the launch, Flight inserted seven "write-in" changes to the rules. One of the last-minute changes, unknown to the crew of Apollo 11, stated that there was no need to abort the landing if the LM's onboard computer experienced a specific series of program alarms.

The book included a long section on "Flight Operations Rules," concerning the overall policy for mission conduct, the treatment of risk by mission phases, and redundancy management. Also included were rules for launch, trajectory and guidance, communications, engine burns, docking, EVA, electrical systems, and aeromedical emergencies. There were rules to cover every conceivable problem and contingency. In each section there was a summary of all the "Go/NoGo" situations (or "Stay/NoStay" situations, in the case of whether to stay or abort immediately after landing on the lunar surface). This was the terminology used by the flight director in a final systems check to make sure that his controllers were confident about proceeding to the next phase of the mission.

Another critical rule—one that became a matter of grave urgency during the descent of Armstrong and Aldrin to the lunar surface—stated that once the warning light came on inside the LM showing a low level of fuel for descent, the astronauts had one minute either to commit to the landing or abort.

So many mission rules were written for Apollo 11 and for the subsequent Apollo missions that they had to be organized according to a numerical code. There was no way any single person could remember all the mission rules; it would have been like memorizing the dictionary. During a spaceflight, flight controllers had to keep their copies of the rules book very close by.

For many of the rules, there was a defined margin, some leeway, a little give. Yet, not until all mission rule requirements were met to the satisfaction of the flight director could any vital decision be made and acted upon. Some mission rules could be interpreted so as to leave an ultimate decision in the hands of the astronauts, but that sort of independent, on-the-spot judgment was not something that NASA managers wanted to encourage.

According to Gene Kranz, one of Mission Control's flight directors, "Buzz was the crewman usually involved in discussing mission rules, demonstrating his knowledge of a variety of subjects, and generally dominating the crew side of the conversations. Neil seemed more the observer than the participant, but when you looked at his eyes, you knew he was the commander and had all the pieces assembled in his mind. I don't think he ever raised his voice. He just saved his energy for when it was needed. He would listen to our discussions, and if there was any controversy, he and Aldrin would try out our ideas in the simulators and then give feedback through Charlie Duke to the controllers. [Astronaut Duke was to serve as the CapCom for Apollo 11 during its landing on the Moon.] Mike Collins used a different tactic. He worked directly with the Trench and system guys." ("The Trench" was the nickname for the men in Mission Control who worked as the flight dynamics team, led by the Flight Dynamics Officer, or FIDO.)

Armstrong generally accepted Kranz's characterization: "It is true that Buzz was talkative and very involved in conversations, and I was probably more reserved. I think that was just our nature."

Almost all the mission rules were written down and formally agreed upon; a very few were not. The most important unwritten rules for Apollo 11 concerned the landing.

"To get a handshake on the unwritten rules for the landing," recalled Kranz, "I had a final strategy session before simulation startup with Neil, Buzz, Mike, and Charlie Duke. It was in this session that I outlined the landing strategy. We had only two consecutive orbits to try to land on the Moon. If we had problems on the first orbit, we would delay to the second. If we still had problems, we would start the lunar descent to buy five additional minutes to solve the problem. If we couldn't come up with answers, we would abort the landing and start a rendezvous to recover the LM, then jettison it and head back home. If problems surfaced beyond five minutes, we would try to land and then lift off from the surface after a brief stay. We would try for the landing even if we could only touch down and then lift off two hours later when the CSM passed overhead in lunar orbit with proper conditions for rendezvous.

"I knew Armstrong never said much," Kranz continued, "but I expected him to be vocal on the mission rule strategy. He wasn't. At that time he was silent. It took time to get used to his silence. As we went through the rules, Neil would generally smile and/or nod. I believe that he had set his own rules for the landing; I just wanted to know what they were. My gut feeling said he would press on, accepting any risk as long as there was even a remote chance to land. I believed we were well in sync, since I had a similar set of rules. I would let the crew continue as long as there was a chance."

Armstrong remembered: "I had high respect for mission rules and how they were developed. . . . But I would admit that if everything seemed to be going well and there was a mission rule that interrupted and said we have to do such-and-such, I would have been willing to use my commander's prerogative on the scene and overrule the mission rule if I thought that was the safest route. After all, aborts were not a very well understood phenomenon—no one had ever done an abort. You were shutting off engines, firing pyrotechnic separation devices, igniting other engines in midflight. Doing all of that in close proximity to the lunar surface was not something in which I had a great deal of confidence. So there is some truth to what Gene is saying, but I would have said, as long as there is a *good* chance of landing, I would proceed."

Like Kranz, a nervous Chris Kraft was also bothered by the fact that he could not be sure what Armstrong might do to override mission rules and force a lunar landing to happen. "In the last month, we'd had Neil in Mission Control to go over the rules for lunar descent, landing, surface operations, and takeoff," Kraft explained. "Mission rules could leave the ultimate decision to the astronaut, but that wasn't something we encouraged. Now I wanted to make certain that all of us understood exactly where we were. We got down to the finest details—descent-engine performance, computer bugs that we knew about, landmarks on the lunar surface, even talking through the most unlikely events we could imagine during the landing.

"The computer and the landing radar got particular attention. We'd be sending last-minute updates to the computer on the lunar module's trajectory, its engine performance, and location over the Moon. Until *Eagle* was about ten thousand feet high, its altitude was based on Earth radars, and its guidance system could be off by hundreds or even thousands of feet. Then the LM's own landing radar was supposed to kick in and provide accurate readings.

"That led to some heated discussion. Neil worried that an overzealous flight controller would abort a good descent, based on faulty information. 'I'm going to be in a better position to know what's happening than the people back in Houston,' Neil said over and over.

"And I'm not going to tolerate any unnecessary risks," Kraft retorted. "That's why we have mission rules."

Arguing about the specifics of the landing radar, Kraft insisted that if the landing radar failed, an abort was mandatory: "I didn't trust the ability of an astronaut, not even one as tried and tested as Armstrong, to accurately estimate his altitude over a cratered lunar surface. It was unfamiliar terrain, and nobody knew the exact size of the landmarks that would normally be used for reference." Finally, Kraft and Armstrong agreed. "That mission rule stayed as written," Kraft recalled. "But I could tell from Neil's frown that he wasn't convinced. I wondered then if he'd overrule all of us in lunar orbit and try to land without a radar system.

"Those conversations came back to me when I saw Neil a few days before the launch," Kraft related. I asked Neil, 'Is there anything we've missed?' 'No, Chris, we're ready. It's all done except the countdown.' He was right. If there was anything undone, none of us could say what it was. We had come to this last point, and for a moment I felt my legs shake."

Because he, too, was worried that the crew might take unnecessary risks in order to make the landing, Dr. Thomas Paine, the NASA administrator, even got into the act. In the week before the launch, he made a point of speaking to Armstrong. According to Neil, Paine told him, "If we didn't get a chance to land and came back, he would give us the chance to go again, on the very next flight. I believe he meant it at that point." The truth was, Paine told every subsequent Apollo crew the very same thing. It was his way of encouraging the crews not to try anything stupid, thinking it would be their only chance. If Apollo 11's landing had been aborted, Armstrong was ready to take the administrator up on his offer.

The crew moved into their astronaut living quarters at the Cape on June 26. Beginning at the stroke of midnight on the twenty-seventh, they participated in a weeklong trial countdown. Simulated launch came on the morning of July 3 precisely at 9:32 A.M., the exact time scheduled for the real launch. Before the trial started, the three men entered a strict physical quarantine that was to last for two weeks before the flight and endure for three more weeks after the flight. The quarantine was invoked in order to limit the astronauts' exposure to infectious organisms. Dr. Charles E. Berry, chief astronaut physician, gave them their last head-to-toe going-over the day the simulated countdown began.

On July 5, the crew of Apollo 11 returned to Houston from Florida for a media day. First up was a press conference. After that came sessions with the wire services, a group of magazine writers, and the three television networks for broadcast that evening. Before it was over, the three astronauts endured a fourteen-hour day answering questions from several hundred international reporters and journalists. Armstrong, Collins, and Aldrin arrived at the morning press conference wearing gas masks. Knowing how silly they looked, the men grinned as they walked onstage. Neil, Mike, and Buzz sat in a three-sided plastic box roughly twelve feet wide. To ensure that no contagion from the journalists circulated into the breathing space of the astronauts, blowers located to the rear of the plastic booth blew air from behind the Apollo crew out into the audience. Once safely within their hygienic box, the astronauts took off their masks and sat down in easy chairs before a desk emblazoned with NASA's emblematic "meatball" and the Apollo 11 seal: an eagle, the symbol of America, coming in for a landing on the lunar surface, its talons bearing an olive branch, a symbol of peace. The atmosphere in the theater felt decidedly strange. Convening to talk about a trip to the Moon still seemed a little fantastical. Naturally the astronauts were a little edgy, too.

As mission commander, Armstrong spoke first. Author Norman Mailer, in the audience on special assignment from *Life* magazine, sensed that Neil was "ill at ease." What Mailer might not have known was that Armstrong often paused in formal conversation, searching for the right words.

"We're here today to talk a little bit about the forthcoming flight, Apollo 11, hopefully the culmination of the Apollo national objective. We are here to be able to talk about this attempt because of the success of four previous Apollo command flights and a number of unmanned flights. Each of those flights contributed in a great way to this flight. Each and every flight took a large number of new objectives and large hurdles, and left us with just a very few additions—the final descent-to-the-lunar-surface work—to be completed. We're very grateful to those large efforts of people here at MSC and across the nation who made those first flights successful, and made it possible for us to sit here today and discuss Apollo 11 with you. I'll ask Mike first to talk about the differences you might see in the command module activities on the flight."

As usual, Neil was brief. Collins talked for a little longer, emphasizing that he was going to be alone in the command module much longer than any previous CM pilot and that rendezvous was going to happen for the first time between a stationary LM, down on the lunar surface, and a CSM "whizzing around the Moon." Last but certainly not least in terms of how much he said came Aldrin. Buzz outlined a complete lunar descent and landing, critical elements of Apollo 11 that involved so much that was new that it did take quite a bit longer to describe them.

In all, the crew responded to a total of thirty-seven questions. Armstrong answered twenty-seven of them. Nine of the newsmen specifically asked for Neil to answer their questions. Twice, Neil turned and asked Buzz to respond to a question directed specifically at Neil; on two other occasions, Buzz, unsolicited, added to Neil's comments. Collins, like Aldrin, answered only three questions directed at him. A few questions called for responses from all three astronauts. It was an overall pattern that would long outlast the Apollo program. People most wanted to hear from the commander, the first man who would step out onto the Moon.

During the conference, Armstrong announced for the first time the nicknames for the Apollo command

and lunar modules: "Yes, we do intend to use call signs other than those you may have heard in simulation. The call sign for the lunar module will be 'Eagle.' The call sign for the command module will be 'Columbia.' . . . Columbia is a national symbol. Columbia stands on top of our Capitol and, as you all know, it was the name of Jules Verne's spacecraft that went to the Moon in his novel of one hundred years ago."

Naturally, the press was curious as to what Armstrong would say when he first stepped onto the Moon. A reporter asked him about it. Not even those few who knew Armstrong personally or who exercised authority over the manned space program had been able to get Neil to disclose his thoughts about the historic first words he would utter on the lunar surface. At one point, the internal pressures inside of NASA had motivated Julian Scheer, the chief of NASA's public affairs office, to write a terse internal memo that asked, in effect, did King Ferdinand and Queen Isabella of Spain tell Christopher Columbus what to say when he reached the New World? In response to the reporter's question about choosing his first words from on the lunar surface, Armstrong simply answered, "No, I haven't." As hard as it may be to believe, that was the truth. "The most important part of the flight in my mind was the landing," Armstrong explained later. "I thought that if there was any statement to have any importance, it would be whatever occurred right after landing, when the engine stopped. I had given some thought to what we would call the landing site. I had also thought about what I would say right at the landing; I thought it was the one that history might note. But not even that was something that I had given a great deal of thought to, because, statistics aside, my gut feeling was that, whereas we had a ninety percent chance of returning safely to Earth, our chances were only even money of actually making the landing."

In fact, Neil had already chosen Tranquility Base as the name of the spot on the Sea of Tranquility where he and Aldrin would land; privately, he had told Charlie Duke about the name, since Duke would serve as CapCom during the landing and Neil did not want Charlie to be caught unawares when Neil used the phrase immediately upon touchdown. No one besides Duke knew about Tranquility Base until *Eagle* landed.

A special high-level government committee had decided that Armstrong and Aldrin should leave three items on the surface as symbolic of humankind's arrival. The first was a plaque mounted on the leg of the LM that held the ladder down which the astronauts would climb. This plaque depicted the Earth's two hemispheres; on it was inscribed the statement, HERE MEN FROM THE PLANET EARTH SET FOOT UPON THE MOON, JULY 1969 A.D. WE CAME IN PEACE FOR ALL MANKIND. The second item was a small disk, less than one and a half inches in diameter, upon which had been electronically recorded a microminiaturized photo-print of goodwill letters from various heads of states around the world. The third item was the American flag. "Some people thought a United Nations flag should be there," Armstrong explained years later, "and some people thought there should be flags of a lot of nations. In the end, it was decided by Congress that this was a United States project. We were not going to make any territorial claim, but we ought to let people know that we were here and put up a U.S. flag."

Reporters tried hard—and mostly in vain—to get Armstrong to philosophize about the historical significance of the Moon landing. "What particular gain do you see in going to the Moon for yourselves as human beings, for your country, and for mankind as a whole?" "Do you think that eventually the Moon will become part of the civilized world just as the Antarctic is now, which was also once a removed and unacceptable place?"

"First, let me repeat something that you have all heard before, but probably addresses itself to your question," Armstrong answered. "That is, the objective of this flight is precisely to take man to the Moon, make a landing there, and return. That is the objective. There are a number of peripheral secondary objectives including some of those you mentioned early in the question that we hope very highly to achieve in great depth. But the primary objective is the ability to demonstrate that man, in fact, can do this kind of job. How we'll use that information in the centuries to come, only history can tell. I hope that we're wise enough to use the information that we get on these early flights to the maximum advantage possible, and I would think that in the light of our experience over the past decade that we can indeed hope for that kind of result."

Nor did the journalists have much luck in provoking Armstrong into giving anything other than unemotional, engineering answers about the grave risks inherent to the flight. "What would, according to you, be the most dangerous phase of the flight of Apollo 11?" "Well, as in any flight, the things that give one most concern are those which have not been done previously, things that are new. I would hope that in our initial statement that we gave to you an idea, at least, of what the new things on this flight are. Now, there are other things that we always concern ourselves about greatly, and those are the situations where we have no alternative method to do the job, where we have only one. You, when you ride in an airliner across the Atlantic, depend on the wing of the airplane to stay on the fuselage; without it, you could not have made the trip, see? We have on recent flights had some of those kinds of situations. In our earlier lunar flights, the rocket engine for the service module must operate for us to return from the Moon. There are no alternatives. Similarly, in this flight, we have several situations like that. The LM engine must operate to accelerate us from the Moon's surface into lunar orbit, and the service module engine, of course, must operate again to return us to Earth. As we go farther and farther into spaceflight, there will be more and more of the single-point systems that must operate. We have a very high confidence level in those systems, incidentally."

"What will your plans be in the extremely unlikely event that the lunar module does not come up off the lunar surface?" "Well," answered Neil tersely, "that's an unpleasant thing to think about and we've chosen not to think about that up to the present time. We don't think that's at all a likely situation. It's simply a possible one, but at the present time we're left without recourse should that occur."

Reporter: "What is the longest time you can wait between the not-firing and the time when Mike Collins would have to go back, the time you would have to work on the LM or fix whatever was wrong or try to fix it?" Neil: "I don't have the numbers. Probably it would be a matter of a couple of days."

Such seemingly passionless answers to questions about the human dimensions of spaceflight and about the historical and existential meanings of going to the Moon piqued Norman Mailer's razor-sharp acumen for disdainful insight. Like other reporters, the Pulitzer Prize–winning author of *The Naked and the Dead* and *Armies of the Night* wanted more from Armstrong. Mailer came to write that Armstrong "surrendered words about as happily as a hound allowed meat to be pulled out of his teeth"; that Armstrong "answered with his characteristic mixture of modesty and technical arrogance, of apology and tight-lipped superiority"; that Armstrong had "the sly privacy of a man whose thoughts may never be read"; that Armstrong, like a trapped animal, seemed to be looking for "a way to drift clear of any room like this where he was trapped with psyche-eaters, psyche-gorgers, and the duty of responding to questions heard some hundreds of times." At the same time, Armstrong was "a professional" who had "learned how to contend in a practical way with the necessary language," always choosing words and phrases that "protected him."

It intrigued Mailer (in his 1970 book on Apollo 11, *Of a Fire on the Moon*, Mailer called himself Aquarius, in reference to the hopeful spirit of a future Age of Aquarius) that Armstrong exuded such an "extraordinarily remote," almost mystical quality that made him appear different from other men. "He was a presence in the room," Mailer noted, "as much a spirit as a man. One hardly knew if he were the spirit of the high thermal currents or that spirit of neutrality which rises to the top in bureaucratic situations, or both. . . . Indeed, contradictions lay subtly upon him—it was not unlike looking at a bewildering nest of leaves: some are autumn fallings, some the green of early spring." Of all the astronauts, Armstrong seemed "the man nearest to being saintly." As a speaker, Neil was "all but limp." Still, the overall impression Armstrong made on Mailer was not unremarkable. "Certainly the knowledge he was an astronaut restored his stature," Mailer realized, "yet even if he had been a junior executive accepting an award, Armstrong would have presented a quality which was arresting. . . . He would have been more extraordinary in fact if he had been just a salesman making a modest inept dull little speech, for then one would have been forced to wonder how he had ever gotten his job, how he could sell even one item, how in fact he got out of the bed in the morning. Something particularly innocent or subtly sinister was in the gentle remote air. If he had been a young boy selling subscriptions at the door, one grandmother might have warned her granddaughter never to let him in the house; another would have commented, 'That boy will go very far.' "

Mailer continued his dogged pursuit of the puzzle-that-was-Armstrong into the press conference organized exclusively for the magazine writers and beyond that into the studio where NBC filmed its interview of the astronauts. As the journalists kept pushing hard for the crew of Apollo 11 to disclose personal feelings and emotions, Mailer watched and listened as Armstrong entrenched himself ever deeper in his engineer's protective cloak, the armor of "a shining knight of technology." Armstrong replied in "a mild and honest voice" to a question about the role of intuition in his flying by remarking that intuition had "never been my strong suit" and by asserting, like a logical positivist, Mailer noted, that the best approach to any problem was to "interpret it properly, then attack it."

Armstrong had mastered "computerese." Instead of saying "we," Neil convoluted the English language and said, "A joint exercise has demonstrated." Instead of saying "other choices," he referred to "peripheral secondary objectives." Rather than "doing our best," it was "obtaining maximum advantage possible." To "turn on" and "turn off" became "enable" and "disable." Mailer, who had rejected in disgust his own college education as an engineer, saw in Neil's vernacular proof not only that "the more natural forms of English had not been built for the computer" but that Armstrong represented "either the end of the old or the first of the new men."

"If not me, another," Neil stated, to Mailer's mind disclaiming "large reactions, large ideas" behind media comparisons of his own journey as commander of Apollo 11 to Christopher Columbus's adventure in 1492. Armstrong's concern was "directed mainly to doing the job," one that could be done by no fewer than ten other astronauts. And hundreds of people were backing up his crew in Houston, at the Cape, at the other NASA centers, and tens of thousands had been working in industrial firms all around the country to enable Apollo to blaze its course. "It's their success more than ours," Neil humbly told the media.

Armstrong was no common hero, Mailer realized. "If they would insist on making him a hero," the author noted, "he would be a hero on terms he alone would make clear."

From Collins and Aldrin, the reporters were able to get a few remarks about family and personal background (Buzz mentioned the family jewelry he was taking with him to the Moon). Nothing of the sort came from Armstrong. "Will you take personal mementos to the Moon, Neil?"

"If I had a choice, I would take more fuel."

"Will you keep a piece of the Moon for yourself?"

"At this time, no plans have been made" came the stiff response.

"Will you lose your private life after this achievement?"

"I think a private life is possible within the context of such an achievement."

Following a comment from Neil about the economic benefits of the space program, a writer jumped in to ask, "So, are we going to the Moon only for economic reasons, only to get out of an expensive hole of a sluggish economy? Don't you see any philosophical reason why we might be going?" "I think we're going to the Moon," Armstrong offered tentatively, "because it's in the nature of the human being to face challenges. It's by the nature of his deep inner soul. We're required to do these things just as salmon swim upstream."

What precisely was in Armstrong's own deep inner soul about the Moon landing, or about anything else that happened in his life—his true feelings about his father, his religious beliefs, the effects of Muffie's death—was hardly laid bare by the remark, or by any other verbal statement he ever made. It was just not his way. Perhaps, his extraordinarily judicious restraint of expression was a deeply inculcated outcome of the avoidance strategy he had developed in childhood. Or perhaps it derived, as his first wife Janet hesitantly came to suggest, from a feeling of social inferiority based on his humble family background in rural Ohio.

What Armstrong on the eve of becoming the First Man did not and would not define or explain about himself, others now sought, almost desperately in the days before the launch, to explain and define for him. All the humanistic and cosmic meanings that he would not fill in, others felt compelled to fill in for him. On the eve of humankind's great adventure to set foot on another heavenly body, Armstrong had become like an oracle of ancient times, a medium, wise, prophetic, mysterious, by which fortunes and misfortunes were told, deities consulted, prayers answered.

Not until he constructed his own myth out of Armstrong could the creative mind of Norman Mailer be satisfied. It did not matter that Mailer would never meet Neil face-to-face, never once talk to him directly, never ask him a single question of his own. Mailer, too, had sat before the oracle, "the most saintly of the astronauts," someone who was "simply not like other men," who was "apparently in communion with some string in the universe others did not think to play." It was up to Mailer, up to Aquarius, to decode Armstrong.

Like Mailer, we were to be the author of our own Moon landing.

Mailer conjured the makings of his own Armstrong while sitting in on NBC correspondent Frank McGee's interview with Neil near the end of the day on July 5. In the interview McGee referred to a story in *Life* by staff writer Dodie Hamblin in which Armstrong told of the recurring boyhood dream in which he hovered over the ground. Mailer had read Hamblin's story when it appeared but dismissed its importance until he heard Armstrong, after a day filled with Neil's engineer-speak, corroborate that, indeed, as a boy, he had such dreams. Mailer was taken with the beauty of the dream: "It was beautiful because it might soon prove to be prophetic, beautiful because it was profound and it was mysterious, beautiful because it was appropriate to a man who would land on the moon." For Mailer, it was a type of epiphany, one by which he could construct "The Psychology of Astronauts" and interpret the entire Space Age: "It was therefore a dream on which one might found a new theory of the dream, for any theory incapable of explaining this visitor of the night would have to be inadequate, unless it were ready to declare that levitation, breath, and the moon were not proper provinces of the dream."

The idea that such a non-whimsical man as Armstrong, as a young boy, dreamed of flight "intoxicated" Mailer, "for it dramatized how much at odds might be the extremes of Armstrong's personality." On the one hand, consciously, Armstrong, the archetypal astronaut-engineer, was grounded in the "conventional," the "practical," the "technical," and the "hardworking." He resided at the very "center of the suburban middle class." On the other hand, what Armstrong and the other astronauts were doing in space was "enterprising beyond the limits of the imagination." Their drive and ambition simply had to have a subconscious element.

It was in this union of opposites, the impenetrable fusion of the conscious and the unconscious, that one found in the modern technological age "a new psychological constitution to man." More than any of the other astronauts, Neil's personality stemmed from the core of that "magnetic human force called Americanism, Protestantism, or Waspitude." He was the Lancelot of the silent majority, "the Wasp emerging from human history in order to take us to the stars." Never mind that Mailer knew almost nothing about Armstrong's family background, personal history, married life, religious beliefs, friends, or genuine psychological state. Aquarius's object was not to understand Armstrong; it was to understand the comings and wrong-goings of humankind in the twentieth century, a century that sought "to dominate nature as it had never been dominated," "create death, devastation and pollution as never before," yet "attack the idea of war, poverty and natural catastrophe as never before." It was also a century "now attached to the idea that man must take his conception of life out to the stars."

There was no denying the brilliance of Mailer's exposé. Yet Mailer really did not care about Armstrong, the man, on a personal level, only as a vessel into which the author could pour his own mental energy and profundity. What Mailer wrote in his chapter "The Psychology of Astronauts" was highly provocative and insightful as social criticism, but as history, biography, or real psychology, it shed considerably more heat than light.

The mythologizing and iconography had only just begun. Fifteen days after the press conference, Armstrong would step onto the Moon. He would no longer be just a man, not for any of us. He would be First Man.

PART SIX

MOONWALKER

He who would bring back the wealth of the Indies, must carry the wealth of the Indies with him.

—INSCRIPTION ON THE FAÇADE OF UNION STATION, WASHINGTON, D.C.

We were told: Save the Moon rocks first. We only have one bag of rocks. We have lots of astronauts.

—MIKE MALLORY, MEMBER OF THE NAVY FROGMAN TEAM THAT RECOVERED THE APOLLO 11 SPACECRAFT UPON ITS SPLASHDOWN INTO THE PACIFIC OCEAN ON JULY 24, 1969

CHAPTER 22

Outward Bound

For Armstrong, Collins, and Aldrin, going into space commenced in the crew quarters three and a half hours before liftoff, shortly after 6:00 A.M., when technicians snapped the astronauts' helmets down onto their neck rings and locked them into place. From that moment on, the crew of the first Moon landing breathed no outside air. They heard no human voice other than that piped in electronically through the barrier of their pressure suits. They saw the world only through the veneer of their face-plates, and could smell, hear, feel, or taste nothing but that which modern technology manufactured for them inside their protective cocoon.

For Armstrong, the isolation was more familiar than it was for his mates. As a test pilot back at Edwards, he had grown accustomed to the confinement of pressurized flight suits. In comparison to the partial pressure suits and headgear he had donned for flying zooms in the F-104 or going to the edge of space in the X-15, the Apollo suit was roomy and easy to maneuver.

Still, as the crew of Apollo 11 left the Manned Spacecraft Operations Building at 6:27 A.M. and paraded in their protective yellow galoshes into the air-conditioned transfer van that was to transport them eight miles to

Launchpad 39A, every tissue of their being acknowledged that they had left the ordinary realm of nature and had entered the totally artificial environment that would sustain them in outer space.

Going into the mission, Neil, Mike, and Buzz possessed great confidence in the Saturn rocket, but one could never be sure about any rocket's performance. "It was certainly a very high-performance machine," Armstrong asserted. "It was not perfect, though." The Saturn V had come to life so quickly. The phenomenally fast pace of its development resulted from the strategy of "all-up testing," a new NASA R&D philosophy championed by Dr. George Mueller, the associate administrator for manned space flight. Mueller put the development of the Saturn V on the fast track by having the rocket tested from the very start with all three of its stages "live" and ready to go, rather than having the stages incrementally tested one at a time and then mating the three of them together only after each had proven itself independently.

Without all-up testing, Kennedy's deadline could not be met. Still, it was not the surest way to assure the development of a sound rocket, particularly one that was such a vast and complex piece of novel machinery—producing an enormous 7.6 million pounds of thrust.

By the time the crew sat on top of the monumentally powerful rocket waiting for it to fire, they were past pondering its dangers. Moreover, there was always the chance that something minor would go wrong at the last minute in one of the several hundred subsystems associated with the rocket, the spacecraft, or the launch complex, and cancel the launch.

The first astronaut into *Columbia* the morning of the launch was not Armstrong, Collins, or Aldrin; it was Fred Haise, Aldrin's backup as lunar module pilot.

"Freddo" preceded the crew into the spacecraft by some ninety minutes in order to run through a 417-step checklist designed to ensure that every switch was set in its proper position. At 6:54 A.M., Haise and the rest of the "close-out crew" gave the spacecraft their thumbs-up. Having taken the elevator up the 320 feet to the level of the waiting spacecraft, Armstrong grasped the overhead handrail of the capsule and swung himself through the hatch. Prior to climbing in, Neil received a small gift from Guenter Wendt, the pad leader: it was a crescent moon that Wendt had carved out of Styrofoam and covered with metal foil. Wendt told him "it's a key to the Moon," and a smiling Neil asked Wendt to hold on to it until he got back. In exchange Neil gave Wendt a small card he had been keeping under the wristband of his watch. It was a ticket for a ride in a "space taxi," reading "good between any two planets."

Inside the command module, Armstrong settled into the commander's seat to the far left. Five minutes later, after a technician hooked up Neil's lines and hoses, Collins, the command module pilot, climbed into the right seat, followed by Aldrin, the lunar module pilot, in the center. (Aldrin was in the center seat because he had trained for that position on Apollo 8. Collins had been out for a while with his neck spur, so rather than retrain Buzz for ascent, NASA left him in the center and trained Mike for the right seat.)

To Neil's left hand was the abort handle. One twist of the handle would trigger the solid rocket escape tower attached to the top of the command module to blast Apollo 11 clear of trouble. In the Gemini program, the spacecraft had possessed ejection seats rather than an escape tower, but the Gemini's Titan booster used hypergolic fuels that could not explode the way the Saturn could, since it was fueled with kerosene, hydrogen, and oxygen. Ejection seats would not have thrown the astronauts far enough away from a Saturn explosion. The Saturn V's ascent was especially suited to Neil's background in research, in that the booster rocket was controllable from the cockpit: "You could not fly the earlier models of the Saturn rocket from the spacecraft. Had there been a failure on the Saturn's inertial system on Apollo 9, for example, McDivitt, Scott, and Schweickart would have had to splash down into the Atlantic or maybe land in Africa, with a high risk of physical injury. For our flight we had added an alternate guidance system in the command module's gear so that if there were a failure of some kind on the Saturn, we could switch to the alternate systems and fly the rocket from the spacecraft." If the autopilot went out, the pilot could fly the booster into orbit manually.

Apollo 11's ascent into orbit involved a number of discrete phases, and within each were discrete changes in abort technique. Armstrong explained, "It was a complete concentration on getting through each phase and being ready to do the proper thing if anything went wrong in the next phase." His most important cues during the fiery ascent came, in his words, from "a combination of looking at the attitude indicator, following the flight performance on the computer, and listening to indications over the radio as to which phase you were in or were about to enter."

In the time it took for a fraction of the heavy automobile traffic to crawl clear of Cape Kennedy's environs, Apollo 11 went around the world one and a half times and was on its way to the Moon. On the front lawn of their Ohio home, Neil's parents had already been interviewed by a small horde of media: "Mr. Armstrong, what did you think of that launch?" and "Mrs. Armstrong, what were your feelings when you saw that rocket disappear into the sky?" Viola exclaimed, "I'm thankful beyond words." Projecting her religious beliefs onto her son as she always did, she asserted, "Neil believes God is up there with all three of those boys. I believe that, and Neil believes that." Steve remarked, "It's a tremendous, most happy time. We'll stay glued to the television for the entire flight." Viola's mother, eighty-two-year-old Caroline Korspeter, remarked before the TV cameras: "I think it's dangerous. I told Neil to look around and not to step out if it didn't look good. He said he wouldn't."

Back on the Banana River, Janet Armstrong and her boys stayed on the yacht listening to transmissions from the spacecraft on a NASA squawk box until the crowds dispersed. Though greatly relieved the launch had gone off smoothly, at Janet's request no bottles of champagne were opened on board, preferring that celebrating be reserved for when the men were home safely. Before departing for home, Janet met briefly with journalists. "We couldn't see the rocket right away," Rick shyly reported, "and I was kind of worried at first. All of a sudden, we could see it, and it was beautiful." Janet told the press, "It was a tremendous sight. I was just thrilled," though her main feeling was simply relief that the launch had gone off safely. "This, too, shall pass" was what she was actually thinking. She had gotten almost no sleep the night before. When she got home to Houston late that afternoon, the press waited in her yard. "I don't feel historic," Janet succinctly told them, ushering her boys into the house. The vigil for the crew had just begun. It would be two and a half more days before the astronauts made it to lunar orbit, a day after that before Neil and Buzz descended to the landing, and four days beyond that before they returned to Earth.

So much could still go wrong.

At 10:58 Houston time, two hours and twenty-six minutes after liftoff, Mission Control gave Apollo 11 the "go" for TLI or "translunar injection"—leaving Earth orbit and heading into deep space. The astronauts fired the Saturn V's third-stage engine, the only stage still attached to the command and service module. This burn, lasting some five and a half minutes, accelerated Apollo 11 to over 24,200 miles per hour, to escape the hold of Earth's gravity.

Although Armstrong reported that the ride had been "beautiful," privately he would have liked it to have been smoother: "In the first stage, the Saturn V noise was enormous, particularly when we were at low altitude because we got the noise from seven and a half million pounds of thrust plus the echo of that noise off the ground that reinforced it. After about thirty seconds, we flew out of that echo noise and the volume went down substantially. But in that first thirty seconds it was very difficult to hear anything over the radio. In the first stage, it was also a lot rougher ride than the Titan. It seemed to be vibrating in all three axes simultaneously." With the burnout of the first stage, the flight smoothed out and quieted down considerably; so much so that the astronauts could not feel any vibration or even hear the engines running. Rocketing upward on the second and third stages of the Saturn proved superior to any stage of the Titan. Mike Collins later described the herky-jerky early ascent of the Saturn V: "It was like a nervous novice driving a wide car down a narrow alley and jerking the wheel back and forth spasmodically." Then on the upper stages the Saturn V turned into "a gentle giant," with the climb out of the atmosphere as "smooth as glass, as quiet and serene as a rocket ride can be."

Out their windows the astronauts could see nothing until three minutes into their ride, when the spacecraft reached sixty miles high. At that altitude, the Apollo 11 crew jettisoned their unused escape rocket and let loose the protective shield that had been covering the command module. Still pointing up, though not straight up following a pitch-over maneuver some three minutes into the ascent, there was nothing for the crew to see except for what Collins called "a small patch of blue sky that gradually darkens to the jet black of space."

Earth orbit was achieved twelve minutes into the flight, when the first burn of the Saturn's single third-stage engine pushed the spacecraft up to the required speed of 17,500 miles per hour. The trio now had an orbit and a half in which to make sure all their equipment was operating properly before they reignited their third-stage engine and committed to leaving the Earth's gravitational field.

According to Armstrong, "The purpose of doing the orbit and a half was twofold. One, it allowed a little more flexibility in launch time, and second, it gave us the opportunity to check out all the principal systems of the spacecraft—the command module, not the lunar module—prior to leaving the Earth's orbit on a translunar trajectory. So systems checkout was the principal reason that we were in this holding orbit, and the responsibility was shared between the crew on board and the people on the ground. The people on the ground could see a good bit more detail of systems operations, and the orbit and a half gave them a long enough time to look at it. If something went wrong on the spacecraft, we would have the time to decide whether we should forget the whole thing and abort."

Initially, the crew found only brief moments to take in the spectacular view of the Earth below. Their first sunrise one hour and nineteen minutes into the flight prompted a hunt for the Hasselblad camera. Fifteen seconds later, Collins found the camera floating in the aft bulkhead.

For the first time since their respective Gemini flights, Armstrong, Collins, and Aldrin again experienced the wonder of weightlessness. In this state, the fluid in the inner ear sloshed freely. Motion sickness could happen more easily in the Apollo spacecraft than in the Gemini because Apollo was more commodious. The mission planners for the subsequent Apollo flights told the crews to move around as slowly and gingerly as possible and not to wiggle their heads back and forth too much, until they got used to being in weightless conditions. Armstrong was intently aware of the potential problem. One hour and seventeen minutes into the flight, he asked Mike and Buzz: "How does zero g feel? Your head feel funny, anybody, or anything like that?" Mike answered, "No. It just feels like we're going around upside down."

According to Neil, "We were very fortunate that none of the crew came down with the malady at any point in the flight. Some of the people that were best known to have an iron gut ended up getting space sickness. No one was sure at the time what exactly was causing it. They were trying various things." He had outgrown his childhood proclivity for nausea, although he could become queasy when doing aerobatics. Curiously, however, space sickness did not correlate to motion sickness on Earth.

Convinced that the ship was ready to leave Earth

orbit, Mission Control gave Apollo 11 the go for TLI some two hours and fifteen minutes after the spacecraft reached orbit. Flight procedures required that the crew, for their protection during the burn, put their helmets and gloves back on; the idea was that if the Saturn third stage, known as the S-IVB, exploded, the astronauts would be inside sealed pressure suits. "The problem with that thinking," Collins said, was that "any explosion massive enough to crack our ship's hull would also result in multiple equipment failures, and we would never get back in one piece. Still, a rule was a rule, so we sat there, helmet and gloves on, ready to be propelled to another planet."

Approaching the point for translunar injection halfway around its second Earth orbit, a preprogrammed sequence fired the Saturn's third-stage engine for one final time, accelerating Apollo 11 to the escape velocity. The TLI burn took just under six minutes. At the moment of ignition, the spacecraft was over the Pacific Ocean; circling more than ninety miles beneath it, a formation of KC-135 aircraft—converted air force tankers carrying a large array of electronic gear—relayed telemetry data from the spacecraft back to Houston. The data indicated that the Saturn V had performed its last job well, speeding away at a rate of six miles per second—faster than a rifle bullet.

The trip out started far busier for Collins than it did for Armstrong or Aldrin. As command module pilot, it was Mike's job (assisted by Neil and Buzz) to separate *Columbia* from the S-IVB and turn the command and service module around. Mike would then maneuver the CSM into a docking with *Eagle,* the lunar module, which, to survive the launch—with its spindly legs, thrusters and antennae stuck out at odd angles, and extremely fragile pressure shell of a body—had flown up to this point tightly secured inside a strong boxlike container attached atop the S-IVB. It was a critical maneuver in the flight plan. "If the separation and docking did not work," Aldrin has explained, "we would return to Earth. There was also the possibility of an in-space collision and the subsequent decompression of our cabin, so we were still in our space suits as Mike separated us from the Saturn third stage."

Neither Aldrin nor Armstrong felt any great apprehension about the maneuver. "Mike did this docking maneuver," Neil related, "as he would need to make a similar docking with the LM after we returned from the lunar surface. This had been done before on both Apollo 9 and 10, so I was pretty confident."

The maneuver came off perfectly. Explosive bolts blew apart the upper section of the large container, giving access to the LM in its garage atop the rocket. Collins controlled rocket thrusters that moved the CSM out and away some one hundred feet from the landing craft. Turning the spacecraft around, he inched forward gently to a successful head-to-head docking. *Columbia* and *Eagle* were now mated; when the time came, Neil and Buzz could enter the LM through an internal tunnel and hatch arrangement. To complete the separation maneuver, the LM had to be released from its mounting points and the CSM/LM stack had to be backed away from the S-IVB. Then all that remained was to slingshot the S-IVB out of the way. A command sent over to the S-IVB from Apollo 11 caused it to dump all of its leftover fuel, resulting in a propulsive reaction that sent the rocket tumbling off on a long solar-orbit trajectory that would keep it far out of Apollo 11's way.

The time was 1:43 P.M. CDT, Houston time, only five hours and eleven minutes into the flight. Apollo 11 was traveling at 12,914 feet per second and approaching 22,000 nautical miles from Earth.

With the separation, docking, and post-TLI maneuvers behind them, the astronauts stripped down and pulled on their considerably more comfortable two-piece white Teflon fabric jumpsuits. In weightlessness, some things were easier to do than in a gravity field, but three men changing out of space suits—in a compartment equivalent in interior space to a small station wagon—was not one of them. Undressing, folding their stiff heavy suits into storage bags, and then stuffing the filled bags under the couch of the spacecraft was a laborious process that, in Aldrin's words, brought about "a great deal of confusion, with parts and pieces floating about the cabin as we tried to keep logistics under control." Collins compared it to "three albino whales inside a small tank, banging into the instrument panel despite our best efforts to move slowly. . . . Every time we pushed against the spacecraft our bodies tended to carom off in some unwanted direction and we had to muscle them back into place."

With their clothes finally off, the crew blissfully removed the gadgets affixed to their private parts. Because the astronauts might need to urinate or have a bowel movement before their suits could be taken off, devices for excreting had been connected to them preliminary to suiting up. Aldrin recalled the nitty-gritty details: "We rubbed our behinds with a special salve and pulled on what were euphemistically called fecal-containment garments." The modified diaper kept the odor of despoiled briefs to a minimum, and the salve kept the men's rear ends from chafing too badly. Urinating was accomplished through prophylactic-like devices from which a connector led to a sack resembling a bikini secured around the hips. To function without leaking, the rubber condom catheter had to fit quite snugly, an uncomfortable reality for male plumbing that the crew joked about privately. Once cleaned up and in their jumpsuits with fresh underwear, going to the bathroom was easier. Feces got stowed in special containers, and urine was vented out of the spacecraft.

Safely on their way Moonward, the astronauts relaxed for the first time. As Collins explained, there was no way to prepare oneself for the novel, Twilight Zone-like experience of cislunar space, the region between the Earth and the Moon: "Unlike the roller-coaster ride of the Earth orbit, we are entering a slow-motion domain where time and distance seem to have more meaning than speed. To get a sensation of traveling fast, you must see something whizzing by: the telephone poles along the highway, another airplane crossing your path. In space, objects are too far from each other to blur or whiz, except during a rendezvous or a landing and in those cases the approach is made slowly, very slowly. But if I can't sense speed out my window, I can certainly gauge distance, as the Earth gets smaller and smaller. Finally the whole disk can be seen."

This cosmic vision of "Spaceship Earth" would deeply move every lunar astronaut. "It was a slowly changing panorama as you went from just the horizon to a large arc, to a larger and larger arc, and finally a whole sphere," Armstrong described. "And depending on what the flight attitude requirements of the vehicle were at any given moment, you may not have been able to see all of that all the time. But we certainly saw the Earth become a sphere. It was a striking event, leaving the planet and realizing that there was no logical reason that you were ever going to fall back to that planet at some point. It was a commitment to excellence, in terms of what you had to do to get back."

Looking back at the "Whole Earth," Armstrong reveled in his knowledge of geography. He radioed at three hours and fifty-three minutes into the flight, "You might be interested that out of our left-hand window right now, I can observe the entire continent of North America, Alaska, and over the Pole, down to the Yucatán Peninsula, Cuba, northern part of South America, and then I run out of window."

In order to make sure that the spacecraft's pipes did not freeze on one side while tank pressures increased from too much heat on the other, Apollo 11 began a slow rotation to ensure that solar rays were absorbed as evenly as possible by all sides of the spacecraft. "We were like a chicken on a barbecue spit," Collins explained. "If we stopped in one position for too long, all kinds of bad things could happen." Visually, the rotisserie action resulted in an incredible panorama, with stunning views of the Sun, Moon, and Earth cycling into the spacecraft's windows every two minutes. Aiding the sightseeing was a simple viewing device called a monocular—half of a set of binoculars. Using it like a magnifying glass, the astronauts took turns getting close looks at different features of their home planet.

It has since become one of the legends of spaceflight that there are only two man-made objects on Earth that can be seen from outer space—the Great Wall of China and the gigantic Fort Peck Dam in Montana. "I would challenge both," Neil stated. In cislunar space, "We could see the continents; we could see Greenland. Greenland stood out, just as it does on the globe in your library, all white. Antarctica we couldn't see because there were clouds over it. Africa was quite visible, and we could see sun glint off of a lake. But I do not believe that, at least with my eyes, there was any man-made object that could be seen."

Whether it was with the naked eye or with the monocular, Neil could not help but contemplate how fragile the Earth looked: "I don't know why you have that impression, but it is so small. It's very colorful, you know. You see an ocean and gaseous layer, a little bit—just a tiny bit—of atmosphere around it, and compared with all the other celestial objects, which in many cases are much more massive and more terrifying, it just looks like it couldn't put up a very good defense against a celestial onslaught." Buzz and Mike felt likewise, with Buzz thinking how crazy it was for the globe to be so politically and culturally divided: "From space it has an almost benign quality. Intellectually one could realize there were wars under way on Earth, but emotionally it was impossible to understand such things. The thought occurred and reoccurred that wars are generally fought for territory or are disputes over borders; from space the arbitrary borders established on Earth cannot be seen."

Next, the astronauts had to eat. They needed to consume sufficient water and between 1,700 and 2,500 calories per day. Already before their first full meal—scheduled for midafternoon of the first day, following the slingshot of the S-IVB and getting out of their suits—the crew had eaten sandwiches made of tubed spreads. The snack pantry contained peanut cubes, caramel candy, bacon bites, and dried fruit.

For the first time in a U.S. lunar spaceflight, the beverage list included not just juice and water but ample servings of coffee. The juice was *not* Tang. Two six-foot flexible tubes attached to spigots supplied cold and hot water. At the end of each tube was a pistol probe with a push button. If the astronaut wanted a cold drink, he held the probe in his mouth, pushed the button, and out came a mouthful of water. If he was preparing food, he stuck the hot water gun into a plastic bag and squirted three blasts into it. Massaging the rehydrating food into an edible form, he ate by sucking the entrées through a tube. Unfortunately, the device designed to ventilate hydrogen from the water did not function well. Considerable gas got into the food, bloating the astronauts and giving them stomach gas. Aldrin jokes that it got so bad, "we could have shut down our attitude-control thrusters and done the job ourselves!"

The fare turned out to be appetizing enough, if bland. A turkey dinner with gravy and dressing was mixed with hot water and eaten with a spoon. "Wet packs" were eaten as they were, including ham and potatoes. Sometimes the crew ate the same meal—as on day two, when they had hot dogs, applesauce, chocolate pudding, and a citrus drink. Other times, they ate individualized meals. Neil's favorite meal was spaghetti with meat sauce, scalloped potatoes, pineapple fruitcake cubes, and grape punch.

Eleven hours into the flight, the crew was ready for its first sleep period. In fact, at 7:52 P.M. CDT, two hours before the scheduled time, Houston wished its tired crew good night and signed off. The urge to sleep had come on the astronauts quite a bit earlier. Just two hours into the flight, before preparations for TLI began, Neil yawned to his mates, "Gee, I almost went to sleep then," to which Collins replied, "Me, too. I'm taking a little rest." For the next nine hours, they fought sporadic drowsiness until it was time to sleep.

Neil and Buzz slept on light mesh hammocks, much like sleeping bags, which were stretched and anchored beneath the left and right couches—the center couch having been folded down, still covering the crews' space suits. "It kept our arms from floating around and from inadvertently actuating switches," Neil explained. The man on watch—Collins the first night out—slept not in a hammock but floated above the left couch, a lap belt keeping him from floating off and with a miniature headset taped to his ear in case Houston called during the "night." "It was a strange but pleasant sensation to doze off with no pressure points falling anywhere on your body," Collins related. It was like being "suspended by a cobweb's light touch—just floating and falling all the way to the Moon." Buzz got to experience the feeling, but Neil did not, as he always slept in his hammock.

With adrenaline levels still fairly high from the excitement of liftoff and TLI, the men slept only five and a half hours that first night. When CapCom Bruce McCandless of Mission Control's Green Team called to wake the crew at 7:48 CDT, all three were already alert. As the astronauts went over their "post-sleep checklist," following updates on the flight plan, McCandless gave them a brief review of the morning news, much of which concerned the world's enthusiastic reaction to their launch.

The first news item concerned the flight of the Soviet Union's Luna 15: according to the story read to the astronauts, the USSR's robotic spacecraft had just reached the Moon and started around it. In a last-ditch effort to steal thunder from America's Moon landing, the Russians had launched the small unmanned spacecraft on July 13, three days prior to Apollo 11's liftoff; its objective was not just to land on the Moon but to scoop up a sample of lunar soil and return it to Earth before Apollo 11 got back. Newspapers in the United States editorialized that the Russians were purposefully trying to upstage the Americans with their "mystery probe" and speculated (inaccurately) that they might also be trying to interfere technically with the American flight.

U.S. space officials worried that Soviet operations and communications with Luna 15 might, in fact, interfere with Apollo—over the years that had happened occasionally when the Russians operated at or near NASA's radio frequencies.

MSC's Chris Kraft telephoned Colonel Frank Borman, the Apollo 8 commander, who was just back from a nine-day tour of the USSR, the first U.S. astronaut ever to visit the country. "The best thing to do is just ask 'em," Borman told Kraft. So, with Nixon's permission, over the famed hotline between Moscow and Washington established by the two superpowers to avert nuclear holocaust following the Cuban Missile Crisis of 1962, Borman sent a message to the head of the USSR Academy of Sciences, Mstislav V. Keldysh, asking for the exact orbital parameters of the Russian probe. Borman was assured that Luna 15's orbit did not intersect the trajectory of Apollo 11.

In fact, nothing about Luna 15 would bother Apollo, and the Soviet mission failed miserably. Luna 15 crashed into the Moon on July 21, the day after Apollo 11's successful landing.

Not confirmed until many years later was the most powerful explosion in the history of rocketry on the launchpad at the Baikonur Cosmodrome in Kazakhstan on July 3, 1969, just nine days before the launch of Luna 15. On that day, the Soviets were test firing a Moon rocket of their own: a mammoth booster designated the N-1. If the unmanned test launch of the N-1 worked, the Soviets were prepared to press on with their clandestine manned lunar program. Seconds after the launch, the N-1 rocket collapsed back onto its pad and exploded—by some estimates with a strength equivalent to 250 tons of TNT. Somehow no one was killed. Not until November 1969 did rumors of the Soviet accident surface in the Western press; by then, American intelligence knew about it through spy satellite photos. The N-1 disaster spelled the end of the Soviet Moon program. Not until after the fall of the Soviet Union in August 1991 did participants in the Soviet Moon program even admit their program existed, let alone that the N-1 disaster had occurred.

The major flight event of day two came at 10:17 A.M. CDT when a three-second burn refined the course of Apollo 11 and tested the CSM engine, which would be needed to get the spacecraft in and out of lunar orbit. At the moment of that slight midcourse correction, Armstrong and his mates were 108,594 miles from Earth—over two-fifths of the way to the Moon—and traveling at a velocity of only 5,057 feet per second. Still in the pull of Earth's gravity, the speed of Apollo 11 would decrease steadily until it was less than forty thousand miles from the Moon—by which point the spacecraft had slowed from its top velocity of approximately twenty-five thousand mph to a mere two thousand mph. Then as the Moon's pull increased, it would speed up again.

Much of the astronauts' time during their midflight coast was taken up with the various minor tasks required to keep the CSM operating properly: purging fuel cells, charging batteries, dumping wastewater, changing carbon dioxide canisters, preparing food, chlorinating drinking water, and so forth. Collins did most of the routine housekeeping so Armstrong and Aldrin could stay focused on reviewing the details of the landing to come—going over checklists, rehearsing landing procedures. "The flight plan to the Moon had several blank pages," Aldrin remembered, "periods in which we had nothing to do. Yet I have no recollection at all of being idle. Everything had to be stowed or sealed away or anchored to one of the many panels by Velcro. We each had little cloth pouches in which we kept various frequently used items, such as pens, sunglasses and, for me, a slide rule. As often or not, one or two of us would be scrambling around on the floor searching for a missing pair of sunglasses, the monocular, a film pack, or a toothbrush."

During rest periods they did relax to some music. It was played on a small portable tape recorder carried on the flight primarily for the purpose of recording crew comments and observations. Neil and Mike requested some specific music be preloaded onto the cassette tape; mostly easy listening. Neil asked specifically for two recordings. One was Antonin Dvorak's 1895 *New World Symphony*. Neil had played the piece when he was in the Purdue concert band, and it also seemed appropriate. The other piece was *Music Out of the Moon* by composer Dr. Samuel Hoffman, the featured instrument of which was the theremin, an unusual device that generated tones electronically by a musician controlling the distance between his hands and two metal rods serving as antennas.

The highlight of day two was the first live television broadcast from Apollo 11, which was scheduled to begin at 7:30 P.M. EDT. Actually, it was the third TV transmission overall from the flight; the first two were conducted to check out camera functions, the picture quality of both interior and exterior shots, and the strength of signal coming into and out of the Goldstone tracking station in California. That way any glitches could be fixed before several million people around the world tuned in to see the broadcast on Thursday evening.

The first fuzzy picture to appear on everyone's screen was a shot looking back at the home planet, which Armstrong described as "just a little more than a half Earth." In plain but wondrous language, Neil pointed out the "definite blue cast" of the oceans, the "white bands of major cloud formations over the Pacific," "the browns in the landforms," and "some greens showing along the northwestern coast of the United States and northwestern coast of Canada." He explained that at their current distance—some 139,000 nautical miles—the depth of the colors was not as great as what they had enjoyed while in Earth orbit or even at 50,000 miles out. For thirty-six minutes, the astronauts put on a show. Aldrin did a few zero-g push-ups, and Neil even stood on his head. Head chef Collins demonstrated how to make chicken stew when traveling at a speed of 4,400 feet per second. The transmission ended emotionally with Neil saying, "As we pan back out to the distance at which we see the Earth, it's Apollo 11 signing off." The crew spent the next three hours on housekeeping items and participating fruitlessly in a telescope experiment during which they were unable to spot a bluish-green laser light being shot at them from the McDonald Observatory near El Paso. None of them fell asleep until after 11:30, this time with Aldrin in the floating "watch" position. The sleep period was scheduled to be a long one, lasting ten hours. Data from the flight surgeon indicated that "the crew slept rather well all night"—so well, in fact, that Mission Control let them have an extra hour before waking them up to perform such chores as charging batteries, dumping wastewater, and checking fuel and oxygen reserves.

In the preliminary flight plan, Aldrin and Armstrong were not scheduled to make their first inspection trip into *Eagle* until Apollo 11 reached lunar orbit around midday of day three, but Aldrin lobbied successfully with the mission planners to enter the LM a day early in order to make sure the lander had suffered no damage during launch and the long flight out. The sojourn began a little after 4:00 P.M. CDT, about twenty minutes into what NASA considered at the time the clearest TV transmission from space ever made. After Collins opened the hatch, Armstrong squeezed through the thirty-inch-wide tunnel and floated in through the top of the LM, followed by Aldrin. Both Neil and Buzz remember the down-up-up-down trip into *Eagle* as one of the oddest sensations of their entire Moon trip, crawling from the floor to the ceiling of the command module only to find themselves descending headfirst from the ceiling of the docked LM.

Though Neil was the first one to take a look inside *Eagle,* it was Buzz as lunar module pilot who began preparing the LM for its separation from *Columbia* that was to come forty-five hours later. Buzz and Neil took the movie camera and television camera along with them, sending back the first pictures from inside the LM. Mission Control knew that the transmission was coming, but it surprised the TV networks, which were not expecting the next pictures from Apollo 11 until 7:30 P.M. EDT, the same time as the previous evening. Scurrying to make the necessary technical connections, CBS went on air with Cronkite and sidekick Wally Schirra at 5:50 P.M. The first images—being broadcast live to the United States, Japan, Western Europe, and much of South America—showed Aldrin taking an equipment inventory in the LM. Later, Buzz gave the international television audience a look at the space suit and life support equipment that he and Neil would wear on the Moon.

No account of the flight of Apollo 11 would be complete without coming to grips with subsequent tales that the crew saw some UFOs. According to the stories, the astronauts saw some things they could not identify, ranging from mysterious lights to actual formations of spaceships. As is true for many a fanciful tale, the stories had a kernel of truth. The first alleged sighting of a UFO by the Apollo 11 crew came early on day one just as the burn for TLI was starting, when the men saw some flashes out of window 5. The astronauts said nothing to the ground about the flashing lights, even though, as Aldrin recalls, they saw the flashes "at least two or three different times," and not just on the outbound flight. The phenomenon of the flashing lights was unusual enough that NASA would brief the next crew about it. And when Apollo 12 went up, they too saw the lights; in fact, they came back and reported, "Guess what? We see them with our eyes closed!" The flashing lights turned out to be a phenomenon that occurred in the especially dark conditions of outer space *inside the human eyeball.* There was an optical threshold tied to a psychological threshold—where a person had to *want* to look and see the flashing lights, or he would not observe them. Experts have since explained that some astronauts have such a sensitive threshold that they see the flashes even when flying in near space below the Van Allen radiation belt.

A second "sighting" took place the evening of the third day—the day of the first sojourn into the LM—shortly after 9:00 P.M. Aldrin apparently saw it first: "I found myself idly staring out of the window of the *Columbia* and saw something that looked a bit unusual. It appeared brighter than any star and not quite the pinpoint of light that stars are. It was also moving relative to the stars. I pointed this out to Mike and Neil, and the three of us were beset with curiosity. With the help of the monocular we guessed that whatever it was, it was only a hundred miles or so away. Looking at it through the sextant we found it occasionally formed a cylinder, but when the sextant's focus was adjusted it had a sort of illuminated 'L' look to it. There was a straight line, maybe a little bump in it, and then a little something off to the side. It had a shape of some sort—we all agreed on that—but exactly what it was we couldn't pin down."

The crew fretted, "What are we going to say about this?" Aldrin remembered. "We sure as hell were not going to talk about it to the ground, because all that would do is raise a curiosity and if that got out, someone might say NASA needed to be commanded to abandon the mission, because we had aliens going along! Our reticence to be outspoken while it was happening was because we were just prudent. We didn't want to do anything that gave the UFO nuts any ammunition at all, because enough wild things had been said over the years about astronauts seeing strange things." At first the crew speculated that what they were seeing was the shell of the Saturn S-IVB that had been slingshot away more than two days earlier. But when Neil radioed Houston about it, the answer was that the S-IVB was about six thousand nautical miles away.

The astronauts scratched their heads. At far closer than six thousand miles, the object in sight could not be the S-IVB, but rather one of the four panels that had enclosed the LM's launch garage. When the LM was extracted for face-to-face mating with the command module, the side panels had sprung off in different directions. Mission Control eventually concluded that it was one of the Saturn LM adapter panels, which would have rotated when ejected from the S-IVB. The reflection of the sun on the panels would be similar to blinking.

The third night out the Apollo 11 astronauts rested more fitfully—they knew that when they awoke, day four was going to be different. Stopping their rotisserie motion and getting into lunar orbit was not automatic. If the spacecraft did not slow down sufficiently, it would sail on by the Moon in a gigantic arc and make a looping return back to the vicinity of the Earth. Arousing the astronauts that morning at 7:32 A.M. CDT, Mission Control again started the day by reading them the morning news. "First off, it looks like it's going to be impossible to get away from the fact that you guys are dominating all the news back here on Earth," said Bruce McCandless, taking his turn as CapCom. "Even *Pravda* in Russia is headlining the mission, and calls Neil 'The Czar of the Ship.' I think maybe they got the wrong mission."

The Sun was now directly behind the Moon, its corona cascading brilliantly around the edges. Only 12,486 miles away, the Moon was a huge dark object completely filling their windows. The Earthshine shone so brightly from behind them that it cast the lunar surface as three-dimensional. Collins later wrote, "The first thing that springs to mind is the vivid contrast between the Earth and the Moon. One has to see the second planet up close to truly appreciate the first. I'm sure that to a geologist the Moon is a fascinating place, but this monotonous rock pile, this withered, sun-seared peach pit out my window offers absolutely no competition to the gem it orbits. Ah, the Earth, with its verdant valleys, its misty waterfalls. I'd just like to get our job done and get out of here."

The vital step in the process of landing was making a very precise burn called "LOI-1," the basic Lunar Orbit Insertion. The burn involved firing the service propulsion system engine for just under six minutes, braking Apollo 11 down to a speed that allowed the Moon's gravity to trap the spacecraft and reel it into orbit. As

Mike Collins said, "We need to reduce our speed by 2,000 mph, from 5,000 down to 3,000, and will do this by burning our service propulsion system engine for six minutes. We are extra careful, paying painful attention to each entry on their checklist." A lot of help with the burn came from the onboard computer and from Mission Control, but it was up to the astronauts to get it right: "If just one digit slips in our computer, and it is the worst possible digit, we could turn around backward and blast ourselves into an orbit headed for the Sun."

It seemed that the LOI burn went well, but Mission Control could not know with certainty until the vehicle swung around the back side of the Moon. Twenty-three minutes later, Houston could once again communicate with it.

"We don't know if all is going well with Apollo 11," Walter Cronkite intoned on CBS's live coverage, "because it is behind the Moon and out of contact with Earth for the first time. Eight minutes ago they fired their large service propulsion system engine to go into orbit around the Moon. We'll know about that in the next fifteen minutes or so. That's when they come around the Moon and again acquire contact with the Earth and they can report. We hope that they are successfully in orbit around the Moon and that the rest of the historic mission can go as well as the first three days."

Inside Mission Control, a few isolated conversations were taking place, but not very many; most people were waiting in silence for the "acquisition of signal," or AOS. On TV, Cronkite accentuated the drama, noting, "It is quiet around the world as the world waits to see if Apollo 11 is in a successful Moon orbit." The anxiety ended when Houston heard a faint, indistinct signal from the spacecraft, at exactly the moment it was expected.

Neil immediately provided Houston with a status report on the burn. Running through a long string of numbers on burn time and residuals,* when Houston then asked to "send the whole thing again, please," Neil exclaimed, "It was like—like perfect!"

Twenty minutes before Apollo came coasting around the Moon and back into contact with Houston, the astronauts thrilled at achieving precisely their intended orbit:

03:03:58:10	Armstrong:	*That was a beautiful burn.*
03:03:58:12	Collins:	*Goddamn, I guess!*†
03:03:58:37	Armstrong:	*All right, let's—Okay, now we've got some things to do . . .*
03:03:58:48	Aldrin:	*Okay, let's do them.*
03:03:59:08	Collins	*Well, I don't know if we're sixty miles or not, but at least we haven't hit that mother.*
03:03:59:11	Aldrin:	*Look at that! Look at that! 169.6* [nautical miles] *by 60.9* [nautical miles].
03:03:59:15	Collins:	*Beautiful, beautiful, beautiful, beautiful!*

* Performing a timed burn with Apollo's SPS rocket engine could produce a substantial acceleration (or deceleration) if the engine shutoff was not precise. It was critically important to know what these differences in speed, or "residuals," amounted to, so as to correct them. Corrections were made, not by firing the main engine again—that could add to the problem—but by briefly firing (the astronauts called it "tickling") the spacecraft's smaller maneuvering thrusters.

† This transcript excerpt as well as the two that follow in the text were from the onboard recorder and were not transmitted to Earth. The astronauts were careful about uttering profanity when they knew that what they were saying was being transmitted home.

		You want to write that down or something? Write it down just for the hell of it: 170 by 60. Like gangbusters.
03:03:59:28	Aldrin:	We only missed by a couple tenths of a mile.
03:03:59:36	Collins:	*Hello, Moon!*

All across the Moon's rocky back side, the part never visible from Earth and densely pockmarked by 4.6 billion years of meteoroid bombardment, Aldrin and Collins had excitedly pointed out one spectacular feature after another, while Armstrong was more restrained in expressing what was his own genuine enthusiasm:

03:04:05:32	Aldrin:	*Oh, golly, let me have that camera back. There's a huge, magnificent crater over here. I wish we had the other lens on but, God, that's a big beauty. You want to look at that guy, Neil?*
03:04:05:43	Armstrong:	*Yes, I see him. . . . You want to get the other lens on?*
03:04:06:07	Collins:	*Don't you want to get the Earth coming up? It's going to be nine minutes.*
03:04:06:11	Aldrin:	*Yes. Let's take some pictures here first.*
03:04:06:15	Collins:	*Well, don't miss that first one. . . .*
03:04:06:27	Armstrong:	*You're going to have plenty of passes.*
03:04:06:30	Aldrin:	*Yes, right.*
03:04:06:33	Collins:	*Plenty of Earthrises, I guess.*

03:04:06:37	Armstrong:	*Yes, we are. Boy, look at that . . . crater. You can probably see him right there. . . . What a spectacular view!*
03:04:08:48	Collins:	*Fantastic. Look back there behind us. Sure looks like a gigantic crater. Look at the mountains going around it. My gosh, they're monsters!*
03:04:09:58	Armstrong:	*See that real big—*
03:04:10:01	Collins:	*Yes, there's a moose down here you just wouldn't believe. There's the biggest one yet. God, it's huge! It's enormous! It's so big I can't even get it in the window. You want to look at that?! That's the biggest one that you ever seen in your life, Neil? God, look at this central mountain peak! Isn't that a huge one?*
03:04:11:01	Aldrin:	*Yes, there's a big mother over here, too.*
03:04:11:07	Collins:	*Come on now, Buzz, don't refer to them as big mothers. Give them some scientific name. . . . Golly damn, a geologist up here would just go crazy.*

In lunar orbit, the crew tried to settle an informal controversy that had arisen from the two previous circumlunar flights. To the Apollo 8 crew, the surface of the Moon appeared to be gray, whereas it looked mostly brown to Apollo 10. As soon as they had a chance, Neil, Mike, and Buzz looked to settle the issue. "Plaster of Paris gray to me," Collins remarked even before they got into orbit. "Well, I have to vote with the 10 crew," Aldrin said shortly after LOI. "Looks tan to me," Armstrong offered. "But when I first saw it, at the other Sun angle, it really looked gray," Buzz continued, and his mates agreed, though they expatiated about the Moon's color throughout several orbits. Ultimately, the controversy was settled in no one's favor. Lighting conditions made all the difference. The color of the Moon shifted almost hourly from charcoal, near dawn or dusk, to a rosy tan at midday.

Armstrong first had a chance to survey his approach to the landing site at 11:55 A.M. Houston time. "Apollo 11 is getting its first view of the landing approach," he reported. "This time we are going over the Taruntius crater, and the pictures and maps brought back by Apollo 8 and 10 have given us a very good preview of what to look at here. It looks very much like the pictures, but like the difference between watching a real football game and watching it on TV. There's no substitute for actually being here." Houston responded: "We concur, and we surely wish we could see it firsthand."

Apollo 11's first television transmission from lunar orbit started at 3:56 P.M. EDT. As it was a Saturday afternoon in July, many Americans tuned to the broadcast after watching the baseball *Game of the Week* on NBC, the Baltimore Orioles against the Boston Red Sox. Given that an orbit circularization burn was scheduled for 5:44 EDT that afternoon, the astronauts were in no mood for a television performance; in fact, if they'd had their druthers, they would not have had a TV show then at all.

The broadcast lasted for thirty-five minutes. Focusing the camera first out of a side window and then out of the hatch window as the spacecraft passed from west to east nearly one hundred miles above the lunar surface, the astronauts took the worldwide TV audience on a guided tour of the Moon's visible side. They talked their way along the path that Neil and Buzz would be taking in the LM in less than twenty-four hours. Neil indicated the "PDI point," where powered descent would be initiated, then Collins and Aldrin spontaneously took turns noting every significant landmark that would be guiding *Eagle* down to its touchdown: the twin peaks of Mount Marilyn, named by Jim Lovell during Apollo 8 after his wife; the large Maskelyne Crater; the small hills dubbed Boothill and Duke Island that would be passed over just twenty seconds into descent; the rilles labeled Sidewinder and Diamondback because they twisted like desert snakes; the Gashes; the Last Ridge; and finally, the landing site on the Sea of Tranquility, which was then barely into the darkness.

It was the first time that the astronauts themselves caught a glimpse of the landing site, as on the previous orbit the spot had lain hidden, beyond the "terminator" line where the astronauts would pass from light into darkness. This time around, the spot was just barely visible, brightened by Earthshine.

Everyone at home and in the spacecraft strained with Neil to take a close look. Collins, for one, did not especially like what he saw, though he kept it to himself: "It is just past dawn in the Sea of Tranquility and the Sun's rays are intersecting the landing site at a very shallow angle. Under these conditions the craters on the surface cast long, jagged shadows, and to me the entire region looks distinctly forbidding. I don't see anyplace smooth enough to park a baby buggy, much less a Lunar Module."

Crossing the terminator, the crew trained its TV camera back through the window for a last look at the landing site before sign-off. "And as the Moon sinks slowly in the west," Collins remarked, "Apollo 11 bids good day to you."

An hour and thirteen minutes later, Apollo 11 fired the SPS engine for the second time that afternoon. Even more than with the first burn, precise timing was critical. "If we over-burned for as little as two seconds," Aldrin explained, "we'd be on an impact course with the other side of the Moon." Concentration was intense as the astronauts, in coordination with Mission Control, made a systematic series of star checks, inertial platform alignments, and navigational calculations with the onboard computer. Collins used a stopwatch to make sure it lasted seventeen seconds, no more and no less. The burn came off perfectly. Apollo 11's orbit dropped and stabilized from an orbit of 168.8 miles by 61.3 nautical miles to one 66.1 by 54.4 miles, close to a perfect ellipse. It was a high degree of precision that excited even the commander:

03:08:13:47	Armstrong:	*66.1 by 54.4—now you can't beat that.*
03:08:13:52	Collins:	*That's right downtown.*
03:08:14:00	Aldrin:	*We're more elliptic now, huh?*
03:08:14:05	Collins:	*That's about as close as you're going to get.*

With Apollo 11 now snug in its orbit, it was time to prepare the LM for its designated job. Powering it up, completing a long list of communications checks, and presetting a number of switches was scheduled to take Neil and Buzz a period of three hours, but it took them thirty minutes less thanks to Aldrin's preparatory work in the module the previous day. By 8:30 P.M. Houston time, *Eagle* was ready. So were the two astronauts, who headed back into *Columbia* for their fourth night's sleep inside the spacecraft, the first in orbit about the Moon. The commander and his lunar module pilot carefully organized all the equipment and clothing they would need in the morning. Then they covered the windows to keep out not only direct light from the Sun but also the Moonshine—far brighter than we see on Earth—and began to settle into their sleeping positions. Knowing of Neil's preference that he sleep before the landing attempt, Buzz eased, as Neil did, into one of the floating hammocks. Dousing the cabin lights, Collins put the punctuation on the day: "Well, I thought today went pretty well. If tomorrow and the next day are like today, we'll be safe."

At three minutes after midnight, the on-duty PAO at Mission Control reported to the press, "The Apollo 11 crew is currently in their rest period." Aldrin recalled sleeping fitfully; Neil remembered sleeping soundly, but not for very long. Houston's wakeup call came at 6:00 A.M. By mid-morning Aldrin and Armstrong would need to be inside the LM ready to separate *Eagle* from *Columbia* for its trip down to the Moon.

CHAPTER 23

The Landing

The critical turning point of the Apollo 11 mission—certainly in Neil Armstrong's life story—was piloting the LM down to the landing.

The day of the first Moon landing was a Sunday for Americans, Europeans, Africans, and some Asians. Of all the lunar landings, only Apollo 11 landed on the Christian Sabbath. Getting up at 5:30 A.M., even earlier than her son did that morning, Viola Armstrong put on a bathrobe and went outside to water her flowers before the reporters could get to her. Then she dressed for a 7:30 church service. She wanted to be home in plenty of time to follow on television her son's separation from the command module.

All around the world, the devout prayed for Apollo 11. The worship service at the Nixon White House was dedicated to the mission, and astronaut Frank Borman read once more from the Book of Genesis. A few days previous to the Apollo 11 launch, H. R. "Bob" Haldeman, the White House chief of staff, had arranged for William Safire, a senior speechwriter for Nixon, to draft statements in case something major went wrong during the mission. One statement covered the hypothetical scenario of the astronauts managing to land on the Moon but then not being able to get off it. Safire recommended that, prior to issuing the statement, the president "should telephone each of the widows-to-be."

At 10:05 A.M. EDT, following a five-minute progress report on the Apollo 11 mission from Walter Cronkite, CBS aired a broadcast featuring a discussion of the religious meaning of Apollo 11. The voice of CBS correspondent Charles Kuralt followed at eleven, beginning the live coverage of what the network called *Man on the Moon: The Epic Journey of Apollo 11.* Providing a voice-over to dramatic pictures that had been taken by the previous Apollo flights, Kuralt also took Genesis as Apollo's spiritual theme yet elucidated its cosmic meaning with insights from modern science. After Kuralt's taped segment, Cronkite—equipped with ten notebooks full of facts about Apollo 11 and the space program—referred to the upcoming Moon landing as "a giant step," not realizing, of course, that, in less than twelve hours, similar words uttered by the First Man would be heard and forever remembered by a worldwide multitude.

Armstrong and Aldrin had been together inside *Eagle* for less than thirty minutes by the time CBS began its comprehensive coverage at 11:00 A.M. Nervous anticipation of what was to come had made it difficult for the astronauts to perform even mundane tasks that morning. Buzz remembered: "The activity of three men in space must, of necessity, be a cooperative venture. By now we had worked out various routines for living together, but in our excitement this particular morning the system came unglued." The rhythm that had been devised for eating, with one man pulling a food packet out, another snipping the packet, and the third liquefying his food with the water gun, got a little out of whack. Attaching new fecal-containment garments, urine catheters, and collection bags prior to suiting up proved especially unpleasant. Nerves frayed as the three men took turns dressing inside the CSM's navigation bay, a space large enough for only one man to change clothes, requiring another man to stand by to help with buttons and zippers. Collins also had to suit up in case something went wrong with the undocking.

Suiting up for spaceflight was always done meticulously, but never more so than on the morning of the landing. Neil and Buzz would need to be in their pressure suits for over thirty hours. The first garment they had carefully squeezed into was liquid-cooled underwear. Resembling long johns, the mesh garments held hundreds of small transparent plastic tubes. On the Moon, cooling water pumped from the astronauts' backpacks would circulate through the tubes, but until the mission progressed to that point the tight undergarments only added to the discomfort of being outfitted for space. Aldrin, in only his underwear, was first into the LM because he wanted to make some initial checks. A half-hour later a fully suited Armstrong crawled into the module. With Neil inside, Buzz returned to the CSM navigation bay to finish suiting up, then immediately reentered the LM. He and Neil sealed their side of the hatch, and Mike did his side.

Inside *Eagle,* Neil and Buzz powered up several more systems preliminary to deploying the LM's spiderlike landing gear. Successful extension of the gear came just before noon EDT. Because a number of communication and equipment checks still had to be made, it took another hour and forty-six minutes before the LM was ready to be detached by a firing of *Columbia*'s engine. Collins and Aldrin did most of the talking over the radio.

"How's the czar over there?" Mike asked from the command module. "He's so quiet."

"Just hanging on—and punching," came Neil's answer, referring to inputs he was making on the LM's primary computer. "You cats take it easy on the lunar surface," Collins told them shortly before he threw the switch to release them. "If I hear you huffing and puffing, I'm going to start bitching at you."

04:04:10:44	Collins:	*We got just about a minute to go. You guys all set?*
04:04:10:48	Armstrong:	*Yes, I think we're about ready. . . . We're all set when you are, Mike.*
04:04:11:51	Collins:	*Fifteen seconds. . . . Okay, there you go. Beautiful!*
04:04:12:10	Aldrin:	*Looks like a good SEP.*
04:04:12:10	Collins:	*Looks good to me.*

His nose pressed against a window, Collins watched them drift away and waited for word from Neil about the efficacy of the two spacecrafts' relative motion. It was a good idea not to drift too far apart until Mike gave the LM a very close visual going-over; it was critical to make sure that all four legs of the landing gear were down and in place. To help Mike with the inspection, Neil performed a little pirouette, turning the vehicle around a full rotation. A few months before launch, Mike had made a special trip to the Grumman factory on Long Island just to see what the LM looked like with its gear properly deployed. In particular, it was important for Collins to take a close look at the six-foot-long touchdown sensor prongs that extended from the LM's left, right, and rear footpads. He also needed to confirm that the footpad on the front gear, the only one without a sensor, was in the correct position. That leg held the ladder down which the astronauts would descend to the lunar surface. Originally that leg, too, had a landing sensor, but it was removed after Armstrong and Aldrin indicated they might trip over it.

04:04:12:59	Armstrong:	*Okay. I've killed my rate, Mike, so you drift out to the distance you like and then stop your rate. . . . Starting my yaw. . . . There's sure a better visual in the simulator.*
04:04:13:38	Collins:	*Okay, I picked up a little roll; I'm going to get rid of it.*
04:04:14:22	Armstrong:	*Okay with you if I start my pitch, or do you think you're not far enough away yet, Mike?*
04:04:14:31	Collins:	*I'd prefer you stand by just a couple of seconds, Neil.*
04:04:14:34	Armstrong:	*Okay, I'll wait for when you're ready, when you think you've got your rates killed perfectly.*
04:04:14:39	Collins:	*Okay. I'm still holding.*
04:04:15:26	Collins:	*Okay, looks pretty good to me now.*
04:04:15:30	Armstrong:	*Okay.*
04:04:16:34	Collins:	*Just like in the simulator, you're drifting off to one side and down below a little bit.*
04:04:16:39	Armstrong:	*Yes.*

04:04:17:06 Collins: *The gear are looking good; I've seen three of them.*

04:04:17:11 Armstrong: *The MESA is not down, right?*

04:04:17:14 Collins: *Say again.*

04:04:17:15 Armstrong: *The MESA's still up?* [The Modular Equipment Storage Assembly, or MESA, folded up against the side of the LM near the front (sensorless) landing leg. The MESA contained a television camera, boxes in which collected rocks would be put, and various tools. Once on the lunar surface, Neil would pull a D-ring that released the MESA and swung it down into an accessible position. Here Armstrong was expressing his concern that the MESA might have come down during the separation maneuver.]

04:04:17:19 Collins: *Yes.*

04:04:17:20 Armstrong: *Good.*

04:04:17:49 Collins: *Now you're looking good.*

04:04:17:59 Armstrong: *Roger.* Eagle*'s undocked. The* Eagle *has wings.*

It was a bird unlike any that had ever flown, and the teasing Collins could not stop himself from poking fun at the LM's appearance: "I think you have a fine-looking flying machine there, *Eagle,* despite the fact that you are upside down."

"*Someone's* upside down," Neil joked back.

Inside the LM, now flying less than sixty-three nautical miles above the Moon, Neil and Buzz stood upright. Without seats, the usable volume of the cabin was greater. Foot restraints with Velcro strips were anchored into the deck of the cabin and a spring-loaded cable and pulley arrangement was clipped to the astronauts' belts. If they needed to brace themselves further, Neil and Buzz could grab handholds and armrests. The astronauts standing erect meant that the triangle-shaped windows could be smaller while giving the astronauts an excellent vantage point from which to peer at the landing zone.

Before *Eagle* could begin its landing approach, Neil and Buzz needed to lower its orbit to the vicinity of fifty thousand feet. Flying feet-first and face-down relative to the lunar surface, they accomplished this by igniting the LM descent engine, its first firing of the mission. This Descent Orbit Insertion burn occurred fifty-six minutes after separation from *Columbia,* at 3:08 EDT. DOI took place while both spacecraft were on the back side of the Moon and out of contact with Earth. Lasting 28.5 seconds, the burn dropped *Eagle* into a coasting path that took it down on the front side of the Moon where the landing would be made. As the descent proceeded, Neil and Buzz checked their range rate so they could return to *Columbia* via the LM's abort guidance system, should the craft's primary navigation system fail or something else major go wrong. *Eagle* was now considerably lower than *Columbia* and thus orbiting at a faster speed. This put the LM out in front of *Columbia*'s orbit by about one minute. As *Columbia* was in a higher orbit and at an angle that brought it in direct line with the Earth, *Columbia*'s carrier signal arrived first in Houston, about three minutes earlier than *Eagle*'s. In both cases, less than a minute after acquisition of signal came the return of voice contact.

04:06:15:02	CapCom:	Columbia, *Houston.* [The CapCom was Charlie Duke.] *We're standing by, over.* [Long pause.] Columbia, *Houston. Over.*
04:06:15:41	Collins:	*Houston,* Columbia. *Reading you loud and clear. How me?*
04:06:15:43	CapCom:	*Rog. Five-by, Mike.* [In communications shorthand, "five-by-five" meant "loud and clear." *How did it* [the DOI burn] *go? Over.*
04:06:15:49	Collins:	*Listen, babe. Everything is going just swimmingly. Beautiful.*
04:06:15:52	CapCom:	*Great. We're standing by for* Eagle.
04:06:15:57	Collins:	*Okay. He's coming along.*

A minute and a half later, Aldrin reported that DOI had come off extremely well, and that it had put *Eagle* into almost the exact, predetermined perilune from which it was to start its final, powered descent. If all went well, in less than thirty minutes, the lunar module would touch down.

Prior to initiating final descent, it was important for Armstrong and Aldrin to check out their onboard guidance and navigation systems. The LM possessed two unique and independent systems. The first was the Primary Navigation, Guidance, and Control System (PNGS), or "pings" for short. This small digital computer, located in the panel in front of and between the astronauts, processed data from a built-in inertial platform—one that was held in a constant position by the action of gyroscopes that sensed movement and kept the platform from tipping in any direction. Finely tuned to the position of distant stars, PNGS flashed yellow-green numbers on a digital display that indicated the LM's position.

The second system was the Abort Guidance System (AGS). Rather than basing its navigation on an inertial platform, the spacecraft itself served as AGS's measuring table, with body-mounted accelerometers providing the flight data. Both PNGS and AGS integrated accelerations that estimated the spacecraft's velocities, with PNGS generally producing considerably more accurate data. Ideally, the mathematics inherent to the two systems—both involving the measurement of angles changing over time—produced the same answers as to where the spacecraft was and where it was heading, but inevitably errors crept into the measurements. If tiny errors were allowed to compound, gross errors in computing the LM's course and location could result.

After DOI and prior to initiating powered descent (PDI), Neil and Buzz ran a number of cross-checks between the two systems. Close agreement was essential to prevent PNGS from initiating an undesired path. The primary cause of error in PNGS was platform drift, a constant concern for any inertial system. Drift needed to be corrected by realignment of the platform via computer-aided celestial navigation, followed by mechanical reorientation by the motors and gears connected to the gyroscope.

During their outbound flight, Apollo 11 performed a number of platform alignments, but these took time and required the spacecraft to be kept relatively still. In the half-orbit prior to DOI when they were busy doing other things, Neil and Buzz made a gross check on the accuracy of their previous alignment. "The way we did that," Armstrong explained, "was by telling the spacecraft to get in an attitude so our sextant was looking directly at the Sun. If our crosshairs were in the center of the Sun, we knew the platform had not drifted. If the crosshairs were an eighth or a quarter out of the Sun's center, we knew the alignment was still okay." Neil performed the Sun check shortly before PDI. Though it had been a few hours since the last alignment, he found that the platform was still aligned satisfactorily, with only a fraction of a degree of drift. "I figured for the next thirty to forty-five minutes, the time it would take us to land, it was probably okay."

Platform drift was not the only worry regarding LM navigation. Both PNGS and AGS had to be "on" during the descent, and this could prove to be a problem if the astronauts did not keep everything about the operation of the two systems straight. While only PNGS could help the astronauts make a successful descent to the lunar surface, AGS had to be waiting and ready to navigate an emergency return to the command module; in the final seconds before touchdown, AGS also could take over from PNGS if it failed. According to Armstrong, "We couldn't land on AGS unless we got right down close to the surface, because you couldn't navigate the trajectory with it." Yet both systems needed to be energized and running, because the crew might have to switch instantaneously from PNGS to AGS. Neil explained, "Both were operating as independent systems, and only one was selected to actually be controlling the spacecraft. We also had information coming out of both that was important to compare."

Another major concern involved the fuel supply.

Recognizing exactly when they were to reach the point where powered descent should begin was extremely important: if Neil and Buzz started down from too high out, *Eagle* would run out of fuel before it was in a position to make a safe landing. The altitude limits, Neil would say, "must have been in the range of plus or minus four thousand feet."

Calculating that altitude was almost as much art as science. A standard altimeter could not tell the astronauts when they reached their perilune because an altimeter determined altitude based on changes in atmospheric pressure, and the Moon has no atmosphere. The LM did have a radar altimeter, but from the cockpit perspective that instrument pointed down and forward. Early in the descent when the LM's vertical axis was nearly horizontal—meaning that the pilots were facing downward—the radar altimeter pointed up and away from the lunar surface and could not provide any landing data. Guesstimating their PDI point from the height of the lunar mountains protruding below them was impossible, because, while Neil and Buzz could roughly figure out the heights of the mountains at the edges of the lunar sphere, they could not judge them in its middle.

The technique devised by Armstrong and MSC engineer Floyd Bennett to determine the PDI point was a relatively simple one. It involved a direct, naked-eye check of the lunar surface combined with what Armstrong called some "barnyard math." "We used the equation $v = r\Omega$," Neil explains, "where r was the altitude that you wanted to know, Ω was the LM's angular rate, and v was the LM's velocity. We knew very well what our velocity was based on radar tracking from Earth and from our own navigation system, so to figure out altitude all we needed was our angular rate. Going into the mission, we knew we could determine that by watching a point on the ground." Early in the descent phase, the LM flew with its windows facing down (the LM's back end forward), which made it easy for Armstrong to spot major landmarks on their way down. The facedown attitude was also useful for solving the altitude equation. On the double-paned window on Neil's side of the LM there was a vertical line with horizontal marks on it. As the LM flew facedown, Neil used a stopwatch to time the number of seconds it took to move from mark A to mark B on the window line. By that he calculated the spacecraft's angular rate. He also had a chart that he used to compare tracking rates with expected values at various positions along the orbit. Differences between his visual observations and the expected values allowed him to estimate both the altitude of the LM's perilune and the time at which they would reach it.

04:06:26:29	Armstrong:	*Our radar checks indicate 50,000-foot perilune. Our visual checks are steadying out at about 53,000.*

A minute and a half later, Houston told *Eagle,* "You're Go for powered descent." Before Mission Control gave them the green light, it made sure that all the pressures, temperatures, and valves checked out. Mission rules allowed the descent to continue even if some minor instrumentation was not functioning with 100 percent efficiency, but generally the flight directors required everything to be working very well before PDI commenced. What ground control could not know very precisely, however, was the LM's altitude. "The barnyard math was something I came up with myself and did on my own. I'm not sure if any of the other astronauts even used it."

Collins relayed the "Go for PDI" to his mates because *Eagle* was still incommunicado. Then even after swinging back around to the front side of the Moon, communications unexpectedly remained broken. "We had a small dish antenna mounted on top of the LM, which was a good antenna," Armstrong noted. "It was steerable, but it had to be pointed very close to Earth to get much signal. We also had an omnidirectional antenna. The omni was just a blade antenna like a person has on his automobile; it was not very accurate or very powerful. It was important to get the dish antenna pointed directly at Earth, but it was not particularly easy to do when you were lying horizontal. If you were off in yaw angle just a little bit, you could easily lose the signal."

It took almost five minutes for Neil and Buzz to make final preparations for powered descent. "We had to get the computer into the right program," Neil explained, "and make sure that all the switches and circuit breakers and everything else was ready to make the systems work and make the engines run for the powered descent." Buzz's focus was exclusively on readouts from the navigational computers, while Neil's was on assuring that everything from engine performance to attitude control was working the way it was supposed to. As *Eagle* started down, the duo activated the sixteen-millimeter film camera that was mounted above the right-hand window next to Buzz; pointing forward and downward, the camera was to film every foot of the historic descent.

Back on planet Earth, tension began to crescendo as network television coverage prepared for PDI and counting down the minutes to touchdown. On air with CBS, Cronkite said to Wally Schirra, "One minute to ignition, and thirteen minutes to landing. I don't know whether we could take the tension if they decided to go around again." In Wapakoneta Viola Armstrong watched Cronkite while clutching a sofa pillow.

PDI came at 4:05 P.M. EDT. Strapped into restraining belts and cables that worked like shock absorbers, neither Neil nor Buzz felt the motion, so they looked quickly at their computer to make sure the engine was in fact firing. For the first twenty-six seconds—or "zoom time"—the two astronauts kept the engine firing at merely 10 percent of maximum thrust. The gentle power gave the guidance computer the leeway it needed to sense when the lunar module was in the proper geometric position to go full-throttle. "Basically, you liked to be at a pretty high thrust for good fuel efficiency," Armstrong noted. "But if something was wrong and you were at too high of a throttle for too long, you couldn't sync to your target. So there was a strategy for the throttle profile—for throttling the engine to start at point A and end up at point B with a relative maximum efficiency."

As power from the engine grew, the motion became noticeable to the astronauts; even though the LM was falling at a rate of thirty feet per second, it seemed no more dramatic than a trip down several floors in a quiet hotel elevator. As they fell, Armstrong watched his instruments for proper readings while Aldrin made sure that the numbers from PNGS and AGS were correlating with preestablished figures written onto a stack of note cards that Buzz had placed between Neil and himself.

By his own admission, Buzz chattered the entire way down, "like a magpie," as he continually read out numbers from the computers, whereas very little was heard from Neil in the minutes leading up to landing.

If Neil had had his way, nothing from either one of them would have been heard by the *outside* world. Late in training Armstrong had asked about the possibility of keeping all the talking in the LM during the last minutes of descent off the radio, so as to minimize distractions. Mission Control quickly rejected the notion because it wanted to hear what was being said; the flight directors wanted their teams at the consoles to be fully informed. The idea was that one of the many experts on the ground might be able to help out the crew even in the very last seconds if a problem popped up. "Whenever I wanted to talk to the outside world," stated Neil, "normally I used the push-to-talk mode," meaning he squeezed a switch. "We had a voice-activated [VOX] position as well, and I think Buzz used that during the descent."

In the first minutes after PDI, while flying with the engine forward and windows down, Neil tracked his surface landmarks in order to confirm *Eagle*'s pathway and its timing down along it. Three minutes into descent, he noticed they were passing over the crater known as Maskaleyne W. a few seconds early:

04:06:36:03	Armstrong:	*Okay, we went by the three-minute point early. A little off.*
04:06:36:11	Aldrin:	*Rate of descent looks good. Attitude—right about on.*
04:06:36:16	Armstrong:	*Our position-checks downrange show us to be a little long.*

Neither Neil nor Buzz could be sure why they were over the crater a little early. They guessed that PDI must have started a little bit late. "Our downrange position appeared to be good at the minus-three and minus-one minute points prior to ignition," Neil reported during Apollo 11's postflight debriefing. On a chart placed in front of them, he had premarked where PDI was supposed to start but, when PDI actually began, things were too hectic for him to pay careful attention to precisely where it had happened. "I did not accurately catch the ignition point because I was watching the engine performance. But it appeared to be reasonable, certainly in the right ballpark. Our cross-range position was difficult to tell accurately because of the skewed yaw attitude that we were obliged to maintain for communications. However, the downrange position-marks on my window after ignition indicated that we were long." From one mark to another represented two or three seconds farther downrange, with every second corresponding to roughly a mile of distance. "The fact that throttle-down essentially came on time, rather than being delayed, indicated that the computer was a little confused as to what our downrange position was. Had the computer known where it was, it would have throttled down later to kill a little velocity. Landmark visibility was very good. We had no difficulty determining our position throughout all the facedown phase of powered descent."*

The reason for the slight delay in starting PDI was not analyzed by NASA until after the mission: what it involved were very small perturbations in the motion of

* Armstrong always discussed the landmark tracking as if he and Aldrin were doing it together, but that was not the case. "I appreciate the 'we,'" Aldrin commented, "but Neil did the tracking, because I wasn't looking out the window. I could have cared less about the landmarks. If it wasn't in the computer displays, I didn't see it." Aldrin quoted in "The First Lunar Landing," *Apollo Lunar Surface Journal,* ed. Eric M. Jones, p. 13.

the lunar module—in engineering terms, small delta-v inputs—that had occurred back at the instant the LM and CSM had separated. Very likely, residual pressure in the tunnel between the two modules had given *Eagle* a little extra "kick," a force resulting, some eighty minutes (and over one orbit) later, in a velocity-induced positional error that put *Eagle* a sizable distance away from where it was supposed to be. Incomplete venting of the tunnel was not considered a serious matter before Apollo 11, but it was afterward. In all subsequent Apollo missions, Mission Control made sure to double-check the status of the tunnel pressure before approving the LM's undocking.

Armstrong had no time to worry that his descent path was taking him a little long, topographically speaking. "It wasn't a sure thing that we were going to be long because we didn't know how accurately the markings on the window would turn out to be. Anyway, it wasn't a big deal as to exactly where we were going to set down. There wasn't going to be any welcoming committee there anyway."

His first indication that *Eagle* might be overflying its landing spot came just as he began to turn the LM into a face-up, feet-forward position. The reason for moving into this unusual position (via a yawing maneuver that took a little more time than expected) was to get the LM's radar antenna pointing down at the Moon. "We needed to get landing radar into the equation pretty soon because Earth didn't know how close we were and we didn't want to get too close to the lunar surface before we got that radar. If we found there was a big difference between where we were and where we were supposed to be, we might have to make some rather wild maneuvers to try and get us back on a proper trajectory, and we wanted to avoid that. So it was a matter of rolling over so that our landing radar was getting contact. It was a Doppler radar that gave three components of velocity and altitude, a pretty unique device." As it turned out, it was good that the radar was working because it showed an altitude of 33,500 feet, some 2,900 feet lower than what PNGS was indicating because PNGS was programmed into the mean surface height, not the actual height above the surface at any one place. Completing the roll, the crew saw in front of them their home planet in all the rare beauty and security it represented. "We got the Earth right out our front window," Buzz said to Neil, looking up from the computer. "Sure enough" was Neil's response.

With a reliable radar reading coming in, Neil prepared for the onboard computer to pitch the LM over so it would be almost upright. As that happened, he would get a great view of the landmarks down below, leading like roadside signs down what the astronauts had been calling "U.S. Highway 1," the pathway to the landing site on the Sea of Tranquility.

It was at that instant—at 04:06:38:22 elapsed time—that a yellow caution light came on and the first of what turned out to be several computer program alarms sounded inside the LM. With only the slightest touch of urgency in his voice, Neil squeezed his comm switch and told Houston: "Program alarm." Three seconds later he added, "It's a 1202." "Give us a reading on the 1202 program alarm," Neil quickly added, not knowing which of the dozens of alarms 1202 represented.

It took Mission Control only fifteen seconds to respond: "We got you . . . we're Go on that alarm." The problem with the computer was not a critical one. *Eagle*'s descent could continue.

"We had gone that far and we wanted to land," Neil asserted. "We didn't want to practice aborts. We were focusing our attention on doing what was required in order to complete the landing."

The 1202 alarm was caused by an overload in the onboard computer incited by the inflow of landing radar data. Fortunately, twenty-six-year-old Steve Bales—the lead specialist in LM navigation and computer software on Flight Director Gene Kranz's White Team—quickly determined that the landing would not be jeopardized by the overflow, because the computer had been programmed to ignore landing radar data whenever there were more important computations to make.

Two more times in the next four minutes, the 1202 flashed on. *Eagle* was only 3,000 feet above the lunar surface. Seven seconds after the third 1202 alarm, the situation grew more intense when a new alarm came on—a 1201.

04:06:42:15 Aldrin: *Program alarm—1201.*

04:06:42:22 Armstrong: *1201!* [Pause] *Okay, 2,000 at 50.* [This meant that the LM was now 2,000 feet above the lunar surface and dropping at a rate of 50 feet per second, which was significantly slower than previously in the descent.]

It took Mission Control only an instant to realize that the 1201 alarm was also not a dangerous problem.

04:06:42:25 CapCom: *Roger, 1201 alarm. We're Go. Same type. We're Go.*

The massive international television audience tuned in to the coverage of Apollo 11 had no idea what the alarms meant. On CBS, Cronkite told his viewers after hearing the crew's reference to the alarm, "These are space communications, simply for readout purposes." Schirra said nothing to correct him. One can imagine how sensational the live coverage would have been if on-air commentators had had an inkling of the alarm call's significance.

For Armstrong, the alarms were mainly a distraction that only endangered the landing slightly by prompting him to turn his eyes away from his landmarks. "We were getting good velocities and good altitudes; the principal source of my confidence at that point was the navigation was working fine. There were no anomalies other than the fact that the computer was saying, 'Hey, I've got a problem.' Everything else was working right and seemed to be calculating fine.

"My inclination was just to keep going ahead as long as everything looked like it was fine. There had never been an abort from this situation, and aborting at this point at rather low altitude would not have been a low-risk maneuver. I didn't want to do that unless I was absolutely out of all other alternatives—and I wasn't out of alternatives at this point. So going ahead looked like the very best thing to me. But I was listening to the ground because I had great respect for the information and help it could provide. When you get that close, why go put yourself intentionally in what was expected to be a dangerous situation—an abort—just because you had a warning light saying you might have a problem."

Armstrong gave no thought at the time as to how worried Aldrin might have been about the alarms. "I don't know if he had the same confidence level that I did that we should keep going."

Neil would have been less distracted by the computer alarms if he had known more about a simulation that had been conducted at Mission Control just a few days before the launch. The mastermind behind the "sim" was Richard Koos, the so-called SimSup, or simulation supervisor, at the Manned Spacecraft Center. A thin guy who wore wire-rimmed glasses, Dick Koos had been with the Army Missile Command at Fort Bliss, Texas, before joining the Space Task Group in 1959. Previously an expert in computer guidance for guided missiles, for Projects Mercury and Gemini Koos became one of Houston's foremost authorities in the computer simulation of spaceflight missions. For Apollo, it was his job to cook up the most intense training sessions imaginable and put every aspect of the vital relationship between a crew in flight and the ground team at Mission Control through trials by fire.

Late in the afternoon of July 5, eleven days prior to launch, Koos had told his technicians to load "Case No. 26" into the simulators. The exercise was not done to educate the astronauts, because the crew sitting inside the LM simulator that afternoon was Dave Scott and Jim Irwin, the backup crew for Apollo 12. The purpose of the simulation was to throw Flight Director Gene Kranz's White Team a wicked curveball. The White Team was the unit that was to be at the consoles inside Mission Control during the landing of Apollo 11, and Koos knew that the only way to train its members for their high-pressure duties was to put them through the wringer. A sly grin on his face, SimSup told his team, "Okay, everyone on your toes. We have never run this case, so it is going to take a helluva lot of precise timing on our part. This one must go by the numbers, so stand by for my call-outs. If we screw it up, I hope you got a bunch of change 'cause we'll end up buying the beer!"

Three minutes into the landing sequence, the devilish SimSup played his wild card: "Okay, gang, let's sock it to them and see what they know about computer program alarms."

The first alarm put to Kranz's team was code 1201, one of the very ones Apollo 11 would ultimately face. Steve Bales, the LM computer system expert, had no idea what it was. Hurriedly paging through a quarter-inch-thick handbook containing a glossary of LM software, Bales read out "1201—Executive Overflow—no vacant areas." What this meant, Bales knew, was that the onboard computer was overloaded with data, but the ramifications of the overload were unknown.

Gene Kranz vividly recalled the thought process that led Mission Control to abort the simulated landing: "Bales had no mission rules on program alarms. Everything still seemed to be working; the alarm did not make sense. As he watched, another series of alarms were displayed. Punching up his backroom loop, Bales called Jack Garman, his software expert. 'Jack, what the hell is going on with those program alarms? Do you see anything wrong?' Steve was counting the seconds, waiting for Garman's response, happy that the crew had not called for an answer. Garman's response did not help. 'It's a BAILOUT alarm. The computer is busier than hell for some reason, it has run out of time to get all the work done.' Bales did not need to consult the rules; he had written every computer rule. But there were no rules on computer program alarms. Where in the hell had the alarm come from? Bales felt naked, vulnerable, rapidly moving into uncharted territory. The computer on the LM was designed to operate within certain well-defined limits—it could do only so much, and bad things could happen if it were pushed to do things it didn't have the time or capacity to do.

"Staring at the displays and plot boards, Steve desperately sought a way out of the dilemma. The computer was telling him something was not getting done, and he wondered what in the hell it was. After another burst of alarms, Steve called, 'Jack, I'm getting behind the power curve, whatever is happening ain't any good. I can't find a damn thing wrong, but the computer keeps going through software restarts and sending alarms. I think it's time to abort!'"

Seconds later, Kranz called the abort. Charlie Duke, who was serving as the CapCom for the simulation just as he would be serving as the CapCom for the actual landing, told astronauts Scott and Irwin inside the LM to carry out the abort, which they accomplished successfully.

The exercise over, SimSup strongly expressed his unhappiness with the outcome in the debrief: "THIS WAS NOT AN ABORT. YOU SHOULD HAVE CONTINUED THE LANDING. The 1201 computer alarm said the computer was operating to an internal priority scheme. If the guidance was working, the control jets firing, and the crew displays updating, then all the mission-critical tasks were getting done." Turning to Bales, Koos told him in a more fatherly tone, "Steve, I was listening to you talk to your back room and I thought you had it nailed. I thought you were going to keep going, but then for some reason you went off on a tangent and decided to abort. You sure shocked the hell out of me!" Then, addressing Kranz, Koos made his last stinging point. "You violated the most fundamental rule of Mission Control. You must have two cues before aborting. You called for an abort with only one!"

Immediately after the debrief Bales pulled his team together in order to figure out where they had gone wrong. Later that evening, he telephoned Kranz at home: "Koos was right, Gene, and I'm damn glad he gave us the run."

The next day, July 6, Koos put them through four additional hours of training exclusively on program alarms. At the end of a thorough analysis of computer performance and response times during a host of different alarm conditions, an enterprise that took until July 11, Bales added a new rule to what was already a long list of reasons to abort the lunar landing. The rule, numbered "5–90, Item 11," read: "Powered descent will be terminated for the following primary guidance system program alarms—105, 214, 402 (continuing), 430, 607, 1103, 1107, 1204, 1206, 1302, 1501, and 1502."

Program alarms 1201 and 1202 did not make Bales's list. In the unlikely event that one or the other popped up during the main event, the lesson from SimSup would not be forgotten.

When Armstrong and Aldrin reported the first program alarm at 4:10 P.M. EST, Bales and his team of LM computer experts were busy in a back room of the control center studying the data just coming in from the landing radar. It took a few seconds before Jack Garman brought the alarm call to Bales's attention. "Stand by, Flight," Bales told Kranz over the flight director's communication loop. Charlie Duke quickly echoed that the alarm was a 1202. Then musing aloud, Duke said almost incredulously, "It's the same one we had in training." Instantaneously, the coincidence dawned on Kranz: "These were the same exact alarms that brought us to the wrong conclusion, an abort command, in the final training run when SimSup won the last round. This time we won't be stampeded."

Mission Control knew that each alarm had to be accounted for, because if an alarm stayed on, the onboard computer could grind to a halt, possibly forcing an abort. But in and of themselves, without additional trouble, neither alarm 1202 nor the later-occurring 1201 required an abort. "We're Go on that alarm," Bales told Kranz as quickly but clearly as he could from the back room after the alarm came up the first time. "He's taking in the radar data." When 1202 came up again, Bales responded even more quickly. "We are Go. Tell him we will monitor his altitude data. I think that is why he is getting the alarm." When the new 1201 alarm popped up, it brought the same speedy response from Bales: "Go . . . same type. . . . We're Go."

Despite Mission Control's decisiveness in keeping the landing going, it would have been helpful to Armstrong and Aldrin if they had experienced a program alarm simulation as part of their training. "We did have some computer alarms in the simulations we were put through, but not these particular ones," Armstrong noted. "I can't tell you how many alarms there were, but there were quite a number—maybe a hundred. I didn't have all those program alarms committed to memory, and I'm glad I didn't." Knowledge of so many alarms would have just cluttered his brain with a type of information he did not absolutely need to know—as long as the guys at Mission Control knew what to do if any of the myriad program alarms sounded.

Still, one would imagine that someone would have thought to brief the Apollo 11 crew about any important results from simulations that occurred after they left Houston for the Cape, or at least have mentioned them informally to both Neil and Buzz. But the astronauts' recollection is that this never happened.

"Neil, in the days before launch, did anyone, perhaps Charlie Duke, tell you about a simulation back in Houston involving LM computer overloads that might happen during the last minutes of the descent?"

"I had heard or remembered somewhere that such failures had been put into the simulator."

"But had you been told that Mission Control in this case had unnecessarily aborted the simulated landing and then figured out afterward that an abort was not commanded if such-and-such a computer program alarm went off but there was no other problem? You don't recall hearing about that?"

"I don't."

"Would that have made a difference in how you reacted to the alarms when they actually occurred in Apollo 11?"

"Well, it would have been helpful to have known that."

Aldrin categorically does not remember hearing anything at all about the last-minute simulation: "I didn't know anything about it until I heard about it a year or two after the flight. That was the first I knew that anyone had experienced that in any training." On the other hand, Buzz felt that Neil must have heard something about it before the launch: "I believe that people briefed Neil on that, so Neil knew that there was something like that that could come up."

"So, Buzz, when the program alarms hit the two of you during the landing, Neil had some recognition that this was a possibility and that it had been worked on in a simulation, but you didn't know anything about it?"

"That's right. I didn't know anything about it. And that was not a good situation. I should have known where the flag was on that. I should have known a few other things. But that's the communication reticence that existed with Neil, and I didn't know how to change that."

"But wouldn't it have been good if you had both known the results of that simulation, so you could both react the most reasonably if such a program alarm actually occurred?"

"I agree, but I didn't find out about it until a year or so after. Then, it was too late to make an issue of it, because it would have brought up things in somebody's methods that were not enhancing, and I sure didn't want to do that."

As it was, the principal effect of the alarms on Armstrong was that he gave more of his time and attention to them than he would have liked: "I had the obligation to make sure that I understood what was happening and that we weren't overlooking something that was important; so in that sense, yes, sure, it was a distraction, and it did take some time. The alarms as they came prevented me from concentrating on focusing on the landmarks. Had I been able to spend more time looking out the window and identifying landmarks, I might have had a better position on just precisely where our landing location was." But never during the alarms did Neil think he might have to abort, because he knew instinctively that such alarms did not command an abort if everything was in order. "In my mind, the operative indicator was how the airplane was flying and the information that was on the panel. If everything was going well, going how you expected it to go, I wasn't going to be intimidated by one computer yellow light."

As Neil shifted his focus to the lunar surface they were fast approaching, he didn't see craters or patterns of craters that he recognized, but, under the circumstances, it wasn't a big concern. For hours on end during training, Neil had studied different maps of the Moon, pored over dozens of Lunar Orbiter pictures of the surface, and scrutinized a score of high-resolution photographs taken by Apollo 10 marking the way, landmark by landmark, down to the Sea of Tranquility. "The landmarks that I was looking at out there were not ones that I had studied or remembered well enough to know just where we were, but I was pragmatic about it. I didn't find it surprising or worrisome that we ended up some other place. Anyway, it would have been surprising on the first try for a lunar landing if we had ended up anywhere very close to where we wanted to be. I didn't count on that at all. From an objective point of view, I didn't particularly care where we landed as long as it was a decent area that wasn't dangerous. It didn't make a lot of difference where it was. I thought we might just have to find somebody's backyard to land in."

Because his attention had been directed toward clearing the program alarms, not until the LM got below two thousand feet was Armstrong actually able to look outside without interruption at the landing area. What he saw as they dropped the next fifteen hundred feet was not good:

04:06:43:08 Armstrong: *Pretty rocky area.*

The onboard computer was taking them right toward the near slope of a crater the size of a football field. Later designated West Crater, it was surrounded by a large boulder field. Some of the rocks in it were the size of Volkswagens.

"Initially, I felt that it might be a good landing area if we could stop just short of that crater, because it would have more scientific value to be close to a large crater. The slope on the side of the big crater was substantial, however, and I didn't think we should be trying to land on a steep slope.

"Then I thought that I could probably avoid the big rocks in the boulder field but, never having landed this craft before, I didn't know how well I'd be able to maneuver in and between them to a particular landing point. Trying to get into a pretty tight spot probably wouldn't be fun. Also the area was coming up quickly, and it soon became obvious that I could not stop short enough to find a safe landing spot; it was not the place where I wanted to be landing. Better to have a larger, more open area without the imminent dangers on all sides."

04:06:43:10	Aldrin:	*Six hundred feet, down at nineteen. Five hundred forty feet, down at thirty. Down at fifteen.*
04:06:43:15	Armstrong:	*I'm going to . . .*

Approaching five hundred feet, Armstrong took over manually. The first thing he did was tip the vehicle over to approximately zero pitch, thereby slowing the descent. By pitching nearly upright, he also maintained his forward speed—some fifty to sixty feet per second—so that, like a helicopter pilot, he could fly beyond the crater.

Now that Armstrong was headed beyond the crater, he needed to pick a good spot to land, a potentially difficult enterprise given the very peculiar lighting conditions affecting the Moon's surface, which there had been no way on Earth to simulate. "It was a great concern," Neil recalled, "that as we got close to the Moon, the reflected light off the surface would be so strong, no matter what angle we came in on, that a lot of our vision would be wiped out, seriously affecting our depth perception."

Fortunately, NASA's mission planners had given plenty of forethought to the photometrics involved. They had concluded that, for optimum depth perception, *Eagle* needed to land at a time of "day" and at an angle that produced the longest possible shadows.

Where there were no shadows, the Moon looked flat, but where shadows were long, the Moon looked fully three-dimensional. An astronaut could then perceive depth on the lunar surface very well: he could detect differences in elevation; he could easily identify the accented shapes and forms of peaks, valleys, craters, ridges, and rims. The ideal condition occurred for the trajectory of the LM when the Sun was 12.5 degrees above the horizon. That was the time when Armstrong and Aldrin would have adequate light over the area and still strong depth-of-field definition.

With Armstrong able to see quite well out into the area beyond the crater, bringing the LM down became a matter of Neil's piloting abilities, pure and simple. It was here that Neil's time in the LLTV really paid off, for he needed to bring the *Eagle* to a touchdown point not simply by hovering and dropping vertically but by sweeping down for another 1,500 feet at a relatively fast speed. "In the Lunar Lander Training Vehicle I had done some of that sort of maneuvering. It was a matter of using those types of techniques and traversing over the ground. If I had had a little more experience in the machine, I might have been a little more aggressive with how fast I tried to get over the crater, but it didn't seem prudent to be making any very large moves in terms of attitude. I just didn't have enough flying experience in the machine in those conditions to know how well it was going to react and how comfortable I would be with it. Fortunately the LM flew better than I expected. So I certainly could have gotten away with being a little more aggressive to moving more smartly over and away from the bad area into the better area, which might've saved us a little fuel."

Normally in flying, "landing long" was not a bad idea, especially when the landing was to take place on a runway where the condition ahead was known for a substantial distance. But when the landing was to occur on the rocky, pitted surface of the Moon, landing long brought in more unknowns than landing "short" in an area where the pilot had already seen the hazards involved. "If you don't like what you see," Aldrin explained, "there are four classes of alternatives: left, right, down or short, or go over. Overwhelmingly, the less traumatic one is to go over, even though there may be some question, 'Well, if I go over, then I don't know where it is. Whereas if I land short, then I know where it is. I'm not on it, I'm in front of it.' As I try to reconstruct it, going right is a hairy thing; going left is a hairy thing; and coming down and stopping short . . . it's just a bad deal." Armstrong concurred: "You might get down there and find, 'Jesus, I've got a terrible situation.'" "So the natural thing to do," Aldrin continued, "is to fly over." "Extend it," added Neil. "We had to pick a spot, and we didn't know how much visibility we would lose as we got closer down to the surface. We wanted to pick a spot that was pretty good while we still had about a hundred and fifty feet of altitude."

04:06:43:46 Aldrin: *Three hundred feet* [altitude], *down three and a half* [feet per second], *forty-seven* [feet per second] *forward. Slow it up. One and a half down. Ease her down.*

04:06:43:57 Armstrong: *Okay, how's the fuel?*

04:06:44:00 Aldrin: *Take it down.*

04:06:44:02 Armstrong: *Okay. Here's a . . . Looks like a good area here.*

04:06:44:04 Aldrin: *I got the shadow out there.*

Seeing the LM's shadow was helpful because it was an added visual cue of how high they were. Buzz estimated that he first saw the shadow at around 260 feet: "I would have thought that, at two hundred sixty feet, the shadow would have been way the hell out there, quite long, but it wasn't. I could tell that we had our gear down and that we had an ascent and a descent stage. Had I looked sooner, I'm sure I could have identified something as a shadow at four hundred feet, maybe higher. Anyway, at the lower altitude, it was a cue that was useful but, of course, you had to have it out your window," which Neil did not. During the final stages of the approach, Armstrong was flying with the LM rotated to the left. As a result, the spacecraft structure over the hatch was blocking his view of the LM's shadow.

Dropping between 200 feet and 160 feet, Armstrong found where he wanted to land, on a smooth spot just beyond another, smaller crater that lay past West Crater:

04:06:44:18	Aldrin:	*Eleven forward. Coming down nicely. Two hundred feet, four and a half down.*
04:06:44:23	Armstrong:	*Gonna be right over that crater.*
04:06:44:25	Aldrin:	*Five and a half down.*
04:06:44:27	Armstrong:	*I got a good spot.*
04:06:44:31	Aldrin:	*One hundred and sixty feet, six and a half down. Five and a half down, nine forward. You're looking good.*

Below him Neil saw a layer of curiously moving lunar dust being kicked up by the LM's descent engine; in fact, the LM's shadow that Buzz was seeing was being cast upon this dust layer rather than on the Moon's surface itself. According to Neil: "We started losing visibility when we got a little below a hundred feet. We started picking up dust—and not just normal dust clouds like we would experience here on Earth. Dust from the lunar surface formed a blanket that moved out and away from the lunar module in all directions. This sheet of moving dust obscured the surface almost completely, though some of the biggest boulders stuck up through it. This very fast, almost horizontally moving sheet of dust did not billow up at all; it just moved out and away in a straight radial sheet.

"As we got lower, the visibility continued to decrease," Neil related. "I don't think the visual altitude determination was severely hurt by the blowing dust, but the thing that was confusing to me was that it was hard to judge our lateral and downrange velocities. Some of the larger rocks were sticking up and out of the moving dust, and you had to look through the dust layer to pick up the stationary rocks and then base your translational velocity decisions on that. I found that to be quite difficult. I spent more time trying to arrest translational velocity than I thought would be necessary.

"Then, after finding the area to land, it was all about lowering the LM down relatively slowly and keeping from inducing any substantial forward or sideward motions. Once I got below fifty feet, even though we were running out of fuel, I thought we'd be all right. I felt the lander could stand the impact because of the collapsible foam inside of the landing legs. I didn't *want* to drop from that height, but once I got below fifty feet I felt pretty confident we would be all right."

From Houston's point of view, the situation was, in fact, critical—the drama at the control consoles over the fuel supply palpable and gripping.

Back at a height of 270 feet, just prior to Buzz's seeing the LM's shadow, Armstrong had asked, "How's the fuel?" When the LM was down to 160 feet, Bob Carlton, Kranz's control systems engineer on the White Team, reported over the flight director loop that the LM's fuel supply had reached "Low Level." This meant that the propellant in the tanks of the LM had fallen below the point where it could be measured, like a gas gauge in an automobile showing empty but the car still running. Kranz asserted later, "I never dreamed we would still be flying this close to empty."

At just under 100 feet, Aldrin had called "Quantity light," indicating that only 5 percent of the original fuel load remained. At Mission Control, this event started a ninety-four-second countdown to a "bingo" fuel call. When "bingo" was called at the end of those ninety-four seconds, Armstrong at his rate of descent would have only twenty seconds to land. If Neil felt he could not land in that amount of time, he would have to abort immediately—something by the time he had gotten to 100 feet he had no thought of actually doing.

At 75 feet, Bob Carlton reported to Kranz that only sixty seconds remained before bingo. Charlie Duke repeated Carlton's call so Neil and Buzz could hear it. As Kranz remembered, "There was no response from the crew. They were too busy. I got the feeling they were going for broke. I had this feeling ever since they took over manual control: 'They are the right ones for the job.' I crossed myself and said, 'Please, God.' "

According to Armstrong, "If we were still at a hundred feet or more, then we would certainly have to abort. But if we were down lower than that, then the safest thing for us to do was to continue. We were very aware of the fuel situation. We heard Charlie make the bingo call and we had the quantity light go on in the cockpit, but we were past both of those. I knew we were pretty low by this time. But below one hundred feet was not a time you would want to abort."

At 04:06:45:07 mission elapsed time, Aldrin read out, "Sixty feet, down two and half. Two forward, two forward. That's good." Armstrong wanted to be moving forward when he landed so he could be sure of not backing into a hole that he had not noticed. "All the way through the final approach, I liked to have a little forward motion because once you were going straight down you couldn't see what was immediately below you. You wanted to get pretty close to the ground so that you knew it was a pretty good area. Then you could stop the forward motion and let it settle."

"Stand by for thirty seconds," came Carlton's next call. In Mission Control, the silence became deafening. Everyone at the consoles and in the observation rooms swallowed hard as they strained to hear what would come next, *Eagle*'s landing or Carlton's next fuel call.

At the controls of the LM, Neil was not terribly worried about his fuel. "Typically in the LLTV it wasn't unusual to land with fifteen seconds left of fuel—we did it all the time. It looked to me like everything was manageable. It would have been nice if I'd had another minute of fuel to fiddle around a little bit longer. I knew we were getting short; I knew we had to get it on the ground, and I knew we had to get it below fifty feet. But I wasn't panic-stricken about the fuel."

04:06:45:26	Aldrin:	*Twenty feet, down a half. Drifting forward just a little bit. Good. Okay. Contact light.*

The contact light went on the instant that one or more of the LM's sensory probes hanging down from three of the four footpads touched the lunar surface.

So focused was Neil on what he needed to be doing to touch down safely that he neither heard Aldrin expressly call "Contact light" nor did he see the blue contact light flash on. His plan had been to shut the descent engine down as soon as the contact light came on, but he did not manage to do it. "I heard Buzz say something about contact. But when he did, we were still over this moving sheet of sand, and I wasn't completely confident at that point that we had really touched. The indicator light might have been an anomaly or something, so I wanted to feel my way down a little closer. We might have actually touched down before I shut the engine down—it was very close anyway. The only danger with that was that if we had gotten the engine bell too close to the surface when it was running, it was possible we could have damaged the engine. We wouldn't have gotten an explosion, so it wasn't something I was concerned about. But looking back on it, I guess there was a possibility that something bad could have happened. If we had landed right on top of a rock with the engine bell still sticking out, that would not have been good."

04:06:45:41 Armstrong: *Shutdown.*
04:06:45:42 Aldrin: *Okay. Engine stop.*

It was a very gentle touchdown, so soft that it was hard for the astronauts to tell when they were actually fully down. "There was no tendency toward tipping over that I could feel," Neil declared. "It just settled down like a helicopter." In actuality, it might have been helpful to land a little harder, as later Apollo crews would purposefully do. "You always try to make a soft touchdown," Neil explained, "but by landing a little harder, engaging the clasps on the landing legs and compressing more foam, the bottom of the LM would have been a little closer to the ground and we wouldn't have had to jump so far up and down the ladder. So there was probably some merit to landing a little bit hard."

04:06:45:58	Armstrong:	*Houston, Tranquility Base here. The Eagle has landed.*

Aldrin knew that Neil was going to call the landing site Tranquility Base, but Neil had not told him when he was going to say it. The same was true for Charlie Duke. Neil had told Charlie in advance of the launch about the name, but when Charlie heard it for the first time at the moment of the landing, the normally smooth-talking South Carolinian turned a little tongue-tied:

04:06:46:06	CapCom:	*Roger, Twan . . .* [correcting himself] *Tranquility. We copy you on the ground. You got a bunch of guys about to turn blue. We're breathing again. Thanks a lot.*
04:06:46:16	Aldrin:	*Thank you.*
04:06:46:18	CapCom:	*You're looking good here.*

In retrospect it seems clear that the matter of *Eagle*'s fuel supply was never quite as dire as Mission Control thought at the time—or as historians have made it out to be. Post-flight analysis indicated that Armstrong and Aldrin landed with about 770 pounds of fuel remaining. Of this total, about a hundred pounds would have been unusable. The remainder would have been enough for about fifty seconds of additional hovering flight. That was some five hundred pounds less usable fuel than what would be left for any of the five subsequent Apollo landings.

Armstrong later said, "The important thing was that we were close enough to the surface that it didn't really matter. We wouldn't have lost our attitude control if we had run out of fuel. The engine would have quit but, from the distance we were at, we would have settled to the ground safely enough."

Touchdown came at 4:17:39 P.M. EDT, Sunday, July 20, 1969 (20:17:39 Greenwich Mean Time). The instant humanity realized that a safe landing had occurred—on TV, Cronkite exclaimed, "Whew, boy! Man on the Moon!"—jubilation broke out. As Cronkite did, people everywhere felt a tremendous emotional release. They sat speechless or wildly applauded. They laughed with tears running down their cheeks. They shouted, whooped, hollered, and cheered. They shook hands and hugged one another, clinked glasses and proposed toasts. The devout offered prayers. In a few parts of the world, some remarked, "Well, the Americans finally did it." Naturally, in the United States, there was a special sense of pride and accomplishment. Even those who were unhappy with their country, as many were in the days of the Vietnam War, felt that the Moon landing was extraordinary.

Inside the LM 240,000 miles away, Armstrong and Aldrin, in the rarefied moments after touchdown, did their best to suppress whatever emotions they felt. The two astronauts gave in only long enough to shake hands and pat each other on the shoulder. It was a defining moment of both men's lives and possibly of humankind in the twentieth century, but for the first two men on the Moon there was no time for enjoying the moment.

"So far, so good" was the only reaction Neil remembered having. Turning back to his checklist, all he said to Buzz was, "Okay, let's get on with it."

CHAPTER 24

One Small Step

Viola Armstrong and her husband offered prayers of thanksgiving with their minister when the men touched down on the Moon. "If I told you that I could feel the power of millions of prayers, you might not believe me, and I could not blame you. But waves of these prayers were coming to me, and I was being gently and firmly supported by God's invisible strength." Outside of Neil's parents' home in Wapakoneta, TV reporters had interviewed Viola and Steve shortly after the landing:

VIOLA: I was afraid that the floor of the Moon was going to be so unsafe for them. I was worried that they might sink in too deep. But no, they didn't. So it was wonderful.

REPORTER: Mr. Armstrong, what were your feelings?

STEVE: I was really concerned the way I understood that Neil had guided the craft to another area. And that would signify that the original was not exactly as they had planned.

REPORTER: What about his voice? Did it sound any different? Or did it sound calm and normal to you?

VIOLA: I could tell that he was pleased and tickled and thrilled. He was much like he always has been.

STEVE: I had the same feeling, that it was the same old Neil.

In El Lago at home with her two boys, Janet Armstrong's experience of the landing was different from her in-laws'. Janet preferred not to watch the television coverage. Instead, she hovered near one of two NASA squawk boxes. She had placed one of the boxes in her living room for all of her houseguests to hear, and the other in her bedroom so she could listen privately. "Watching TV was not something that I did during the flight. Now it's true that we watched TV during the landing, for the landing, and while the men were walking on the lunar surface, because that was a good way to hear and see, because they had the cameras there. The speculation by the TV commentators—the drama of things that could happen if there was a problem along the way—I didn't need to hear all that. That just drove me nuts." Janet's house was packed full of neighbors and guests, not all of them invited. Her sisters were there, one with her husband and their children. Inside were the ubiquitous folks from *Life* magazine; outside was a swarm of reporters. Neil's brother Dean and his wife and their son were there. A local priest was there at Janet's invitation. Janet's mother had come to Houston for the launch, but afterward returned home to Southern California. People came and went throughout the day. Janet stuck a clipboard on her front door with a sign-up sheet and a ballpoint pen hanging from it. Otherwise, if people arrived, she would not have known it. "I was paying attention to the flight, and that was most important. It was not a social occasion."

As always, the other astronauts and their wives came around to lend the families of the crew their emotional support. Janet was quite savvy about the mission. In her bedroom, she kept maps of the Moon and other technical material that Neil had given her. She studied graphs indicating the stages of the powered descent, and pencil in hand she checked off landmarks as radio communications made clear that *Eagle* was passing over them. After NASA named the Apollo 11 crew she had taken some pilot training, in part so that when her family was flying in the Beech Bonanza that Neil had just bought a part-ownership in she would have a better idea of how to bring the plane down in an emergency. She also sought to better understand and communicate what her husband was doing before the press and to her boys.

"Rick was twelve, five years older than Mark. He was interested, but Mark was too young. Mark doesn't remember much about it—any of it." At the time the little boy repeated, "My daddy's going to the Moon. It will take him three days to get there. I want to go to the Moon someday with my daddy."

Janet remembered "talking to Neil just before he left for the Apollo flight, asking him to talk to the boys and explain to them what he was doing. I said to Neil, there is a possibility you might not come back. It was right in front of the boys when I said that. I said, 'I'd like *you* to tell the boys.' I don't think that went very far."

For the boys on the day of the landing, with their house so full of people, it was a big party. Because her sisters and sister-in-law were there, Janet had help. "I had people in the house. I didn't really have to worry too much about the boys. They would go swimming, and people would watch them. They had some friends over. I tried to keep life as normal as possible for them, but that day probably wasn't very normal."

During the crew's TV transmissions during the outbound journey, she would urge, "Mark, hurry up. We're going to see Daddy." When Neil's arm came into view on the screen, she quickly pointed it out: "That must be Daddy right there. There he is!" Rick was attentive, but Mark, being young, was distracted by other things. With all the preparations for houseguests, Janet had barely slept the night before the landing. No doubt the pressure of it all wore on her, and she lit cigarette after cigarette to ease the tension. The afternoon before the landing, when Mission Control seemed to be a little late in reporting the acquisition of *Columbia*'s signal, Janet banged her fist on a coffee table.

By the time PDI came, it had already become a very long day in the Armstrong home. For the terrors of the landing, Janet again needed to be alone, so she retired to the privacy of her bedroom. Bill Anders decided to join her. Bill and Janet together had given Pat White the bad news that awful night in January 1967 when her husband, Ed, died in the Apollo fire, and Bill felt he should stay with Janet right through the touchdown. Rick, a very intelligent and sensitive boy, also wanted to be with his mother. She and Rick had been following the NASA flight map step by step, now with Anders's help. Rick settled on the floor near the squawk box, and Janet hunched down on her knees next to Rick, putting her arm around her son tightly as *Eagle* dropped its final 250 feet.

Janet recalled issuing a big sigh of relief at the moment of landing. Other people came in, hugged her, kissed her, and offered her congratulations. Returning to the living room, she and her entire company enjoyed a celebratory drink. Yet the worry was far from over. "I really wasn't too concerned about the landing. I felt Neil could do that, if at all possible. But, God, you didn't know if that ascent engine was going to fire the next day. If you listened to the TV, as I did later that evening, the drama was on the landing. Well, forget the landing! Are they going to be able to get off of there?!"

In retrospect, two items may seem curious about Apollo 11's technical situation immediately following touchdown. First, no one in NASA knew exactly where *Eagle* had landed. "One would have thought that their radar would have been good enough to pinpoint us more quickly than it did," remarked Neil. When a spacecraft was in a trajectory or when it was in orbit, with all the optical and radar measurements being taken, both the ground and the crew had a pretty good idea of where the flight vehicle was, but it was a different problem when the object was sitting in one spot and all that anyone was getting was the same single measurement over and over again. "There was an uncertainty in that that was bigger than I would have guessed it would have been."

Up in *Columbia,* which was passing over Tranquility Base at a height of sixty miles, Collins peered hard through his sextant trying to spot the LM. Over his radio he had heard the whole thing and rightfully felt he shared in the achievement. "Tranquility Base, it sure sounded great from up here," Mike had radioed to his mates. "You guys did a fantastic job." "Thank you," Neil replied warmly. "Just keep that orbiting base ready for us up there." "Will do," answered Collins. With his right eye straining through his eyepiece, Mike had tracked them as long as he could during their descent until they disappeared from his view as a "minuscule dot" about 115 miles from the landing site. Now even with the ground sending up tracking numbers for him to input on his DSKY (display-keyboard) unit so that the command module's guidance computer could accurately point his sextant, it frustrated Mike that he could not see them.

04:07:07:13 Collins: [To Houston] *Do you have any idea whether they landed left or right of centerline? Just a little bit long—is that all you know?*

04:07:07:19 CapCom (Charlie Duke): *Apparently that's about all we can tell, over.*

The limited information provided by Houston was no help to Mike: "I can't see a darn thing but craters. Big craters, little craters, rounded ones, sharp ones, but no LM anywhere among them. The sextant is a powerful optical instrument, magnifying everything it sees twenty-eight times, but the price it pays for this magnification is a very narrow field of view, only 1.8 degrees wide (corresponding to 0.6 miles on the ground), so that it is almost like looking down a gun barrel. The LM might be close by, and I swing the sextant back and forth in a frantic search for it, but in the very limited time I have, it is possible to study only a square mile or so of lunar surface, and this time it is the wrong mile."

Collins never did locate *Eagle* down on the surface,

not on any of his passes, which was more of a concern to Mike than it was to anyone else. The main concern at Mission Control over the LM's exact location did not come from the geologists—they were happy enough that Apollo 11 had landed anywhere in the mare. "They just wanted us to get out there and get some stuff!" Yet the question of where exactly the LM had come down did bother Mission Control, as Neil explained: "A lot of people were interested in where we landed, particularly those people who were involved in the descent guidance trajectory controls. After all, in later flights, we were going to try to go to specific spots on the surface and we needed to get all the information we could regarding methods that might help precision. However, not knowing exactly where the LM had landed did not affect what we did very much. Nor did people on the ground think that this was a disastrous occurrence. But the fact was, they didn't know exactly where we were and they did want to know if they could."

Related to the question of where exactly they had landed was the mystery of how mascons might have affected *Eagle*'s pathway down to the surface. Though NASA had figured out how perturbations caused by mascons in the vicinity of the Moon's equator might affect a spacecraft, at the time of Apollo 11, as Armstrong noted, NASA was "trying to reduce the error from these uncertainties to the point that we could have increasing confidence about going to a particular point on the surface."

Much more pressing was the question of whether Neil and Buzz should stay on the surface for any time at all. There was always a chance that some spacecraft system was not operating properly, requiring a quick takeoff by *Eagle*'s ascent stage. "If we had problems that indicated that it was not safe to continue staying on the surface," Neil related, "we would have had to make an immediate takeoff."

Within the lifetime of the electrical power system on the LM, there were three early times that the LM could lift off and get into a satisfactory trajectory to rendezvous with the command module. The first of these times, designated T-1, came a mere two minutes after landing. T-2 followed eight minutes later, with T-3 not coming until *Columbia* completed another orbit in two hours' time. If there was an emergency that absolutely forced *Eagle* to leave at any other time than these three, it would be up to Armstrong and Aldrin in the LM and Collins in the CSM to find some way, any way, to get in a decent position for joining up.

From a quick initial look at the LM systems, everything seemed to be okay. Gene Kranz's White Team quickly assented to a "Stay/NoStay" decision, which Charlie Duke passed on to Neil and Buzz.

04:06:47:06	*CapCom:*	Eagle*, you are Stay for T-1.*
04:06:47:12	Armstrong:	*Roger. Understand. Stay for T-1.*

Five minutes later, after more checks of spacecraft systems had been made, Duke relayed to *Eagle,* "You are Stay for T-2." The astronauts were going to remain on the Moon at least until the final Stay/NoStay decision.

A major technical concern in the first minutes after landing was the possibility that too much pressure was building up in the LM's fuel lines due to the high daylight temperature on the lunar surface. "Those fuel lines were not a new subject," Armstrong remembered. In the last days before launch, hydraulics experts had discussed with the crew what could happen if the tanks became too hot and lines overpressurized. "If we closed all valves and trapped fluid in certain lines," Neil explained, "then we were sitting on a two-hundred-degree surface of sunlight with a lot of reflected heat coming up towards the bottom of the LM, and it's heating up the pipes. The fluid pressure might really build in that line, and then we'd have a problem. It was a thing we talked about before launch in terms of optimum procedures, and we knew it was something to pay attention to when we landed, but it wasn't an uncontrollable situation. We had a couple of options of how to handle the situation, and we knew the guys on the ground were going to be doing their job, so we were not too concerned about it."

Just as predicted, immediately after engine shutdown, there was a sharp rise in pressure inside the fuel lines of the LM's descent engine. "Within two minutes after landing," as Neil told it, "we vented both fuel and oxidizer tanks as planned. But the pressure still subsequently rose, probably due to evaporation of residual propellant in the tank as a consequence of the high surface temperature. Then we vented again. The ground was getting a different reading than we were, due to a different transducer location; I think theirs was in a trapped line. It was my view that the worst that could happen was a line or a tank could split open. As we would no longer be using the descent stage, it was less than a serious problem, in my opinion. I wasn't too worried about it."

Houston considered the situation dangerous, however. If fuel sprayed onto what was still a hot descent engine, a fire could result, though unlikely in a vacuum. Fortunately, the venting eased the pressure and the problem was resolved.

For Armstrong and Aldrin there certainly was no time to savor the landing. Even after getting the stays, and even before they could take their first close look at the lunar landscape, they had to go through a complete dress rehearsal for the next day's takeoff from the lunar surface. According to Neil, "The intention was to go through all the procedures for a normal takeoff and find out if they all worked okay. This required aligning the LM platform, which was a first because no one had ever done a *surface* platform alignment before. We used gravity to establish the local vertical and a star 'shot' to establish an azimuth; in that way we got the platform aligned and ready for takeoff. Even though everyone considered it a simulation, we still went through all the systems checks just the way we would have if we had been going to make a real takeoff."

As Neil looked at it, the simulation run time allowed Mission Control to make a thorough evaluation of mission progress. "Our data resources on the lunar surface were limited. If we found there was some problem, we needed to maximize the time available for the people at Mission Control to work the problem and figure out what we might do about it. So I think it was a good strategy to get that simulated takeoff out of the way, first thing."

Only after Collins passed overhead a second time and Buzz and Neil could cease their simulated countdown following the Stay for T-3 did the two LM crewmen breathe easier. During the first two hours on the Moon, while Aldrin was painstakingly communicating back to Earth a variety of measurements and alignments that he was making for navigational purposes, Armstrong took his first opportunities to describe what he saw outside:

04:07:03:55	Armstrong:	*The area out the left-hand window is a relatively level plain cratered with a fairly large number of craters of*

		the five-to-fifty-foot variety, and some ridges that are small, twenty, thirty feet high, I would guess, and literally thousands of little, one- and two-foot craters around the area. We see some angular blocks out several hundred feet in front of us that are probably two feet in size and have angular edges. There is a hill in view, just about on the ground track ahead of us. Difficult to estimate, but might be a half a mile or a mile.
04:07:04:54	CapCom:	*Roger, Tranquility. We copy, over.*
04:07:05:02	Collins:	*Sounds like it looks a lot better than it did yesterday at that very low Sun angle. It looked rough as a* [corn] *cob then.*
04:07:05:11	Armstrong:	*It really was rough, Mike. Over the targeted landing area, it was extremely rough, cratered, and large numbers of rocks that were probably—some, many—larger than five or ten feet in size.*
04:07:05:32	Collins:	*When in doubt, land long.*

Neil returned to documenting the Moon's color: "I'd say the color of the local surface is very comparable to that we observed from orbit at this Sun angle—about ten degrees Sun angle. It's pretty much without color. It's gray, and it's very white, chalky gray, as you look into the zero-phase line. And it's considerably darker gray, more like ashen gray, as you look out ninety degrees to the Sun. Some of the surface rocks in close here that have been fractured or disturbed by the rocket-engine plume are coated with this light gray on the outside, but where they've been broken, they display a very dark gray interior, and it looks like it could be country basalt." According to the flight plan, the takeoff simulation was followed by meal time and then, officially, by a four-hour rest period. Aldrin recalled, "It was called a rest period, but it was also a built-in time pad in case we had to make an extra lunar orbit before landing, or if there was any kind of difficulty which might delay the landing. Since we landed on schedule and weren't overly tired, we opted to skip the four-hour rest period. We were too excited to sleep anyway."

The idea of skipping the rest period had actually been fully discussed and strategized about prior to the launch. "From our early discussions of how we would organize our time line of activities," Neil related, "we concluded that the best thing to do, if everything was going well, was to go ahead outside as soon as we could and do the surface work before we took our sleep period. We recognized that the chances for even getting down safely—having things go well enough with all the systems to allow a landing—were problematical. If we scheduled the surface activity immediately for as soon as we could after *Columbia*'s first revolution and after the practice takeoff and so on—immediately after that—and then didn't make it on time, the public and the press would crucify us. That was just the reality of the world. So we tried to finesse things by saying that we were going to sleep and then we would do the EVA.

"But we never had any plan to do it that way. We had discussed it with Slayton and Kraft—and a few other people. My recollection was that they all thought it was a reasonable thing to do. And so everyone agreed we'd do it that way if we could. We knew it would create a change that people weren't expecting, but we thought that was the better of the two evils."

With everything in order, at 5:00 P.M. Eastern time, Armstrong radioed a recommendation that they plan to start the EVA in about three hours at 8 PM, earlier than scheduled. Aware of the prearranged deal, Charlie Duke approved the change.

They did eat a meal as scheduled, but not before Aldrin first reached into his Personal Preference Kit, or PPK, and pulled out two small packages given to him by his Presbyterian minister back in Houston. One package contained a vial of wine, the other a wafer. Pouring the wine into a small chalice that he also pulled from his kit, he prepared to take Holy Communion.

At 04:09:25:38 mission elapsed time, Buzz radioed, "Houston, this is the LM pilot speaking. I would like to request a few moments of silence. I would like to invite each person listening in, wherever or whoever he may be, to contemplate the events of the last few hours and to give thanks in his own individual way." Then, with his mike off, Buzz read to himself from a small card on which he had written John 15:5, traditionally used in the Presbyterian communion ceremony.

It had been Buzz's intention to read the passage to Earth, but Slayton had advised him not to do it and Buzz agreed. Apollo 8's Christmas Eve reading from Genesis had generated sufficient controversy to make the space agency shy away from overt religious messages. Madalyn Murray O'Hair, the celebrated American atheist, had sued the federal government over the Bible reading by Borman, Lovell, and Anders. By the time of Apollo 11, O'Hair had added a complaint that NASA was purposefully withholding "facts" about Armstrong being an atheist. Though the U.S. Supreme Court eventually rejected O'Hair's lawsuit, NASA understandably did not want to risk getting embroiled in another battle of this type. Regrettably to NASA, the word of Aldrin's religious ceremony had made its way to the press. Cronkite passed advance word to his viewers: "Buzz Aldrin did take something most unusual with him today, and it has become public—made public by the pastor at his church outside of Houston. He took part of the Communion bread loaf, so that during his evening meal tonight he will, in a sense, share communion with the people of his church, by having a bit of that bread up there on the surface of the Moon. The first Communion on the Moon."

Characteristically, Neil greeted Buzz's religious ritual with polite silence. "He had told me he planned a little celebratory communion," Neil recalled, "and he asked if I had any problems with that, and I said, 'No, go right ahead.' I had plenty of things to keep busy with. I just let him do his own thing."

After eating a meal and performing a few housekeeping chores, the astronauts turned all their attention to gearing up for the EVA. No matter how much they had practiced inside the LM mockup, doing it for real was much more difficult and time-consuming. "When you do simulations of EVA Prep," Neil explained in NASA's technical debriefing following the mission, "you have a clean cockpit and you have all the things that you're going to use there in the cockpit and nothing else. But in reality, you have a lot of checklists, data, food packages, stowage places filled with odds and ends, binoculars [actually a monocular], stopwatches, and assorted things, each of which you feel obliged to evaluate as to whether its stowage position is satisfactory for EVA and whether you might want to change anything from the preflight plans. We followed the EVA preparation checklist pretty much to the letter, just the way we had done during training exercises—that is, the hookups and where we put equipment—and the checks were done precisely as per our checklist. That was all good. It was these other little things that you didn't think about and didn't consider that took more time than we thought."

It took an hour and a half before Buzz and Neil were ready to start the EVA prep procedures and then three hours to do the preps, which were expected to take two hours. Much of the time involved getting their backpacks on, donning helmets and gloves, and getting everything configured for going outside. One of the main reasons why it took so long was because it was so cramped inside the LM. Aldrin recalled: "We felt like two fullbacks trying to change positions inside a Cub Scout pup tent. We also had to be very careful of our movements. Weight in the LM was an even more critical factor than in the *Columbia.* The LM structure was so thin, one of us could have taken a pencil and jammed it through the side of the ship."

Neil explained, "It was pretty close in there with the suits inflated. It was certainly a larger cockpit than the Gemini, so there was more room than I was used to. Nevertheless, you had to be very careful and move slowly. It was very easy to bump things. That backpack was sticking out behind you almost a foot and it had a hard surface; if you made a quick motion, you could easily bang into something." And things were banged into. For example, the outer knob of an ascent engine-arming circuit breaker broke off, which Buzz was able to depress prior to liftoff with a felt-tipped pen.

Proceeding with great care, the two men used all the estimated time for suiting up and then some. Then it also took longer than anticipated to get the cooling units in their PLSS backpacks operating and even more time than expected to depressurize the LM for egress. According to Neil, "We had to depressurize the cabin and we wanted to protect the lunar surface from Earth germs so we had filters on all the vents. We had never done the tests with the filters on, and it took a much longer time to depressurize the cabin than we had anticipated." They were ready to swing open the hatch and for Neil to step out onto the surface an hour later than estimated, though that was still five hours ahead of the original schedule.

Opening the hatch proved to be a chore. "It was an effort in patience more than anything else," Neil explained. "It was a pretty good-size hatch—five or six hundred square inches or something like that. So when we got the cabin pressure down to a very low psi, it took something like two hundred pounds of pressure to open that up. You can't put two hundred pounds of pressure into pulling on a handle very easily—not in those cumbersome suits. So we had to wait until we got to a very low-pressure difference between the inside of the door and the outside of the door before it would break free. We tried a number of times to open it up, but we didn't want to bend or break anything. Mostly it was Buzz doing the pulling because the door opened his direction; it was easier for him to pull toward himself than it was for me to push."

The hatch finally opened, Neil began backing through a fairly tiny opening. Peering down and around, Buzz helped navigate. According to Neil, "Egress required you go through the hatch backward, feetfirst. The technique was to get the door wide open and face the rear of the lunar module cabin, then kneel down and slide backwards, allowing your feet to go out through the hatch first. Then you had to get around the backpack. The backpack extended quite a long way above your back. You needed to get quite low but then you also had things on the front of you that you didn't want to damage. So it was a matter of doing that kind of awkward procedure with as much care as possible so as not to damage."

So intent was Armstrong on his egress technique that when he got out onto the small porch of the LM, he forgot to pull the lanyard just north of the ladder rigged to deploy the swing-action Modular Equipment Storage Assembly. The MESA lanyard also activated the television camera that was to transmit to Earth images of Neil's descent down the ladder and his first step onto the lunar surface. Quickly noticing the omission, Houston reminded him about it, and Neil moved back a bit to pull the deployment handle.

The television camera was black and white. "We did have a color camera in the command module," explained Neil, "but it was quite big and bulky, and for the LM we were very concerned about weight. Principally, weight and electrical power were the factors that required the much smaller black-and-white-image orthicon TV camera." Essentially, the orthicon was a pickup tube that used a low-velocity electron beam to scan a photoactive mosaic.

"When I first exited the lunar module out onto the porch and pulled the handle to release the MESA table, as I remember it, Buzz turned a circuit breaker powering the camera. I asked Houston if they were getting a picture and they said, yes, they were, but it was upside down. I was the most surprised guy probably of anybody listening to that conversation, because I did not expect them to get a picture [none had been obtained during any preflight simulation]."

Standing at the top of the ladder seemed not at all precarious. "You are so light up there and you fall so slowly that, if you have anything to hold on to anywhere, you are going to be able to control yourself. So I was not ever concerned about falling from the ladder."

In the CapCom seat, astronaut Bruce McCandless had taken over from Owen Garriott for the EVA:

04:13:22:48	McCandless:	*Okay, Neil, we can see you coming down the ladder now.*
04:13:22:59	Armstrong:	*Okay, I just checked getting back up to that first step, Buzz. The strut isn't collapsed too far, but it's adequate to get back up.*
04:13:23:10	McCandless:	*Roger. We copy.*
04:13:23:25	Armstrong:	*It takes a pretty good little jump* [to get back up to the first rung].
04:13:23:38	Armstrong:	*I'm at the foot of the ladder. The LM footpads are only depressed in the surface about one or two inches, although the surface appears to be very, very fine-grained as you get close to it. It's almost like a powder.* [The] *ground mass is very fine.*
04:13:24:13	Armstrong:	*I'm going to step off the LM now.*

Every one of the global millions who watched what next happened on their television sets will never forget the moment that Armstrong took his first step out onto the surface of the Moon. Watching the shadowy black-and-white TV pictures coming back from a quarter of a million miles away, it seemed like an eternity before Neil, his right hand on the ladder, finally stepped off onto the Moon, leading with his booted left foot.

The historic first step took place at 10:56:15 P.M. EDT, which was 02:56:15 Greenwich Mean Time. In terms of mission elapsed time, the step came, according to NASA's official press statement, at four days, thirteen hours, twenty-four minutes, and twenty seconds.

In the United States, the largest share of the television audience, including everyone at the Armstrong homes in Wapakoneta and El Lago, were watching CBS and listening to Cronkite, who for one of the very few times in his broadcasting career was virtually speechless. Having taken his eyeglasses off, and rubbing tears from his eyes, Cronkite declared, "Armstrong is on the Moon! Neil Armstrong, a thirty-eight-year-old American, standing on the surface of the Moon! On this July twentieth, nineteen hundred and sixty-nine."

What also so impressed Cronkite, as it did everybody else, was that the world was watching something that was happening so far away, at a place no human being had ever been before, via a live television feed. "Boy! Look at those pictures!" the veteran newsman exclaimed. "It's a little shadowy, but he [Neil] said he expected that in the shadow of the lunar module."

Television pictures afforded the audience the virtual sensibility of being there with Armstrong when he stepped out onto the Moon. Without them, the human experience of the First Man's first step would still have been meaningful, yet surely very different. As Neil later said, "The pictures were surreal, not because the situation was actually surreal, but just because the television technique and picture quality gave it sort of a superimposed unreal image." Even given all the ridiculous conspiracy theories over the past four decades about the Moon landing having been a faked telecast from a remote movie studio location somewhere out in the desert, Armstrong confessed, "I have to say that it almost looked contrived.

"That certainly wasn't planned. Had we had the ability to make a much clearer picture, we certainly would have opted to do so.

"From a technical standpoint, the TV was still valuable," Armstrong recalled, "to various individuals in and around NASA." But no piece of information carried a greater worth, or was more closely guarded, than what words Armstrong would say when he stepped out onto the lunar surface. No one knew, not even his crewmates. Buzz recalled: "On the way to the Moon, Mike and I had asked Neil what he was going to say when he stepped out on the Moon. He had replied that he was still thinking it over."

Armstrong always maintained he spent little time thinking about what he would say until sometime after he had successfully executed the landing.

At 04:13:24:48 mission elapsed time, which was a few seconds before 10:57 P.M. EDT, Neil spoke his eternally famous words:

> *That's one small step for man, one giant leap for mankind.*

In El Lago, Janet reportedly said as Neil was coming down the ladder, "I can't believe it's really happening," then when Neil stepped off, "That's the big step!" As he began to walk upon the Moon, she coaxed him, "Be descriptive now, Neil." In Wapakoneta, Viola, clutching the arms of her chair ever so tightly, thanked God that her son was not sinking into the lunar dust, a fear that many people harbored even after the LM had landed. Janet kept telling her company that she had absolutely no idea what her husband would say when he stepped onto the Moon. An hour earlier, she had jested, as everyone grew more impatient for Neil and Buzz to begin the EVA, "It's taking them so long because Neil's trying to decide about the first words he's going to say when he steps out on the Moon. Decisions, decisions, decisions!"

Janet's joke was not too far from the truth, as Neil would explain: "Once on the surface and realizing that the moment was at hand, fortunately I had some hours to think about it after getting there. My own view was that it was a very simplistic statement: what can you say when you step off of something? Well, something about a step. It just sort of evolved during the period that I was doing the procedures of the practice take-off and the EVA prep and all the other activities that were on our flight schedule at that time. I didn't think it was particularly important, but other people obviously did. Even so, I have never thought that I picked a particularly enlightening statement. It was a very simple statement."

Then there was the matter of the missing "a"—the fact that Neil fully intended to say, "That's one small step for *a* man," but, in the rush of the moment, forgot to say, or just did not say, the "a." Or said the "a" but said it in such a way that it was inaudible or indistinguishable as a separate word to his global audience.

In terms of memory, "I can't recapture it. For people who have listened to me for hours on the radio communication tapes, they know I left a lot of syllables out. It was not unusual for me to do that. I'm not particularly articulate. Perhaps it was a suppressed sound that didn't get picked up by the voice mike. As I have listened to it, it doesn't sound like there was time there for the word to be there. On the other hand, I think that reasonable people will realize that I didn't intentionally make an inane statement, and that certainly the 'a' was intended, because that's the only way the statement makes any sense. So I would hope that history would grant me leeway for dropping the syllable and understand that it was certainly intended, even if it wasn't said—although it actually might have been."

When asked how he prefers for historians to quote his statement, Neil answered only somewhat facetiously, "They can put it in parentheses.

"As for what I did say on the Moon, I took a small step—so that part of it came real easy. Then it wasn't much of a jump to say what you could compare that with." Some people would come to think that Neil came across the idea for his statement while reading J. R. R. Tolkien's *The Hobbit.* (In the book, the protagonist Bilbo Baggins jumps over a villain in "not a great leap for a man, but a leap in the dark.") But Neil did not read *The Hobbit* until well after Apollo 11. (Encouraged by their two sons, who were Tolkien fans, Neil and Janet would use the name "Rivendell" for the farm that became their home in Lebanon, Ohio, in 1971.) A far less chimerical theory was that a high NASA official gave him the idea. This hypothesis is based on the existence of an April 19, 1969, memorandum from Willis Shapley, an associate deputy administrator at NASA Headquarters to Dr. George Mueller, head of the Office of Manned Space Flight. Early in the memo, in talking about what sort of message the Moon landing should present to the world, Shapley wrote: "The 'forward step for all mankind' aspect of the landing should be symbolized primarily by a suitable inscription to be left on the Moon and by statements made on Earth." As the story goes, Mueller passed this memo on to Deke Slayton, who shared it with Armstrong. However, Armstrong had absolutely no recall of the memo, or ever hearing about it. It seems to be another example of a similar statement having been made independently.

"So, in your mind, Neil, there was never any particular context for your coming up with the phrase? It did not connect back to any other quotation or experience?"

"Not that I know of or can recall. But you never knew subliminally in your brain where things come from. But it certainly wasn't conscious. When an idea runs for the first time through your own mind, it comes out as an original thought."

For the first few minutes after stepping off the LM, Armstrong kept his exploring close to the ladder. He was intrigued by the peculiar properties of the lunar dust. He told Houston: "The surface is fine and powdery. I can kick it up loosely with my toe. It does adhere in fine layers, like powdered charcoal, to the sole and sides of my boots. I only go in a small fraction of an inch, maybe an eighth of an inch, but I can see the footprints of my boots and the treads in the fine, sandy particles." As was expected, motion posed no problem. "It's even perhaps easier than the simulations of one-sixth g that we performed in the various simulations on the ground. It's absolutely no trouble to walk around." Still in close proximity to the LM, Neil saw that the descent engine had not left a crater of any marked size. "It has about one foot clearance on the ground. We're essentially on a very level place here. I can see some evidence of [exhaust-induced erosion] rays emanating from the descent engine, but a very insignificant amount."

He was anxious to have the mission's photographic camera, a seventy-millimeter Hasselblad, sent down to him. To do that, Buzz, just inside the hatch, needed to hook the camera to a device known as the Lunar Equipment Conveyor, or LEC. The astronauts nicknamed it the "Brooklyn Clothesline" because it worked pretty much the same way as the line in New York apartment buildings used to hang out and dry wash. The idea for the LEC came along not so much to solve the problem of bringing the camera and other things down from the LM but for taking things back up from the lunar surface at the conclusion of the EVA. Neil explained, "We had done some practice sessions on the final segment of the lunar surface work where we brought all the rock boxes, cameras, and various equipment that needed to go inside. It was very cumbersome. We found it very difficult to manhandle all that stuff around and get it in the proper position so that the other person—the top man—could pick it up. I think it was my suggestion that we try the clothesline technique. So we did that, and it seemed to work out all right."

Unhooking the heavy camera from the LEC, Armstrong set it in the bracketing framework of the Remote Control Unit, or RCU, which was built according to his own design right into the front of his suit. As soon as he got the camera mounted, Armstrong was so intent on taking a few pictures that he neglected to scoop up the contingency sample of lunar dirt, a higher priority item that he was supposed to accomplish first in case something went wrong and he quickly needed to get back into the LM. NASA did not want to get all the way to the Moon and then not be able to bring back any lunar sample for scientific study. Houston had to remind Neil, never one to be rushed, a couple of times to get the sample. "It was going to take somewhat more effort to get . . . the equipment and the container for that sample than it was to get a few pictures. My thought was just that I was going to get a few quick pictures—a panoramic sequence of the LM's surroundings—while I was there, and then I was going to get the sample."

In the technical debrief following the mission, Neil explained his conservative reasoning for changing the order of doing these first two things. He said that at first he was standing in the shadow of the LM, where good pictures could be taken. To do the sample, he would have to stow the LEC and go ten or more feet into the non-shadowed area, and thus he changed the order. He also had to assemble a pooper-scooper-like device with a collapsible handle with a removable bag at the end. After he scooped up a small amount of soil sample, he deposited the bag inside a strap-on pocket on his left thigh. Digging into the top surface was no problem at all, as the soil was very loose. Though the sample did not require him to take anything from any real depth, he did try digging an inch or more into the surface, only to find that it quickly became very hard. He also made sure to get a couple of small rocks into his bag before closing it up. Finally, he conducted a little soil mechanics experiment by pushing the handle-end of his sampler down into the surface from four to six inches.

His sample completed, Neil took a moment to gaze out at the lunar landscape. "It has a stark beauty of its own," he reported. "It's much like the high desert of the United States. It's different, but it's very pretty out here." Then, still thinking about what he could do to experiment, he detached the ring that had been holding the collection bag on to the contingency sample and threw it sidearm to see how far it would go. "Didn't know you could throw so far," Aldrin teased, watching out his window. Chuckling, Neil answered, "You can really throw things a long way up here!"

Sixteen minutes into the EVA, it was time for Aldrin to egress, something he was itching to do.

Standing southwest of the ladder, Neil used the Hasselblad to snap a series of remarkable photographs of Buzz slowly emerging from the hatch, studiously coming down the ladder, kneeling on the porch, moving down to the last rung, jumping down to the footpad, and hopping off onto the lunar surface. These are the pictures that people would later see and forever remember in terms of the first human stepping onto the Moon: Buzz doing it rather than Neil, for whom no photographs from below could be taken because he went out first. Actually, Buzz climbed down to the last rung twice before stepping off—the first time just as rehearsal.

04:13:41:28	Aldrin:	*Okay. Now I want to go back up and partially close the hatch.* [Long pause] *Making sure not to lock it on my way out!*
04:13:41:53	Armstrong:	[Laughing] *A particularly good thought.*

Not that the two men were actually worried that they could lock themselves out, as the hatch could be opened from the outside, if necessary. Aldrin's reason for partially closing the hatch was apparently to prevent radiative cooling of the LM cabin.

As a matter of fact, though Buzz and Neil did not think of it at the time, there *was* a way that they could have locked themselves out, if the hatch's pressure valve had somehow gone awry and started repressurizing. "Did we really ever investigate that problem?" Aldrin asked. "It probably would have been a good idea to use a brick or a camera to keep it from closing. Somebody must have thought through that. We had a handle [on the outside] to unlatch it, but, considering the difficulty we had, if you had a couple of psi [in the cabin], you'd never get it open. Well, you'd get it open, but you'd never get the bent hatch closed again!"

Down on the surface, it was at this moment that Buzz referred to the Moon's unique beauty as "magnificent desolation." Leaning toward Buzz so close that their helmets almost touched, Neil clapped his gloved hand on his mate's shoulder. According to Buzz's autobiography, Neil then said to him, "Isn't it fun?" But Neil later insisted that " 'fine' is definitely what I said," in reference to the fine powder that the two astronauts were examining.

After that, they moved off their separate ways and began testing their mobility. Though their substantial number of practice hours in one-sixth g had not been spent moving very far or very fast, inside the LM they had been standing, bending, and leaning and, in Neil's words, had "a pretty good appreciation of what the one-sixth g environment felt like before we ever got out." What they were not accustomed to were major and very rapid body movements. In ground simulations and in the one-sixth g airplane, they had practiced a number of different possible lunar gaits. In one of the ground simulations, Neil remembered, "You were suspended sideways against an incline plane and walked sideways while hooked to an assembly of cables." Although a truer feeling came in the one-sixth g airplane—a converted KC-135, flying parabolas—that only gave them a few seconds each flight to polish their techniques.

During the EVA, it was Aldrin's job to test all the different lunar gaits. These included a "loping gait" (Neil's preference) in which the astronaut alternated feet, pushed off with each step, and floated forward before planting the next foot; a "skipping stride," in which he kept one foot always forward, hit with the trailing foot just a fraction of a second before the lead foot, then pushed off with each foot, launching into the next glide; as well as a "kangaroo hop," which few Apollo astronauts ever employed, except playfully, because its movements were so stilted.

With their big backpack and heavy suit on, the astronauts would have weighed 360 pounds apiece on Earth; on the Moon, in one-sixth gravity, they each weighed merely sixty pounds. Since they felt so lightweight, special care in all movements did need to be taken, primarily because of their backpacks, whose mass effects on their balance, they quickly discovered, pitched their walk slightly forward. When looking out in any direction toward the horizon, both men felt a bit disoriented. Because the Moon was so much smaller a sphere than Earth, the planetoid curved much more visibly down and away than they were accustomed to. Also, because the terrain varied a good bit relative to their ability to move over it, they had to be constantly alert. "On Earth, you only worry about one or two steps ahead," Buzz recalled. "On the Moon, you have to keep a good eye out four or five steps ahead." Mostly, the two astronauts, trained as they were to be very conservative in their EVA mobility, walked flat-footed, with one foot always firmly planted into the lunar surface.

Armstrong did try making some fairly high jumps straight up off the ground. What he found was a tendency to tip over backward upon landing. "One time I came close to falling and decided that was enough of that." After he and Buzz stretched out the TV cable so the television camera could be moved to its position some fifty feet away from the LM, Neil also tripped over the cable. "The TV cable was coiled in storage, so when we stretched it out we had a spiral on the ground that was lifting up, and with the low gravity that was accentuated a little bit. It was very easy to trip over that cable, which I did a few times." Exacerbating the problem was the fact that the astronauts really could not see their feet very well. "Because of our suits, it was hard to see anything right below you. It was hard to see your feet; they were pretty far down there." The fact that the cables got dusty almost immediately also contributed to the problem.

Sewn to each man's left gauntlet was an ordered checklist of EVA tasks. Even though Neil and Buzz, through repeated simulations, knew from memory the order of events, they still used the checklists consistently, as professional pilots did, no matter how well they knew the procedures.

The astronauts' next task (another late addition) was unveiling the commemorative plaque that was mounted on the ladder leg of the LM. "For those who haven't read the plaque," Neil said to the world at 04:13:52:40 elapsed time, "we'll read the plaque that's on the front landing gear of this LM. First, there's two hemispheres, one showing each of the two hemispheres of the Earth. Underneath it says, 'Here Men from the Planet Earth first set foot upon the Moon, July 1969 A.D. We came in peace for all mankind.' It has the crew members' signatures and the signature of the president of the United States."

Another item that was not on their checklist but

that NASA wanted accomplished fairly early during their EVA was the planting of the American flag. As discussed earlier, the decision to erect an American flag on the Moon had been controversial. Armstrong remembered: "There was substantial discussion before the flight on what the flag should be. It was questioned as to whether it should be an American flag or a United Nations flag." Once it was decided (with no input from the crew) that it should be the American flag, Neil, a former Eagle Scout, did give some thought as to *how* the flag should be displayed. "I thought the flag should just be draped down, that it should fall down the flagpole like it would here on Earth. It shouldn't be made to stand out or put into any rigid framework, which it ultimately was. I soon decided that this had gotten to be such a big issue, outside of my realm and point of view, that it didn't pay for me to even worry about it. It was going to be other people's decision, and whatever they decided was okay."

While he and Buzz had trained in minute detail to execute virtually every other assigned task during the EVA, they had done no training at all for the flag ceremony, as it, like the unveiling of the plaque, was a late addition. As it turned out, planting the flag (some thirty feet in front of the LM) took a lot more effort than anyone had imagined—so much more that the whole thing nearly turned into a public relations disaster.

First there was difficulty with the small telescoping arm that was attached as a crossbar to the top end of the flagpole; its function was to keep the flag (measuring three feet by five feet) extended and perpendicular in the still, windless lunar atmosphere. Armstrong and Aldrin were able quickly enough to lock the arm in its 90-degree position, but as hard as they tried, they could not get the telescope to extend fully. Thus, instead of the flag turning out flat and fully stretched, it had what Buzz has called "a unique permanent wave." Then, to the dismay of the two men, fully aware that the whole world was watching them through the TV camera they had just set up, they could not get the staff of the flag-pole to penetrate deeply enough into the soil to support itself in an upright position. "We had trouble getting it into the surface," recalled Neil. "It ran into the sub-surface crust." With the pole sticking barely six inches into the Moon, all the two men could think about was the dreaded possibility that the American flag might collapse into the lunar dust right in front of the global television audience.

Fortunately, the pole, with its funny curly flag, stayed standing. With his camera, Neil took the memorable picture of Aldrin saluting the flag. According to Aldrin, he and Neil were just about to change positions and transfer the camera so that Buzz could take a picture of him when Mission Control radioed that President Nixon was on the line and wanted to talk to them. This distracted them from taking the picture, Buzz related, so a photo of Neil never got taken. However, the sequence of events as evidenced in the NASA communications transcript shows that the first word of Nixon's call did not come to the astronauts until well after Neil took the picture of Aldrin saluting the flag; the picture was taken during a break in communications very shortly after 04:14:10:33 elapsed time whereas the news that Nixon wanted to talk to them came at 04:14:15:47. During most of that five-minute-and-fourteen-second interval, the two men were no longer even together. Following the flag planting, Armstrong moved back to the LM, the camera still with him. There, at the MESA, he prepared to collect his first rock samples. Aldrin moved out westward from the LM a distance of some fifty feet before rejoining Neil at the MESA. Next Houston told the men that President Nixon was calling in from the Oval Office. Nixon came on the line and congratulated them, saying that the country and the whole world was proud of them.

There is no question that President Nixon's phone call came as a surprise to Aldrin. In his autobiography, Buzz recalled: "My heart rate, which had been low throughout the entire flight, suddenly jumped. Later Neil said he had known the president might be speaking with us while we were on the Moon, but no one had told me. I hadn't even considered the possibility. The conversation was short and, for me, awkward. I felt it somehow incumbent on me to make some profound statement, for which I had made no preparation whatsoever. I took the handiest possible refuge. Neil was the commander of the flight, so I let him do the responding. I conveniently concluded that any observation I might make would look as though I was butting into the conversation, so I kept silent."

Armstrong later explained, "Deke had told me shortly before the flight that we might expect some special communication. He didn't say it would be the president necessarily, but just to expect some special communication that would come through the CapCom. It was just a heads-up, to tell me that something might come through that seems unusual, but Deke didn't tell me exactly what it was. I didn't know it was going to be the president, and I'm not sure Deke knew exactly who or what it was going to be, either."

Aldrin would in later years take issue with the fact that he hadn't been alerted that the President might call, as if Neil had been told specifically that Nixon would call. Without question, these two men who had to work so closely together to be the first to get to another world and explore it had a highly unusual relationship.

Consider the fact that, while Armstrong took dozens of wonderful photographs of Aldrin, Buzz took not a single explicit picture of Neil. The only pictures of Neil were one with a reflection of him in Aldrin's helmet visor in a picture Neil took, or a very few where Neil was standing in the dark shadow of the LM with his back to the camera or only partially shown. One image would have provided a good view of Neil at the MESA if the exposure had been more appropriate.

It is one of the minor tragedies of Apollo 11 that posterity benefits from no photos of the First Man on the Moon. Not of him saluting the American flag. Not of him climbing down the ladder. Not of him stepping on the Moon. Not of him directly anywhere. Sure, there are the grainy, shadowy, black-and-white TV pictures of Armstrong on the Moon, and they are memorable. There are also a number of frames from the 16mm movie camera. But, very regrettably, there are no high-resolution color photographic images of the First Man with the spectacular detail provided by the Hasselblad.

Why not? The answer, according to Aldrin, was that he simply did not think to take any—except at that moment when they were planting the American flag and President Nixon's call allegedly ended what would have been a Buzz-at-Neil photo shoot.

In his autobiography, Aldrin excuses what he failed to do. "As the sequence of lunar operations evolved, Neil had the camera most of the time, and *the majority of the pictures taken on the Moon that include an astronaut are of me* [author's emphasis]. It wasn't until we were back on Earth and in the Lunar Receiving Laboratory looking over the pictures that we realized there were *few pictures of Neil.* My fault perhaps, but we had never simulated this during our training."

"We didn't spend any time worrying about who

took what pictures," Armstrong graciously recalled. "It didn't occur to me that it made any difference, as long as they were good.

"I don't think Buzz had any reason to take my picture, and it never occurred to me that he should. I have always said that Buzz was the far more photogenic of the crew."

At the same time, Armstrong offered some real clarification of the situation pertaining to cameras and the photographic plan for surface activities during Apollo 11. "We always had a plan for when we were going to transfer the camera. He was going to take some pictures, and I was going to take some. And I think roughly we did it approximately like the plan called for in terms of the camera transfer. I had the camera for a large fraction of the time and I had more assigned photographic responsibilities, but Buzz did have the camera some of the time and did take pictures. It was in the flight plan."

Besides the Hasselblad that Neil mounted on his chest bracket, another Hasselblad was kept in the LM as a spare, but it was never brought out. The only other still-photo camera that was used on the surface was the Apollo Lunar Surface Close-Up Camera (ALSCC), a stereoscopic camera—often called the "Gold camera," as its proponent was Dr. Thomas Gold, a prominent Cornell University astronomer. Specifically designed for close-ups of the lunar surface, the Gold camera was solely Neil's responsibility. But Buzz definitely took a number of pictures with the EVA Hasselblad. This means Neil painstakingly took the camera off his chest bracket and handed it directly and carefully over to Aldrin. Buzz took two complete 360-degree panoramas; pictures of the distant Earth, and of the LM. He took the famous shots of his footprints in the lunar dust, but he took no purposeful shots of Neil. To be fair, all of the photos Buzz took were planned photo tasks of his. Not even Apollo 11 crewmate Mike Collins realized it until well after the mission. "We came back, the pictures got developed—they came back from the NASA photo lab. I loved them. I thought they were terrific. Never once did it occur to me, 'Which one of them is that?' It's just some guy in a pressure suit. It was not until later that people said, 'That's Buzz,' and 'That's Buzz,' and 'That's Buzz,' and the only Neil was the one where he was in Buzz's visor. But even then, I attributed it to technical stuff—you know, the timeline, who was carrying what piece of equipment, what they were supposed to be doing at a given time, experiments they were running on the surface, and so forth." Flight Director Gene Kranz only shook his head sadly trying to come up with an answer: "I don't have an explanation. In recent years I have been speaking to about 100,000 people a year. I do sixty to seventy public appearance engagements. And the only picture I can put up on the screen of Neil is his reflection in Buzz's facemask. I find that shocking. That's something to me that's unacceptable." According to Chris Kraft and others involved in Apollo 11's mission planning, "There were all kinds of scientific reasons to take pictures and all kinds of plans to take pictures of the lunar landscape, but I don't think there was ever any game plan to have them take a picture of each other like you would do at the beach. I don't recall that ever being discussed." Gene Cernan sees it similarly. "Certainly Neil realized the significance of the moment, but he was not going to be so arrogant as to say, 'Here, Buzz, take a picture of me.' What I can imagine Neil thinking was, 'Oh well, we don't have time to take a picture of me, so I'll take a few pictures of Buzz to show everyone we were here.'

"Myself, if I had been in Neil's place, I would have said, 'Buzz, take a picture of me—quick.'"

At the conclusion of the telephone conversation with President Nixon, Armstrong immediately returned to the MESA to gear up for his primary geological work. Up to this point the only lunar material that had been collected was the contingency sample. Now he needed to get to work on the bulk sample—enough for scientists around the world to share in—and also a variety of rock forms. He needed to collect enough samples for scientists around the world.

Over a period of about fourteen minutes, Armstrong made some twenty-three scoops. This took longer than anticipated because the vacuum-packed containers were difficult to seal. In addition, the area in which Neil was working was in deep shadow, making it hard to see. More significantly, in one-sixth gravity he couldn't apply as much force as he had in training on Earth.

In all, Apollo 11 brought back nearly 48 pounds (21.7 kilograms) of rock and soil samples, the great majority scooped up by Armstrong. Overall in the Apollo program, 841.6 pounds (381.69 kilograms) of Moon rock were returned. Understandably, given the unknowns of the first landing mission, the load brought back by Apollo 11 was the lightest of all the landing missions.

Mostly the rocks Armstrong collected were basalt: a dense, dark-gray, fine-grained igneous rock composed chiefly of calcium-rich plagioclase feldspar and pyroxene; on Earth, basalt is the commonest type of solidified lava. The oldest basalts brought back by Apollo 11 had been formed some 3.7 billion years ago. Later flights brought back a greater variety of specimens, including lighter-colored igneous rocks that were even older, called gabbros and anorthosite.

Some critics in the years following Apollo 11 were disappointed that the Moon rocks did not unlock secrets of the universe, but not Armstrong. "I am persuaded that they produced an extraordinary proof of the constituency of the regolith, the layer of loose rock atop the lunar mantle. They also demonstrated the different kinds of rock types, while confirming their plutonic character, their deep igneous or magmatic origin. Many of the rock types also revealed evidence of valuable metallic ores." By 1975, the 2,200 distinct samples collected by the six Apollo Moon landings had been further split into some 35,600 specimens. As of 2015, only 17 percent of the Apollo lunar material had been provided to researchers around the world for study. Of the remaining 83 percent, the majority remains in storage at NASA's Johnson Space Center in Houston and Brooks Air Force Base in San Antonio, Texas, with less than 5 percent on loan to museums and educational institutions or presented to foreign countries and the U.S. states as goodwill gifts.

Besides the rock sampling, the astronauts had a number of experiments to conduct, and precious little time to conduct them, as surface activity for Apollo 11 was limited to two hours and forty minutes. There were six experiments in all, each one selected by a NASA scientific panel after rigorous peer review.

The most generic was a soil mechanics investigation with core samples (taken primarily by Aldrin) measuring soil density, grain size, strength, and compressibility as a function of depth. Near the end of the EVA, Buzz hammered a couple of core tubes into the surface, the Moon's tightly locked soil grains yielding only about six inches. The objective was not just to improve scientific knowledge, but to provide engineering data toward the design of an astronaut-carrying lunar vehicle, the later Lunar Rover that first went to the Moon with Apollo 15 in late June 1971.

The Solar Wind Composition Experiment was designed to trap evidence of the flux of electrically charged particles emitted by the Sun. With Armstrong's help, it took Aldrin five minutes to deploy the solar wind instrument (a small banner of thin aluminum foil 11.7 inches [30 centimeters] wide by 54.6 inches [140 centimeters] that unrolled downward from a reel to face the Sun) early in the EVA, right after he and Neil had unveiled the plaque on the LM ladder leg. Exposed on the lunar surface for seventy-seven minutes, the foil collector entrapped ions of helium, neon, and argon, expanding scientists' knowledge of the origin of the solar system, the history of planetary atmospheres, and solar wind dynamics.

The other five experiments came as part of EASEP, the Early Apollo Scientific Experiment Package. EASEP consisted of two units about the size of small backpacks. The PSEP, or Passive Seismometer Experiment, deployed by Aldrin, was designed to analyze lunar structure and to detect moonquakes. Attached to the back of the PSEP was a lunar dust detector experiment that monitored the effects of lunar dust on the experiments.

At the same time Aldrin was deploying the seismic experiment (from 04:15:53:00 to 04:16:09:50, a duration of roughly seventeen minutes), Armstrong assembled the LRRR, or "LR-cubed." Designed to measure precisely the distance between the Moon and Earth, the LRRR device consisted of a series of corner-cube reflectors, essentially a special mirror that reflected an incoming light beam back in the direction it came—in this case from a laser aimed at the Sea of Tranquility from inside a large telescope at the University of California's Lick Observatory, east of San Jose. Though the laser beam remained tightly focused over a very large distance, by the time it traveled the quarter of a million miles from Earth, its signal was widely dispersed, to a signal about two miles in diameter. To maximize reception, Armstrong had to align the reflector quite accurately.

Neil recalled: "We wanted to make sure that all the mirrors were pointed at Earth, and we wanted to make certain that the reflector was mounted on a fairly stable surface where it wouldn't be likely to get shifted later. We aligned it with the local vertical by means of a circular bubble—like the bubble in a level, except it was in a circle—so once you got the bubble in the middle of the circle the platform was level. Then we also had to align the whole platform by turning it until the mirrors were pointed directly at the Earth." For that he used a shadow stick—a gnomon—where the shadow made by the stick created the alignment. On Earth the concave bubble was fairly stable, but in lunar gravity it kept circling.

Mysteriously, the bubble finally did stabilize. It ended up being one of the most scientifically productive of all the Apollo experiments, deployed on Apollo 14 and 15 as well. Together, the three LRRR instruments deployed by the Apollo missions produced many important measurements. These included an improved knowledge of the Moon's orbit, of variations in the Moon's rotation, of the rate at which the Moon is receding from Earth (currently 1.5 inches or 3.8 centimeters per year), as well as of the Earth's own rotation rate and precession of its spin axis. Scientists have used data from the laser reflectors to test Einstein's theory of relativity.

Armstrong recalled the decision against utilizing the large S-band dish antenna, which was stowed in LM Quad 1 to the right of the ladder. "We didn't have to erect it as the signal for the LM antenna was strong enough to transmit the TV to Earth." From a mission efficiency viewpoint, Neil was happy that the S-band antenna, which was roughly eight feet across, did not need to be deployed. It took about twenty minutes to assemble, and he and Buzz were already running thirty minutes behind schedule. On the other hand, "It was really fun putting that thing together. I would have enjoyed doing it if I had had to, and finding out if it really worked. I'd done that quite a few times on the ground, and I was always amazed watching that thing bloom like a flower."

According to Armstrong, the overall plan for the entire EVA was well conceived. "We had a plan. We had a substantial number of events to complete that were all in a proper order. We had built that plan based on the relative importance of the different events and the convenience and practicality of doing them in a certain order. We'd gone through a lot of simulations and developed the plan over a period of time. We knew it forwards, backwards, and blindfolded. That wasn't going to be any trouble. I didn't feel any restriction against violating a plan or drifting away from a plan somehow if the situation warranted."

The most noteworthy change in the plan came late in the EVA when Armstrong decided he wanted to go over and take a look at the sizable crater about sixty-five yards east of the LM (known today as East Crater). "When I went over to look at the crater, that was something that wasn't on the plan, but I didn't know the crater was going to be there. I thought seeing and photographing it was a worthwhile addition, although I did have to give up some documented-sample time to do that. But it looked to me like that could be a piece of evidence that people would be interested in." There were guidelines but no specific mission rules as to how far away from the LM a crew member could go. If he or Buzz strayed too far from the LM, Mission Control would have definitely reined them back in. "In fact, I had some personal reservations in taking the time to go over and snap a picture of the crater. But I thought it was of sufficient interest that it was worth getting."

With EVA time running out, Neil hustled to get to the crater and back. Based on subsequent analysis of the TV footage showing him running there and back (he used a loping, foot-to-foot stride), his speed appears to have been about 2 miles (3.2 kilometers) per hour. In all, Neil's expedition took three minutes and fifteen seconds. While there, he took eight shots showing various features of East Crater, including outcroppings in its sidewalls that he felt would be of interest to geologists.

The moment Armstrong headed toward the crater, Houston informed Buzz that it was time for him to start thinking about heading back into the LM. Neil would follow Aldrin up the ladder some ten minutes later. Before either headed up the ladder, though, they needed to finish up and cap the final core samples, and Neil, with a pair of long-handed tongs, had to complete his final rock sampling. Everything had to be brought to the ladder, including the camera film magazines, the solar wind experiment, and all the rock boxes.

As Armstrong explained in a post-mission press conference, "There was just far too little time to do the variety of things that we would have liked to have done. There were rocks in a boulder field that we had photographed out of Buzz's window before going out that were three to four feet in size. Very likely they were pieces of lunar bedrock. It would have been very interesting to go over and get some samples of those. There were just too many interesting things to do.

"When you are in a new environment, everything around you is new and different and you have the tendency to look a little more carefully at 'What is this?' and 'Is this important?' or 'Let me look at it from a different angle,' which you would never do in a simulation. In a simulation, you just picked up the rock and threw it in the pot.

"So it doesn't surprise me that it took us somewhat longer to get through things. We didn't have that presidential call either—that was never in our simulations. And there were questions coming from the ground. We were responding to those, which took a little extra time. No one was asking us questions when we went through this in our practice sessions.

"It would have been nice from our point of view to have had more time to ourselves so that we could have gone out and looked around a little bit. But a lot of people had needs based on whatever discipline they belonged to, and these people had spent a lot of time getting ready to have their experiments done. I felt that we had a substantial obligation to try to honor those needs as best we could, and in a most timely fashion. I didn't mind breaking the rules if it seemed like the right thing to do.

"I do remember thinking, 'Gee, I'd like to stay out a little longer, because there are other things I would like to look at and do.' It wasn't an overpowering urge. It was just something that I felt, that I'd like to stay out longer. But I recognized that they wanted us to go back in." Back on Earth, it was approaching 1:00 A.M. EDT, and they were told to go up the ladder. Armstrong was supposed to dust off Buzz's suit before Buzz went back inside the

LM, but that was forgotten, perhaps because it seemed pointless. "The dust was so fine that you couldn't have got rid of all of it," explained Neil.

Armstrong's last tasks on the lunar surface were labor-intensive and physically demanding. To prevent contamination, the NASA contractor that built the rock boxes had cleaned their hinges rather than leaving them lubricated. To close their lids, Neil had to apply thirty-two pounds of force. After struggling to close the bulk sample box, it took "just about everything I could do, an inordinate amount of force," to close the documented sample, his second box. Low gravity made for an added difficulty: the boxes tended to skid away. In order to close the boxes, Neil placed them on the MESA table, a surface that was not very rigid. Just holding the box securely enough in place to apply the necessary force on the sealing handles caused him considerable trouble. Then he had to carry the rock boxes one by one over to the LEC, hook them to the "Brooklyn clothesline" running from the porch of the LM up to the hatch, and, with Buzz's help, hoist them up.

In Houston, a cardiac monitor showed that Neil's heart rate rose during the EVA close-out period to 160 beats per minute, the typical heart rate of an Indy car driver at the start of the Indianapolis 500. Five minutes before he was to head up the ladder, Houston made a disguised request for Neil to slow down for a moment, by asking him to report on the status of the tank pressure and oxygen in his EMU.*

---

* In recent years an American physician by the name of William J. Rowe, who has spent over two decades researching the impact of spaceflight on human physiology, particularly its vascular complications, has published a series of papers on what he calls the "Neil Armstrong Syndrome." In a nutshell, Dr. Rowe has argued, based on the medical data from Apollo 11, that Neil, during the last 20

minutes of his lunar EVA, "suffered severe dyspnea"—that was, difficult or labored breathing—with him twice verbally notifying Mission Control about it during a four-minute interval, and that Neil also experienced "severe tachycardia with a heart rate up to 160/min" [Rowe, "Neil Armstrong Syndrome," *International Journal of Cardiology*209 (2016) 221-222].What caused "Armstrong's Moon Cardiac Scare" [Rowe in *Spaceflight* 58 (Feb. 2016): 56–57]? According to the research physician, "Since catecholamine levels [i.e., chemically related neurotransmitters such as epinephrine and dopamine] in Space are twice the supine [i.e., inactive, lying on the back] levels on Earth, it should not be surprising that space flight is conducive to catecholamine cardiomyopathy, a form of acute temporary heart failure," and that Armstrong likely suffered from it late in his EVA. "In addition to high catecholamines," Rowe has underscored, "there are low magnesium ions in Space and vicious cycles between the two," which can quickly take an astronaut's heart into tachycardia and trigger an "oxidative stress" that so intensifies "endothelial dysfunction" (the endothelium being the layer of smooth, thin cells lining the heart and blood vessels) that "it may be fatal." In Rowe's view, it was likely that it was this condition that also explains astronaut James Irwin's cardiovascular complications on Apollo 15, when Irwin became seriously dehydrated when his in-suit water device did not function and he had no access to any water during his three lunar excursions. In Neil's case, "Armstrong's lunar heart rate of 160 was conducive to oxidative stress; yet, while still in microgravity, approximately 30 minutes before splashdown into the Pacific, his heart rate dropped all the way down to 61. This significant reduction can best be explained this way: during the three days back to Earth, despite the reduction in thirst in microgravity, he replenished his very depleted plasma volume, thereby reducing the gradient at the site of protrusion of the septum into the ventricle" (Rowe, "Neil Armstrong Syndrome," 221). Dr. Rowe's complete explanation of what he calls the "Neil Armstrong Syndrome" is too complex to be explained fully here; his bottom line is that "A major problem with space flight is dehydration, conducive to heart failure.

It should be added that Dr. Rowe does not believe that the Neil Armstrong Syndrome is, from a cardiovascular standpoint, peculiar to Armstrong; Rowe, now living in retirement in Virginia, believes that it is a matter of human cardiovascular physiology generally that has, and will continue, to impact the health of all astronauts in space—and even the health of future human beings on Planet Earth if global warming continues unabated.

Dr. Rowe's analysis and views on the issues of human physiology in space are controversial and would become even more

More concerned about getting every necessary object inside the LM, the astronauts almost forgot to leave a small packet of memorial items on the lunar surface. Aldrin recalls the near-oversight: "We were so busy that I was halfway up the ladder before Neil asked me if I had remembered to leave the mementos we had brought along. I had completely forgotten. What we had hoped to make into a brief ceremony, had there been time, ended almost as an afterthought. I reached into my shoulder pocket, pulled the packet out and tossed it onto the surface." The packet contained two Soviet-made medals, in honor of deceased cosmonauts Yuri Gagarin, the first human to orbit the Earth, who died in a MiG-15 accident in March 1967; and Vladimir Komarov, killed a month after Gagarin at the conclusion of his Soyuz 1 flight when his spacecraft's descent parachute failed to open. Also in the packet was an Apollo 1 patch commemorating Gus Grissom, Ed

---

controversial if they were more widely known, especially by those proposing human trips to Mars. For that long-duration spaceflight mission, Dr. Rowe states in his article "Genetic gifts and a Mars mission" [*Spaceflight* 59 (Aug. 2017): 303-304], that "even without considering unknown radiation, our best chance of surviving a 20-month roundtrip to Mars is to take advantage of genetic gifts. Recently, a Kenyan ran a marathon in 2 hours and 26 seconds. Similarly, Man's best chance of surviving a trip to Mars is by educating a group of young Bushmen, capable of running for 2 days across the Kalahari desert without water, and in their twenties, send them to Mars with return before age 30, [as] the vascular repair mechanism is incomplete over this age."

Neil Armstrong's heart issues as they related to his health and cause of death will be discussed later in this book. At this point the reader may only be reminded that Neil died in 2012 after coronary bypass surgery; also, he suffered, and recovered, from a heart attack in 1991. Astronaut James Irwin suffered three heart attacks, the first occurring two years after Apollo 15, when Irwin was 43. He suffered his second in 1986 and died following his third attack, in 1991.

White, and Roger Chaffee. Also inside was a small gold olive-branch pin, symbolic of the peaceful nature of the American Moon landing program. The token was identical to the pins that the three Apollo 11 astronauts were carrying as gifts for their wives. Aldrin's packet landed just to Neil's right. Armstrong straightened it out a little by moving it with his foot.

Immediately after doing that, at 1:09 A.M. EDT (04:15:37:32 elapsed time), Neil climbed onto the LM footpad, put his arms on the ladder arms, and, pushing with his legs and pulling with his arms, jumped all the way up to the third rung of the ladder.

"The technique I used was one in which I did a deep knee bend with both legs and got my torso down absolutely as close to the footpad as I could. I then sprang vertically up and guided myself with my hands by use of the handrails. That's how I got to the third step, which I guess was easily five or six feet above the ground."

Characteristically, the engineer was experimenting, not showboating. "It was just curiosity. You could have really jumped high if you didn't have that suit on. But the suit's weight. . . . You didn't really feel the weight of the suit because it was pressurized from the inside, so the interior pressure was holding most of the weight of the suit up. But when you jumped you had to carry that, and our lunar weight was sixty-two pounds or something like that. So if you are a sixty-two-pound man, how high can you jump? If you are unencumbered in a real stiff suit, you can probably jump pretty high. I just wanted to get an idea of how high could you go if you took a good leap up."

Armstrong's leap up the ladder probably stands as a lunar record, as subsequent Apollo astronauts were usually carrying something in their hands or arms when they ascended. If Neil had missed the step while making the jump—and the steps were slippery from lunar dust—there was only a slight chance he could have hurt himself. With his hands on the rails, he could have easily guided himself to a soft landing. In addition, if Neil had fallen, he would have had no trouble getting up, having practiced that in the water tank back at the Manned Spacecraft Center.

Aldrin's ingress a few minutes earlier had proven relatively easy, considering that the bulky PLSS meant he had to arch his back to get inside. Navigating solo, Aldrin first brought his knees inside the cockpit, then he moved from a kneeling to an upright position. Before turning around, he had to ensure adequate allowance for the switches and other equipment immediately behind him. Neil's ingress, which took one minute and twenty-six seconds from the time he climbed on the LM footpad, benefited from Aldrin's guidance:

04:15:38:08 Aldrin: *Just keep your head down. Now start arching your back. That's good. Plenty of room. Okay now, all right, arch your back a little, your head up against* [garbled]. *Roll right just a little bit. Head down. Getting in in good shape.*

The time between the hatch opening and its closing was two hours, thirty-one minutes, and forty seconds. Earth time upon closure was 1:11 A.M. EDT. Humankind's first direct sojourn onto the surface of the Moon was over in less time than it took to watch a football game.

On CBS, Eric Sevareid and Walter Cronkite summed up the momentous events. "Man has landed and man has taken his first steps. What is there to add to that?" Cronkite asked. Sevareid answered: "I don't know what one can add now. We've seen some kind of 'birth' here. . . . When they moved around, you sensed their feeling of joy up there. I never expected to see them bound, did you? Everything we've been told was that they would move with great care. Foot after foot with great deliberation. We were told they might fall. And here they were, like children playing hopscotch."

"Like colts almost," Cronkite interjected.

"But I never expected to hear that word 'pretty.' He said it was 'pretty.' What we thought was cold and desolate and forbidding—somehow they found a strange beauty there that I suppose they can never really describe to us."

Cronkite: "It may not be a beauty that one can pass on to future beholders, either. These first men on the Moon can see something that men who follow will miss."

Sevareid: "We're always going to feel, somehow, strangers to these men. They will, in effect, be a bit stranger, even to their own wives and children. Disappeared into another life that we can't follow. I wonder what their life will be like, now. The Moon has treated them well, so far. How people on Earth will treat these men, the rest of their lives, that gives me more foreboding, I think, than anything else."

One of the gaps in the record of Apollo 11 concerns the personal items and mementos Armstrong and his crewmates took with them to the Moon. All three men had a Personal Preference Kit stowed on board for them at launch. A PPK was a beta-cloth pouch about the size of a large brown lunch sack, with a pull-string opening at the top, coated with fireproof Teflon.

Exactly how many PPK pouches were taken by each

Apollo 11 astronaut to the Moon is unknown. At least one belonging to each stayed for the entire flight in the lower equipment stowage compartment of the command module; these CM PPKs could weigh no more than five pounds per astronaut. At least two other PPKs, one for Neil and one for Buzz, were stowed inside the LM. These LM PPKs—it is likely there were only two of them, one for each of them—were limited to half a pound per astronaut. Neil, Mike, and Buzz agreed to authenticate all items on board Apollo 11 as "carried to the Moon," whether they went to the lunar surface or stayed in the CSM, so as not to devalue the symbolic importance of the items carried by Collins only in lunar orbit.

None of the three astronauts has ever shared an inventory of the souvenirs that were in those six bags. (The astronauts also took another PPK for frequently used personal items, such as pens and sunglasses.) What is known about them has been based solely on what the astronauts over the years have said or written and on what they have released and identified from their private holdings for sale or display. In Armstrong's case, this has amounted to almost nothing, since Neil never spoke about what he took to the Moon—and, unlike Buzz and Mike, never put any of his items up for auction.

All attempts to discover the contents of the PPKs have failed. Even before the launch of Apollo 11, rumors circulated, but NASA refused to shed any light upon the subject. Janet Armstrong admitted that Neil had taken something to the Moon for her, but she refused to reveal what it was. NASA's policy was to keep the astronauts' personal belongings strictly confidential.

So cautious was NASA about releasing what its astronauts carried as souvenirs that it is not known with certainty even today what Apollo 11 was carrying in its OFK, or Official Flight Kit. An OFK manifest for Apollo 11 was never released publicly, and none has ever been located.

Apollo 11's OFK might not even have been an actual bag; OFK items may have been stowed in cabinets inside the command module. A NASA document from 1972 would later indicate that "the total weight of this kit shall not exceed 53.3 pounds per mission." Clearly, the contents of the OFK comprised a much larger stash of souvenir items than what went into the astronauts' PPKs. As official NASA souvenirs, OFK items were meant for distribution, either by the astronauts or by leading NASA officials, to VIPs and organizations. None of these items were transferred to the LM prior to its separation in lunar orbit. Thus, the only items taken to the surface of the Moon were whatever was in the PPKs that had been stowed in the LM—and we do not know what was in them. To wit:

- Four hundred fifty silver medallions that had been minted by the Robbins Company of Massachusetts. These had been divided equally between the three astronauts and stowed in their PPKs. How many of the medallions were taken to the surface is unknown.
- Three gold medallions, also minted by the Robbins Company, one for each astronaut. One can assume that these medals were in the LM PPKs.
- An unknown number of miniature (4 x 6–inch) flags of the United States; of the fifty U.S. states, District of Columbia, and U.S. territories; of the nations of the world; and of the United Nations. According to a NASA press release of July 3, 1969, "These flags will be carried in the lunar module and brought back to Earth. They will not be deployed on the Moon." As part of the OFK kept in-

side the command module, there were a great many additional miniature American flags. There were also two full-size (5 x 8–foot) American flags, which were to be presented to the two houses of Congress upon return to Earth. These exact flags had been flown over the U.S. Capitol prior to the Apollo 11 mission and were to be flown again there after the mission. Aldrin carried miniature U.S. flags in his PPK, some of which he later sold. It is not certain whether they went to Tranquility Base or just stayed in orbit.

- A commemorative Apollo 11 envelope issued by the U.S. Post Department. On it was a newly issued ten-cent stamp also commemorating Apollo 11. It is not known whether these items were in Neil's or in Buzz's LM PPK. While on the lunar surface, they were supposed to cancel the cover, but they forgot to do so. (That was not done until July 24 when the crew was together in the quarantine facility. Nonetheless, the cancellation read July 20.) In the command module, in either Collins's PPK or the OFK, the crew also brought along the die from which the commemorative stamp had been printed. In his CM PPK, Aldrin carried 101 philatelic covers on behalf of the Manned Spacecraft Center's Stamp Club. Another 113 envelopes, perhaps more, were also carried onboard the command module. Each member of the crew signed all of the covers carried. In later years, Aldrin and Collins initialed some of their covers, in the upper left corner, and some were put up for sale. Armstrong never initialed one.
- An unknown number of Apollo 11 "beta cloth" patches, so named by their manufacturer, Owens-Corning Fiberglass of Ashton, Rhode Island, because they were made of tightly woven and fireproof glass fiber. Each astronaut may have had a small number of beta patches in his PPK, but how many of them went to the surface, if any, is unknown.

- An unknown number of embroidered Apollo 11 patches. Most of these were likely part of the OFK, but a few might have been taken in PPKs, though few if any of them traveled to the lunar surface.
- Three gold olive-branch pins, exact replicas of the gold olive branch in the packet that Aldrin tossed down at the last minute to the lunar surface during the EVA. After the flight, the crew presented the pins to their wives as gifts. Presumably, each LM astronaut carried his own wife's pin in his respective PPK, with either Neil or Buzz carrying the pin that Collins was to give his wife, Patricia.
- A vial filled with wine and a miniature chalice, in Aldrin's LM PPK.
- Pieces of jewelry for his wife and family, in Aldrin's LM PPK.

Armstrong never released any information about the contents of his PPK. He agreed to do so for publication in this book, but reported that he was unable to find the manifest among his many papers. All he had to say about what he took with him to the Moon was, "In my PPK I had some Apollo 11 medallions, some jewelry for my wife and mother [simply the gold olive branch pin for each], and some things for other people." He was most clear about, and most proud of, the pieces of the historic Wright Flyer that he took to the Moon. Under a special arrangement with the U.S. Air Force Museum in Dayton, he took in his LM PPK a piece of wood from the Wright brothers' 1903 airplane's left propeller and a piece of muslin fabric (8 x 13 inches) from its upper left wing.

Armstrong also took along his college fraternity pin from Purdue, which he later donated for display at Phi Delta Theta's headquarters in Oxford, Ohio. Contrary to published stories, he did not take Janet's Alpha Chi Omega sorority pin.

"I didn't bring anything else for myself," Neil would declare. "At least not that I can remember." As for Janet, the only thing taken to the Moon for her was the olive branch pin. "He didn't ask me if I wanted to send anything."

Perhaps surprisingly, Armstrong took nothing else for family members—not even for his two boys, a fact that still distresses Janet. "I assumed he had taken things to give to the boys later, but I don't believe he has ever given them anything. Neil can be thoughtful, but he does not give much time to being thoughtful, or at least to expressing it."

Another loved one that Neil *apparently* did not remember by taking anything of hers to the Moon was his daughter, Karen. What could have made the first Moon landing more meaningful "for all mankind" than a father honoring the cherished memory of his beloved little girl, by taking a picture of the child, dead now over seven years (she would have been a ten-year-old), one of her toys, an article of her clothing, a lock of hair, her baby bracelet? Astronaut Gene Cernan, just before he left the lunar surface on Apollo 17, had written the initials of his nine-year-old daughter, Tracy, in the dust. Buzz Aldrin carried photos of his children to the Moon. Charlie Duke left a picture of his family on the surface.

What if Neil did something for Muffie but never told anyone about it, not even Janet, because it was of such an intensely personal nature? How much more would posterity esteem the character of the First Man? It could have elevated the first Moon landing to an even higher level of significance. Among those who feel so are Neil's sister June, who knew her brother as well as anyone.

“Did he take something of Karen with him to the Moon?” was June’s rhetorical question.

“Oh, I dearly hope so.”

Perhaps the mystery will be solved when humankind returns, as it surely will, to Tranquility Base.

CHAPTER 25

Return to Earth

Neil had always worried most about the final descent to the Moon landing. "The unknowns were rampant. The systems in this mode had only been tested on Earth and never in the real environment. There were just a thousand things to worry about in the final descent. It was hardest for the systems, and it was hardest for the crew. It was the thing I most worried about, because it was so difficult.

"Walking around on the surface, on a ten-point scale, I deemed a one. The lunar descent on that scale was probably a thirteen."

Somewhere in between was what it took to pilot the ascent stage of the LM back to a reunion with Mike Collins. That tender piece of flying, though perhaps only a five or six on the scale of difficulty, was a ten-plus on the scale of the mission's ultimate success. If that ascent to docking, for whatever reason, did not work out, nothing about Apollo 11's remarkable performance up to this point, or about the dedicated efforts of four hundred thousand talented people who had endeavored to get Apollo 11 to the Moon, could be regarded as anything but a tragedy. The first Moon landing would have happened, but the astronauts who accomplished it would never return home.

Back inside *Eagle* with the hatch closed, Armstrong and Aldrin repressurized their cabin, doffed their PLSSs, and looked at control panel readings to ensure the safety of the LM. They started to fill up a trash bag with unnecessary gear to be left on the lunar surface to save weight. The astronauts again hooked up to the LM's environmental control system, and they took their helmets and visors off, so the two tired and hungry men could eat.

Before their meal, they used up the rest of their film. The EVA Hasselblad had been purposefully left outside after retrieving its finished film magazines. They trained the spare Hasselblad through the portals, snapping pictures of the American flag, the TV stand, and the faraway Earth. (The IVA [intra-vehicular] Hasselblad photos are distinctive in the absence of grid-patterned reseau crosses.) Buzz finally got around to taking two pictures of Neil, showing the commander, tired and relieved, in what the astronauts had come to call their "Snoopy cap," the stretchable black and white cap with foam ear covers that looked like the dog from *Peanuts*. Neil took five shots of Buzz.

While they were eating, a delighted Slayton sent his congratulations:

04:18:00:02	Slayton:	*Just want to let you guys know that, since you're an hour and a half over your timeline and we're all taking a day off tomorrow, we're going to leave you. See you later.*

04:18:00:13	Armstrong:	*I don't blame you a bit.*
04:18:00:16	Slayton:	*That's a real great day, guys. I really enjoyed it.*
04:18:00:23	Armstrong:	*Thank you. You couldn't have enjoyed it as much as we did.*
04:18:00:26	Slayton:	*Roger.*

To achieve jettison, the astronauts had to depressurize their cabin once again, refitting their helmets so they could open the hatch. It was almost like performing another EVA prep, though this took less than twenty minutes and did not involve any hose-swapping or donning of the PLSSs.

In an act some might regard as lunar littering, they threw all the garbage out. First came the PLSSs, their cooling water drained into a plastic bag that was then stowed. "We could get down far enough in our pressurized suits to reach the backpacks with our gloves and then tossed them rather than kicked them out as later crews did," Neil offered. On TV, both backpacks could be seen tumbling down. The exact moment they hit the ground was detected back on Earth thanks to the seismometer experiment Buzz had put out during the EVA. Neil tossed out both pairs of dust-covered boots, a bagful of empty food packages, and the spare Hasselblad, minus the exposed film. He jettisoned the lithium hydroxide canister that he and Buzz had changed out.

Though less cluttered, the cockpit was hardly clean. It was incredible how much dust the men had picked up while on the surface. When they returned to zero g, some of it began to float around inside the cabin. It even affected the way they sounded due to the particles they had breathed in. Neil remembered, "We were aware of a new scent in the air of the cabin that clearly came from all the lunar material that had accumulated on and in our clothes. I remember commenting that we had the scent of wet ashes."

After answering a few questions from Mission Control, Neil was ready to take a break. When asked for a more lengthy and detailed description of the geology they had observed, Neil said, "We'll postpone our answer to that one until tomorrow. Okay?"

At 2:50 A.M. CDT, July 21, Mission Control finally signed off and told the men to get a good night's sleep. Up in *Columbia,* Collins had fallen soundly asleep shortly after hearing his mates had gotten back into the LM okay. Armstrong and Aldrin had been up for nearly twenty-two hours. Neil and Buzz were relieved. "There are always some regrets that you didn't do more or accomplish everything that you wanted, but we had gotten a pretty fair share of stuff done. There is always the great satisfaction of getting things behind you and getting things accomplished. That satisfaction outweighed any regrets we might have had. Also, we were thinking, 'That's another couple hundred pages of checklist items we no longer have to remember and worry about.' "

It was their first and only night's sleep in the LM, and it was not at all pleasant. According to Neil, "The floor was adequate for one person—not to stretch out, but to lay halfway between a fetal position and a stretched-out position. That was where Buzz slept. The only other place to rest was the engine cover, which was a circular table some two and a half feet in diameter. To support my legs from it we configured a sling from one of our waist tethers. We attached that to a pipe structure that was hanging down. It was a good structure to hang a sling from, so I stuck my legs in there and kept the center part of my body on the engine cover. That kept my legs suspended. Behind the cover there was a flat shelf where I could sort of rest my head. It was a jerry-rigged operation and not very comfortable."

Neither man slept well. Compounding their uncomfortable sleeping positions was the fact they were sleeping with their helmets and gloves on to protect their lungs from all the dust they had brought in. Then there was the temperature. Even though it was over 200 degrees Fahrenheit outside the LM, it was quite cold inside, about 61 degrees F. "When we put the window covers on so that it would be relatively dark inside," Armstrong explained, "the temperature got quite brisk in the cockpit." Also disrupting their sleep was light from the control panels, and noise from a loud water pump. Scheduled to sleep for seven hours, Neil may have gotten two restful ones at the very end. As the commander struggled to slumber, he thought through the geology question he had promised to answer. He did not overly worry that lack of rest might affect his flying of the LM the next day. "What was painfully obvious was that I didn't really have any choice. The schedule was there, and I had to perform. I had to do it." The obstacle was hardly a first. "One night. Most people can get by with a low amount of sleep—for several nights, actually," he told himself. "I had relaxed and slept well generally in the command module. Mike said things like, 'This part of the flight is easy. All these other guys have done it and haven't had any trouble. So just relax and enjoy it and save yourself for when you need to be bright-eyed.' And I took it to heart."

Ron Evans, the capsule communicator for the night shift, made the wake-up call to the LM crew at 9:32 A.M. CDT. Liftoff from the Moon, after a stay totaling twenty-one hours, was scheduled to occur shortly after noon.

Most of the intervening time was taken up going over checklists in preparation for the ascent, taking star sightings, establishing the proper state vector for the flight up, inputting computer code, and tracking the command module for one last hack on the LM's precise landing location. The only significant change to the checklist was Houston wanting the LM's rendezvous radar turned off during the ascent. As CapCom Evans told the crew, "We think that this will take care of some of the overflow of program alarms that you were getting during descent."

The science experts on the ground were also anxious to hear more from Neil and Buzz about what they had observed on the lunar surface. Neil was now ready to tell them. "I was excited about the experience myself and was honored and willing to share it with the guys, who I knew were really interested in what was going on. This was a very exciting day for some of those guys; they had been working for many years on what might be found. All of a sudden they had a chance to get real information. It was important to them."

Armstrong's observations that morning impressed everyone with their incisiveness and clarity. "I don't remember writing notes. I think it was just so fresh in my memory that it wasn't hard to re-create what I'd just seen."

"Houston, Tranquility Base is going to give you a few comments with regard to the geology question of last night.

"We landed in a relatively smooth crater field of elongate secondary . . . [correcting himself] circular secondary craters, most of which have raised rims, irrespective of their size. That's not universally true. There are a few of the smaller craters around which do not have a discernible rim. The groundmass throughout the area is a very fine sand to a silt. I'd say the thing that would be most like it on Earth is powdered graphite. Immersed in this groundmass are a wide variety of rock shapes, sizes, textures—round and angular—many with varying consistencies. As I've said, I've seen what looked to be plain basalt and vesicular basalt. Others with no crystals, some with small white phenocrysts, maybe one to less than five percent.

"And we are in a boulder field where the boulders range generally up to two feet with a few larger than that. Now, some of the boulders are lying on top of the surface, some are partially exposed, and some are just barely exposed. And in our traverse around on the surface—and particularly working with the scoop—we've run into boulders below the surface; it was probably buried under several inches of the groundmass.

"I suspect this boulder field may have some of its origin with this large, sharp-edged, blocky-rim crater that we passed over in final descent. Now, yesterday, I said that was about the size of a football field, and I have to admit it was a little hard to measure coming in. But I thought that it might just fit in the Astrodome as we came by it. And the rocks in the vicinity of this blocky-rim crater are much larger than these in this area. Some are ten feet or so and perhaps bigger, and they are very thickly populated out to about one crater diameter beyond the crater rim. Beyond that, there is some diminishing, and even out in this area [around the LM], the blocks seem to run out in rows and irregular patterns, and then there are paths between them where there are considerably less surface evidence of hard rocks. Over."

Heading into the countdown for lunar liftoff, Neil's mind-set was that of a typical test pilot: pragmatic and hard-nosed. "The LM's ascent engine was a single chamber. The tanks and the propellants and the oxidizer were what they were. We did have various means of controlling the circuitry to the valves—opening the flow of propellants to the engine. So that was an alternative. I had proposed that we just put a big manual valve in there to open those propellant valves rather than, or in addition to, having all the electronic circuitry. But management didn't think that that was up to NASA's standards of sophistication. So I really knew that circuitry very well. But it wasn't really a problem, because if we fired the engine and it didn't fire, we weren't out of time. We had a lot of time to think about the problem to figure out what else we could do. When pilots really get worried is when they run out of options and run out of time simultaneously.

"The ascent was a very simple trajectory. We were on PNGS. If we had PNGS malfunction we could have gone to AGS and got into a safe orbit—at least in that point in time we thought we could. How could Houston help? Maybe if PNGS was acting up or there were questions, they would certainly have been able to do more analysis of the problem down there than we could. We were in a pretty good position. We were on the eastern side of the Moon and we were moving west, so during that ascent phase we were going right through the center of the Moon and should be getting pretty good data from Earth's radars there. Maybe they could tell us that we needed to switch to AGS. But other than that, there was not a lot that they could do. They were going to be watching other things, too—systems problems, batteries, environmental systems, and various things. I'm sure if they saw anything funny they would want to know about it and we would have to work out what should be done. But the ascent trajectory itself was pretty straightforward. All through our rendezvous we were calculating the different trajectory changes—the burns—we needed to make. They were doing the same thing, using different sources of information on Earth."

At 05:04:04:51 elapsed time, Ron Evans cleared them for takeoff. "Understand," Aldrin answered. "We're number one on the runway." Some seventeen minutes later, at 12:37 P.M. CDT, it was time for that single, nonredundant engine to fire for its first time. Next to the landing itself, there was no more tense moment in the entire Apollo 11 mission—correct that: in the history of the entire U.S. manned space program.

On CBS, Cronkite uttered to Schirra, "I don't suppose we've been this nervous since back in the early days of Mercury." Both Neil's mother and wife had the same fears.

05:04:21:54	Aldrin:	*Nine, eight, seven, six, five, Abort Stage, Engine Arm, Ascent, Proceed.*
05:04:22:00		LAUNCH OCCURS
05:04:22:07	Aldrin:	[Static] [Garbled] *shadow. Beautiful.*

In his autobiography, Buzz eloquently described the liftoff: "The ascent stage of the LM separated from the descent stage with its chunky body and spindly legs, sending out a shower of brilliant insulation particles which had been ripped off from the ascent of the ascent engine."

05:04:22:09	Aldrin:	*Twenty-six, thirty-six feet per second up. Be advised for the pitchover.*
05:04:22:14	Armstrong:	*Pitchover.*

Again from Buzz: "There was no time to sightsee. I was concentrating intently on the computers, and Neil was studying the attitude indicator, but I looked up long enough to see the American flag fall over. Seconds after liftoff, the LM pitched forward about forty-five degrees, and though we had anticipated it would be an abrupt and maybe even a frightening maneuver, the straps and springs securing us in the LM cushioned the tilt so much and the acceleration was so great it was barely noticeable."

05:04:22:15 Aldrin: *Very smooth. Balance couple, off. Very quiet ride. There's that one crater down there.*

05:04:23:04 Evans: *One minute and you're looking good.*

05:04:23:10 Aldrin: *Roger.* [Pause] *A very quiet ride, just a little bit of wallowing back and forth. Not very much thruster activity.*

05:04:23:31 Evans: *Roger. Mighty fine.*

05:04:23:37 Aldrin: *Seven hundred* [feet per second horizontal velocity], *150* [feet per second vertical velocity] *up. Beautiful. Nine thousand feet* [altitude]. *AGS agrees* [with PNGS] *within a foot per second.*

05:04:23:59 Evans: Eagle, *Houston. You're looking good at two* [minutes]. . .

05:04:24:06 Aldrin: *And that's a thousand. One hundred seventy up. Beautiful. Fourteen thousand. And a foot per second again . . .*

Neil's mother was hardly the only person in tears as she heard Cronkite exclaim. "Oh, boy! Their words 'beautiful' . . . 'very smooth' . . . 'very quiet ride.' Armstrong and Aldrin, just short of twenty-four hours on the Moon's surface, on their way back now to rendezvous with Mike Collins orbiting the Moon."

For the past six months Mike Collins's "secret terror" had been that he might have to leave his mates on the Moon and return to Earth alone. "*Columbia* has no landing gear; I cannot help them if they fail to rise from the surface, or crash back into it." If either tragedy happened, Mike was coming home, but he knew he would be a marked man for life. "It would almost be better not to have that option," he sometimes thought.

The ascent stage had to fire for slightly over seven minutes to achieve the requisite altitude and speed to reach orbit. In the command module, Collins followed their progress very carefully. More than anyone, he knew the precariousness of "rendezvous day." As soon as he awoke that morning, there had been a "a multitude of things to keep me busy," including approximately 850 separate computer keystrokes, "eight hundred fifty chances for me to screw it up." If all went well with *Eagle,* then he would just serve "as a sturdy base-camp operator and let them find me in my constant circle. But if . . . if . . . if any one of a thousand things goes wrong with *Eagle,* then I become the hunter instead of the hunted." At the instant of LM liftoff, Mike was "like a nervous bride." He had been flying for seventeen years, had circled the Earth forty-four times in Gemini X, but had "never sweated out any flight" like he was sweating out the LM.

As *Eagle* rose upward to meet him, Collins knew "One little hiccup and they are dead men. I hold my breath for the seven minutes it takes them to get into orbit." A Gemini veteran, he was "morbidly aware of how swiftly a rendezvous could turn sour. A titled gyro, a stubborn computer, a pilot's error—ah, it was that last one that troubled me the most. If Neil and Buzz limped up into a lopsided orbit, would I have enough fuel and enough moxie to catch them?" Next to him in the CM was a notebook outlining eighteen different variations of what he could try to intersect the LM if the module did not manage to get up to him on its own.

At *Eagle*'s controls, Armstrong drew not just on his Apollo training but also on his Gemini experience to fly to the proper rendezvous point. In terms of the piloting he needed to do and the thruster activities required, flying the LM to its rendezvous with *Columbia* was similar to what he had done in Gemini VIII: the same relative strategy and techniques, the same size velocity changes. "That was one of the main reasons we felt comfortable in the situation."

The ascent was very different from the descent to landing. During much of the descent, the cockpit of the LM faced up; the crew was not able to see the lunar surface. Now, they were staring right at it. "Yes, we were looking at it now from very close range and going over it facedown where we could look at things very closely. The ascent also had a characteristic unlike any other portion of the flight. The attitude control rockets were being used to position the proper attitude of the lunar module. Normally to pitch up—to pitch the nose upward—you would fire the forward rockets upward and the aft rockets downward, both of which tended to rotate the vehicle upward. But in the ascent phase, any rockets that were pointed forward and firing actually slowed you down and fought against the action of the main ascent engine. So the forward-firing rockets were disabled for the ascent engine. To pitch the vehicle we only used half of the rockets—only the ones that were pointing downward. The result of that, since the center of gravity was never quite on center, was those rockets would fire and move the vehicle upward. Then they would shut off, because the CG was offset, and we would be pushed the other way. Then they would kick in again. The whole thing was like a rocking chair going up and down through the entire ascent trajectory.

"That was different than Gemini. We had tried to implement the experience of this motion in the LM simulator, but because the simulator was a stationary object, you had no sensation of that rocking motion. That was quite an unusual characteristic. I didn't remember having it reported to me by the previous crews that had fired the ascent engine on Apollo 9 or 10. If they did report it, I had overlooked it somehow."

As was typical for Neil when he was piloting, he spoke sparingly during the ascent. Heading westward over the same landmarks they had been trying to identify when coming down, Neil remarked, "We're going right down U.S. 1." His only other comment was, "It's a pretty spectacular ride."

At one o'clock in the afternoon Houston time, July 21, a NASA public affairs officer reported that *Eagle* had achieved lunar orbit, one with an apolune of 47.2 nautical miles and a perilune of 9.1 miles. To move from this orbit below *Columbia* to a docking with it would take almost another three hours. Neil, Buzz, and Mike would all be busy with a long and detailed series of rendezvous procedures, navigational maneuvers, and backup checks. "Three hours may seem like a long time," Buzz remarked, "but we were too busy to notice." Mike recalled that his hands were full with the "arcane, almost black-magical manipulations" called for by his notebook full of rendezvous procedures.

Eagle needed to make three separate maneuvers to catch up with *Columbia.* The first, occurring at 1:53 P.M. CDT, took place on the back side of the Moon. Firing the LM's reaction control system (RCS) engines, Armstrong brought the spacecraft into a higher orbit that was just fifteen miles below the command module. An hour later, a second burn put the LM even more in plane with its target, reducing the altitude variations as it incrementally overtook the CSM.

Collins recalls their coming up the rest of the way. "The LM is fifteen miles below me now, and some forty miles behind. It is overtaking me at the comfortable rate of 120 feet per second. They are studying me with their radar and I am studying them with my sextant. At precisely the right moment, when I am up above them, twenty-seven degrees above the horizon, they make their move, thrusting toward me. 'We're burning,' Neil lets me know, and I congratulate him, 'That-a-boy!' We are on a collision course now; our trajectories are designed to cross 130 degrees of orbital travel later (in other words, slightly over one third of the way around in our next orbit). I have just passed 'over the hill,' and the next time the Earth pops up into view, I should be parked next to the LM. As we emerge into sunlight on the back side, the LM changes from a blinking light in my sextant to a visible bug, gliding golden and black across the crater fields below."

So close yet still so far away, the "amiable strangers" jested over the radio about their manner of reconvening:

05:07:22:11	Collins:	*Well, I see you don't have any landing gear.*
05:07:22:15	Armstrong:	*That's good . . . You're not confused on which end to dock with, are you?*

Continuing to close in, even the conversation between Neil and Buzz became more lighthearted:

05:07:25:31 Armstrong: *One of those two bright spots is bound to be Mike.*

05:07:25:36 Aldrin: *How about picking the closest one?*

05:07:25:44 Armstrong: *Good idea.*

For Neil, the image of the command module passing so closely overhead brought back memories of his days as a fighter pilot:

05:07:28:23 Armstrong: *Looks like you're making a high-side pass on us, Michael.*

Buzz got his first good look at *Columbia* as well:

05:07:32:25 Aldrin: *Okay. I can see the shape of your vehicle now, Mike.*

05:07:32:42 Armstrong: *Oh, yes. . . . Got your high-gain* [antenna] *in sight. Your tracking light . . . whole vehicle shows. I see that you're pointed at me. Now, you're turning a little bit. Great.*

05:07:33:49 Collins: *Are you burning yet?*

05:07:33:50 Armstrong: *We're burning.*

"All that remains," Collins recalled, "was for them to brake to a halt using the correct schedule of range versus range rate. . . . While they are doing this, they must make certain they stay exactly on their prescribed approach path, slipping neither left nor right nor up nor down. . . . The sextant is useless this close in, so I close up shop in the lower equipment bay, transfer to the left couch, and wheel *Columbia* around to face the LM."

Peering out through his docking reticle, Mike marveled at the steady, centered approach of the LM as Neil and Buzz brought it home:

05:07:43:43	Collins:	*I have 0.7 mile and I got you at thirty-one feet per second. Look good . . .*
05:07:44:15	Aldrin:	*Yes, yes. We're in good shape, Mike. We're braking . . .*
05:07:46:13	Armstrong:	*Okay, we're about eleven feet a second coming in at you.*
05:07:46:43	Collins:	*That's good . . .*

Bigger and bigger the LM appeared in Collins's window, and it was hard for him to hold back the feeling of exultation. "For the first time since I was assigned to this incredible flight six months ago, for the first time I feel that it *is* going to happen." Inside *Eagle,* however, the commander and the lunar module pilot were nervously entertaining what still needed to be done—and what still could go wrong.

05:07:47:05	Aldrin:	*Hope we're not going to get a pitch straight down.*
5:07:47:16	Armstrong:	*We've got a pitch down and then a yaw to do . . . It flies good . . . Okay, if I pitch over, I'm going to be looking right into the Sun.*
05:07:50:09	Aldrin:	*I hope you know how to roll.*

05:07:50:11 Armstrong: *Yes, I do.*

05:07:50:23 Aldrin: *You want to end up with that window opposite his right window, so you don't want to roll right. Right?*

05:07:50:32 Armstrong: *Yes.*

05:07:50:34 Aldrin: *The only trouble is, it's towards—towards* [a] *ninety* [degree roll], *isn't it? . . . You could . . . You . . .*

05:07:50:58 Armstrong: *If I roll 120—it'll roll left.*

05:07:51:06 Aldrin: *Ninety, huh? . . . Sixty?*

05:07:51:21 Armstrong: *Well, why don't I start to roll . . .*

05:07:51:24 Aldrin: *Yes, I think if you roll up sixty . . .*

05:07:51:29 Armstrong: *I'll be looking into his left window when I pitch up.*

05:07:51:32 Aldrin: *I don't think so. If you did it right now you'd . . .*

With the LM only fifty feet away, technically the rendezvous was over. Neil having turned the LM around, *Eagle*'s drogue directly faced *Columbia*'s docking port. Collins could not contain his emotions when he caught a gorgeous view of Earthrise:

05:07:51:36 Collins: *I got the Earth coming up already. It's fantastic!*

Houston broke in at this crucial moment to learn what was going on:

05:07:52:00 Evans: Eagle *and* Columbia, *Houston. Standing by.*

05:07:52:05	Armstrong:	*Roger. We're station keeping.*

Neil's succinct answer and sharp tone made it clear the unwelcome intrusion was barely tolerable.

05:07:52:24	Aldrin:	*Pitch up . . . Pass right up a little. You got a better view . . . Bottom side . . . Move back.*
05:07:52:45	Collins:	*That's right.*
05:07:53:08	Armstrong:	*Okay, I'm getting about into the right attitude, I think . . .*
05:07:53:18	Aldrin:	*Yes.*
05:07:53:21	Armstrong:	*That roll's pretty far. I just don't know how much . . . So that's . . . Oh, it's going to go* BLOCK!

Although the alignment between *Eagle* and *Columbia* looked good, as the vehicles came together they experienced a potentially nasty phenomenon known as gimbal lock. Put simply, two of the three pivoting gimbals that were located between the inertial platform of the LM's guidance system and the spacecraft itself accidentally got into alignment and temporarily could not move, resulting in the loss of the platform's stability and the firing of some attitude jets. Armstrong recalled how it happened. "The docking technique was to have the lunar module stabilize itself in the vicinity of the command module and maneuver to a point where it would be convenient for the command module to go ahead with the docking. Then Mike would do the actual command module motion to engage the docking mechanism. In a way it's similar to how the Gemini spacecraft had docked with Agena, because Mike's position in the command module was just like the position of the commander of a Gemini. He's looking out his front window and through his docking reticle, a device that helped him make sure the vehicles were properly aligned. We, on the other hand, were looking up. The docking hatch was in the roof of the lunar module so we are looking upward through a small flat window in the roof.

"In trying to achieve the best attitude for the lunar module so that Mike could make an easy docking, I was looking through the top window and making the attitude corrections relative to the command module. Unfortunately, I neglected to be looking at the attitude indicator, which would have told me that we were getting close to gimbal lock. In the process of flying through the top window I flew it right into gimbal lock.

"Now the consequence of that was not very bad, particularly since we were finished flying the lunar module at that point. We weren't going to be in it anymore; we were going to leave it behind. There were alternatives for stabilizing the system and [we] were approximately in the right spot at that point for Mike to complete the docking.

"It's not something that you would do intentionally. But we didn't have any substantial motions or [tumbling] resulting from it."

Perhaps because Collins was controlling the actual docking from his end, and perhaps because Mike had waited so long, all alone, to master this critical final maneuver, his reaction to the gimbal lock was more extreme. As soon as the two spacecraft were engaged by the small capture latches, he flipped a switch that fired a nitrogen bottle to pull the two vehicles together. As soon as he flipped it, he got what he later called "the surprise of my life": "Instead of a docile little LM, suddenly I find myself attached to a wildly veering critter that seems to be trying to escape." Specifically, the LM yawed to his right, instigating a misalignment of about fifteen degrees. Working with his right hand to swing *Columbia* around, there was nothing he could do to stop the automatic retraction cycle designed to pull *Eagle* into a deep embrace. "All I can hope for is no damage to the equipment, so that if this retraction fails, I can release the LM and try again."

Wrestling with his controller, the two vehicles veered back into proper alignment. The docking was sealed. Later, when Neil and Buzz reentered the command module, Mike sought to explain. "That was a funny one. You know, I didn't feel a shock, and I thought things were pretty much steady. I went to retract there, and that's when all hell broke loose." Armstrong offered Mike his own explanation: "It seemed to happen at the time I put the plus-X thrust to it, and apparently it wasn't centered, because, somehow or other, I accidentally got off in attitude and then the attitude-hold system started firing." "I was sure busy there for a couple seconds," Mike declared.

It was 4:38 P.M. CDT. It then took well over an hour for Armstrong and Aldrin to disable the specified LM systems (some were left on), snare and stow floating items, and get *Eagle* into configuration for its final jettisoning.

At 5:20 P.M., Collins opened the hatch mechanism from the other side, and Neil and Buzz, still very dusty, made their way up, down, and into *Columbia*'s cockpit. "The first one through is Buzz, a big smile on his face," Collins noted. "I grab his head, a hand on each temple, and am about to give him a big smooch on the forehead, as a parent might greet an errant child; but then, embarrassed, I think better of it and grab his hand, and then Neil's. We cavort about a little bit, all smiles and giggles about our success, and then it's back to work as usual, as Neil and Buzz prepare the LM for its final journey."

On CBS, Cronkite brought the historic thirty-two hours to an end with the following thoughts:

> *Man has finally visited the Moon after all the ages of waiting and waiting. Two Americans with the alliterative names of Armstrong and Aldrin have spent just under a full Earth day on the Moon. They picked at it and sampled it, and they deployed experiments on it, and they packed away some of it to pack with them and bring home.*
>
> *Above the men on the Moon, satellite over satellite, orbited the third member of the Apollo team, Michael Collins. His bittersweet mission was to guide and watch over the Command Service Module whose power and guidance system provided the only means of getting home . . .*
>
> *With this flight, man has really begun to move away from the Earth. But with this flight, some new challenges for mankind. A challenge to determine yet, whether in coming to the Moon, we turn our centuries-old friend in the sky into an enemy, that we invaded, conquered, exploited, and perhaps someday left as a desolate globe once more. Or will we make the most of it, as perhaps a way station on beyond the stars. Apollo 11 still has a long way to go—and so do we.*

Thus concluded the longest continuous scheduled broadcast in the history of television.

Back in lunar orbit, Collins helped his mates transfer their equipment, film, and rock boxes into the mother ship.* That done, they tried to clear *Columbia* of Moon

* In January 2015, two and a half years after Armstrong's death in August 2012, Carol Armstrong, Neil's widow, informed curators at

dust. They extracted from storage a small vacuum cleaner head, as directed by the microbe people. "The

---

the National Air and Space Museum (NASM) that she had found in the back of one of Neil's closets "a white cloth bag filled with assorted small items that looked like they may have come from a spacecraft." The bag turned out to be what was known as a "McDivitt purse," named after the Apollo 9 commander Jim McDivitt, who first had the idea to include such a bag aboard the Apollo spacecraft. Inside the bag were Neil's waist tether, utility lights and their brackets, equipment netting, an emergency wrench, the optical sight that had been mounted above Neil's window in the lunar module, and the 16mm data acquisition film camera (DAC) that had recorded the now iconic footage of the lander's final approach as well as Neil's descent down the ladder to take his "small step" onto the Moon. As NASM space historian Dr. Allan Needell stated at the time, "The 16mm DAC, given the images that it captured, ranks as enormously important," though all of the items in the bag were significant given their connection to Armstrong. It is not known why Neil kept the bag or that he even recollected that he had it. But, as space memorabilia expert Robert Pearlman explained, "To be clear, the bag was not something Armstrong snuck home from the Moon. After returning to lunar orbit, the bag and its contents were moved from *Eagle* to the command module *Columbia.* Neil mentioned the purse to Michael Collins as it was being transferred from one craft to the other; what Neil said exactly was "That [is] just a bunch of trash that we want to take back—LM parts, odds and ends." Collins later called Mission Control to note where the bag was being placed for the trip home and about how much it weighed. Pearlman continued his assessment ("Neil Armstrong's purse: First moonwalker had hidden bag of Apollo 11 artifacts," collectSpace.com, February 6, 2015): "It is not known how the purse came to be in Armstrong's possession after the mission, but it wasn't unusual for the astronauts to retain small spent parts of their capsules as souvenirs. In September 2012, one month after Armstrong died, President Barack Obama signed into law a bill that confirmed the Mercury, Gemini, and Apollo astronauts had legal title to their mementos." Presently, the purse and its contents are on extended loan from the Armstrong estate to the Smithsonian; two of the artifacts, the data camera and waist tether, went on display immediately at NASM as part of the exhibition, "Outside the Spacecraft: 50 Years of Extra-Vehicular Activity." In addition to the McDivitt purse, the Armstrong family also donated to the Smithsonian an extensive collection of Neil's personal items and memorabilia.

vacuum didn't take off much of the dust at all," Buzz stated. "We got more off by dusting each other by hand, but even that didn't do the job." Before closing the hatch, Neil and Buzz tidied things up in the LM. It was very hard to say good-bye to the machine. *Eagle* had done absolutely everything it had been asked to do, and then some.

At 6:42 P.M. CDT, it was time to send the LM on its way. On subsequent Apollo lunar missions, the LM would be impacted into the Moon for seismographical measurements, but *Eagle* floated for several years until it crashed into the lunar surface. Buzz and Neil were both glad it was Collins who flipped the switches to release it. During the subsequent meal, Collins started throwing question after question at his mates: "How did liftoff feel? . . . Do the rocks all look the same? They're different? Good, great. I'm glad to hear it. . . . Luckily, you were able to get a little bit of everything."

At 11:10 P.M. Houston time, Monday, July 21, Mission Control gave *Columbia* the go-ahead for Trans-Earth Injection. Collins later called TEI "the get us out of here, we don't want to be a permanent Moon satellite" maneuver. It was a two-and-a-half-minute burn of the service propulsion engine that was to send them home by increasing their velocity to 6,188 miles per hour, the speed necessary to escape lunar orbit. If TEI did not go well, as Neil explained, "we would have been in for a long, lonely ride."

TEI took place on the back side of the Moon, out of contact with Earth. Along with Earth reentry, it was the only truly nervous moment left to face. As complicated as the whole mission had been, what the astronauts had to make absolutely sure of when they did deorbit the Moon was that they were pointed in the right direction. They relieved the tension with humor:

05:15:14:12	Collins:	*I see a horizon. It looks like we are going forward* [laughter].
05:15:14:26	Armstrong:	*Shades of Gemini.*
05:07:14:29	Collins:	*It is most important that we be going forward* [more laughter]. *There's only one really bad mistake you can make there.*
05:15:14:50	Aldrin:	*Shades of Gemini retrofire. Are you sure we're* [laughter] . . . *no, let's see, the motors point this way and the gases escape that way, therefore imparting a thrust that-a-way.*
05:15:15:03	Collins:	*Yes, horizon looks good.*

Actually, there was a very remote chance that the astronauts could have shot themselves off in the wrong direction. "I wouldn't put it at zero," Armstrong admitted. "There was certainly a chance—particularly when you are in the dark, without external references, and dependent on your instruments. Is it possible to get that attitude wrong? I would say it's possible. It's something that Mission Control always worried about because they can't see you on the far side and they don't have any data."

As soon as the spacecraft peeked around the disk of the Moon half an hour later, Houston wanted to know what had happened:

05:15:35:14	Duke:	*Hello, Apollo 11. How did it go? Over.*
05:15:35:22	Collins:	*Time to open up the*

		LRL [Lunar Receiving Laboratory] *doors, Charlie.*
05:15:35:25	Duke:	*Roger. We got you coming home. It's well stocked. . . . All your systems look real good to us. We'll keep you posted.*
05:15:36:27	Armstrong:	*Hey, Charlie boy, looking good here. That was a beautiful burn. They don't come any finer.*

As Collins recalled, all three men next took turns with the cameras, pointing them alternately at the Moon and the Earth. "The Moon from this side is full, a golden brown globe glorying in the sunshine. It is an optimistic, cherry view, but all the same, it is wonderful to look out the window and see it shrinking and the tiny Earth growing." Not only seeing it from this distance but knowing they were coming back home to it made the sight "unforgettable."

The remainder of the two-and-a-half-day trip home was relatively routine. The first night's sleep after the reunion was the deepest and most satisfying of the entire trip, and lasted some eight and a half hours, until noon Houston time, Tuesday, July 22. The spacecraft passed the point where Earth's gravity took over and began drawing the astronauts progressively homeward—a point 38,800 nautical miles from the Moon and 174,000 from the Earth—shortly after they woke up. Later that afternoon, they made their only midcourse correction, slightly adjusting their flight for the best trajectory for a return to Earth. By midafternoon of the next day,

Columbia reached the midway point of the journey home, 101,000 nautical miles from splashdown. So relaxed was the crew and so uneventful their duties that they created a little mischief by playing over their radio to Houston a special tape of sound effects they had brought along. On it were sounds of dogs barking and of a speeding diesel locomotive.

What everyone back on Earth most remembered about the return home were the two evening primetime color television transmissions. In the final TV transmission from Apollo 11, each astronaut explained what the Moon landing meant to him within the grander scheme of things. At 7:03 P.M. EDT, Armstrong opened the broadcast:

> *Good evening. This is the commander of Apollo 11. A hundred years ago, Jules Verne wrote a book about a voyage to the Moon. His spaceship,* Columbia, *took off from Florida and landed in the Pacific Ocean after completing a trip to the Moon. It seems appropriate to us to share with you some of the reflections of the crew as the modern-day* Columbia *completes its rendezvous with the planet Earth and the same Pacific Ocean tomorrow. First, Mike Collins.*

At Mission Control, Janet and her boys, Pat Collins and her youngsters, as well as one of the Aldrin children, were taking in the show from the viewing room.

Mike Collins:

> *Roger. This trip of ours to the Moon may have looked, to you, simple or easy. I'd like to assure you that has not been the case. The Saturn V rocket which put us into orbit is an incredibly complicated piece of machinery, every piece of which worked flawlessly. This*

computer up above my head has a 38,000-word vocabulary, each word of which has been very carefully chosen to be of the utmost value to us, the crew. This switch, which I have in my hand now, has over 300 counterparts in the command module alone. . . . We have always had confidence that all this equipment will work, and work properly, and we continue to have confidence that it will do so for the remainder of the flight. All this is possible only through the blood, sweat, and tears of a number of people. First, the American workmen who put these pieces of machinery together in the factory. Second, the painstaking work done by the various test teams during the assembly and retest after assembly. And finally, the people at the Manned Spacecraft Center . . . . This operation is somewhat like the periscope of a submarine. All you see is the three of us, but beneath the surface are thousands and thousands of others, and to all those, I would like to say, thank you very much.

In Buzz's time on camera, he presented the first of his many future statements on behalf of the spirit of exploration:

Good evening. I'd like to discuss with you a few of the more symbolic aspects of the flight of our mission, Apollo 11. As we've been discussing the events that have taken place in the past two or three days here on board our spacecraft, we've come to the conclusion that this has been far more than three men on a voyage to the Moon; more still than the efforts of a government and industry team; more even than the efforts of one nation. We feel that this stands as a symbol of the insatiable curiosity of all mankind to explore the unknown. Neil's statement the other day upon first set-

> *ting foot on the surface of the Moon, "This is a small step for a man, but a great leap for mankind," I believe sums up these feelings very nicely. We accepted the challenge of going to the Moon; the acceptance of this challenge was inevitable. The relative ease with which we carried out our mission, I believe, is a tribute to the timeliness of that acceptance. Today, I feel we're fully capable of accepting expanded roles in the exploration of space. . . . Personally, in reflecting on the events of the past several days, a verse from Psalms came to mind to me: "When I consider the heavens, the word of Thy fingers, the moon and the stars which Thou hast ordained, what is man that Thou art mindful of him."*

The man of the fewest words, Commander Armstrong, closed the broadcast eloquently. His mood was as reflective as it would ever be in public:

> *The responsibility for this flight lies first with history and with the giants of science who have preceded this effort; next with the American people, who have, through their will, indicated their desire; next, to four administrations and their Congresses, for implementing that will; and then to the agency and industry teams that built our spacecraft: the Saturn, the* Columbia, *the* Eagle, *and the little EMU, the space suit and backpack that was our small spacecraft out on the lunar surface. We would like to give a special thanks to all those Americans who built the spacecraft, who did the construction, design, the tests, and put their hearts and all their abilities into those crafts. To those people tonight, we give a special thank-you. And to all the other people that are listening and watching tonight, God bless you. Good night from Apollo 11.*

For everyone who was watching at home in their living rooms that midsummer night's eve, these were proud moments. Wrapping up the broadcast on CBS, Cronkite called the crew's closing statements "a heartwarming vote of appreciation from those three astronauts who have done the incredible—gone to the Moon and walked upon it." Total success for Apollo 11 now hinged upon reentering the Earth's atmosphere, splashing down, and being safely recovered.

But back on Earth unforeseen danger threatened the final moments of Apollo 11. A bad storm was brewing over the Pacific that a couple of fast-thinking meteorologists saw was moving right over the splashdown point. NASA therefore changed the landing site. Early on the morning of Thursday, July 24, the prime recovery ship, the USS *Hornet,* a carrier built in 1943, and with President Nixon aboard, was ordered to move northwesterly a distance of some 250 miles to a calmer area. *Columbia* then changed its inbound trajectory. Otherwise, the entry part was routine.

At 11:35 Houston time on the twenty-fourth, Apollo 11 started down through the Earth's atmosphere. It slammed into the first fringes of air at some four hundred thousand feet when the spacecraft was northeast of Australia. Collins, at the controls, graphically detailed the reentry: "We are scheduled to hit our entry corridor at an angle of six and a half degrees below the horizon, at a speed of 36,194 feet per second, nearly twenty-five thousand miles per hour. We are aimed at a spot eight miles southwest of Hawaii. We jettison our Service Module, our faithful storehouse still half full of oxygen, and turn around so that our heat shield is leading the way. Deceleration begins gradually and is heralded by the beginnings of a spectacular light show. We are in the center of a sheath of protoplasm, trailing a comet's tail of ionized particles and heat shield material. The ultimate black of space is gone, replaced by a wispy tunnel of colors: subtle lavenders, light blue-greens, little touches of violet, all surrounding a central core of orange-yellow." Dropping fast but feeling as if they are in a state of suspended animation, the three astronauts see the first earthly forms, a big bank of gorgeous stratocumulus clouds. Then their three huge main chutes blast open, "beautiful orange-and-white blossoms of reassurance." Soon the astronauts were able to make out the wide expanse of ocean below. At 08:03:09:45 elapsed time, Air Boss, the head of the interservice recovery team, radioed it had visual contact with the descending capsule. Dawn was just breaking over the southwestern Pacific.

Eight minutes and thirty-three seconds later, at 11:51 A.M. CDT, the spacecraft hit the water like a ton of bricks, forcing a grunt out of each astronaut. Armstrong radioed to Air Boss, "Everyone okay inside. Our checklist is complete. Awaiting swimmers." Air Boss verified an on-target landing, 940 nautical miles southwest of Honolulu and 230 miles south of Johnston Island. The *Hornet* was only thirteen miles away. Navy helicopters were in the immediate area. Armstrong and his mates had each taken an anti-motion-sickness pill before reentry, only to discover they should have taken two, given the wave height and the fact that the command module was floating small-end down. Mike owed Neil a beer, payment for Neil betting that the module would topple over. Technically, it was called the "stable two" position—the CM's hatch underwater and the astronauts hanging from their straps. Armstrong remembered, "It was unusual being upside down looking into the water while hanging from the straps. Everything looked completely different because gravity had now established an orientation that had been missing for a long time. All of a sudden you had a gravity vector that you could identify with, but it was not like anything you'd ever seen before! Everything looked like it was in the wrong spot."

Quickly the crew started the motorized pumps to inflate three small airbags that changed the spacecraft's buoyancy center of gravity and turned it back over big end down. It took almost ten minutes for the float motors to fill the bags. Waiting for the team of three navy frogmen, they sat in silence, willing themselves not to be seasick, especially Aldrin. "It was one thing to land upside down," Buzz later remarked, "it would be quite another to scramble out of the spacecraft in front of television cameras tossing our cookies all over the place." The swimmers attached the orange flotation collar, then opened the spacecraft's hatch; it was 12:20 P.M. CDT, 6:20 in the morning Hawaii time. The astronauts sensed they had been in the water for eternity, but only twenty-nine minutes had elapsed. Into the command module were thrown the Biological Isolation Garments, or BIGs. Grayish-green in color, these were the rubberized, zippered, hooded, and visored containment suits meant to save the world from "Moon germs." The astronauts labored to put on their BIGs inside the cramped module. Dealing with gravity for the first time in eight days, they were so light-headed, and their feet and legs so swollen, they could barely stand, especially against eighteen-knot winds.

The BIGs finally donned, the astronauts squeezed through the small hatch; as the commander, Neil came out last. Before they were escorted one by one into the raft bobbing alongside, frogmen sprayed them with a precautionary disinfectant against lunar microbes. Once inside the dinghy, they were then given cloths and two different doses of chemical detergent to continue the scrub-down. When they were finished, the frogmen tied the cloths to weights and dropped them into the ocean. Supposedly the BIGs were airtight, but within minutes moisture began seeping into them. Virtually nothing was said by the astronauts during any of this, mostly because the visors and headgear of their BIGs made it almost impossible to be heard, especially with four helicopters beating their rotors overhead.

Again they sat, for fifteen minutes, until a helicopter got the order to drop down and pick them up. The *Hornet* was now in view, less than a quarter of a mile away. With TV cameras on board a couple of the helicopters, every moment of the recovery was being broadcast live around the world. Waiting for them inside the helicopter was Dr. William R. Carpentier, their flight surgeon from the Manned Spacecraft Center. They gave him the thumbs-up sign as they entered.

At 12:57 P.M. CDT, the helicopter landed on the *Hornet*'s flight deck. A brass band was playing. Cheering sailors crowded on deck. A big grin on his face and his hands crossed atop a rail, President Nixon stood on the bridge along with Secretary of State William P. Rogers and NASA Administrator Dr. Thomas O. Paine, who were accompanying the president on a twelve-day, round-the-world trip that included a stop in Vietnam. The astronauts could barely see the hoopla. Still inside the chopper and in their BIGs, they rode one of the ship's elevators down to the hangar deck. Disembarking, they walked down a newly painted line through a cheering crowd of seamen and VIPs into the mobile quarantine facility—a thirty-five-foot-long modified house trailer—in which they were to remain until they arrived at the Lunar Receiving Laboratory in Houston on July 27.

Neil remembered what it was like landing on the ship

and getting up on his feet. "We all felt pretty good. We didn't have any seasickness kinds of problems." They were able to go right into the quarantine trailer where they immediately sat down in easy chairs to undergo the microbiology sampling and a preliminary medical exam by Dr. Carpentier.

They had time only for a quick shower before seeing the president. "There were the Nixon ceremonial activities to attend to," Neil reflected. "We needed to do [that] and get it behind us so that we could celebrate." Following the playing of the National Anthem, President Nixon, nearly dancing a jig of pleasure, addressed the astronauts via intercom at 2:00 P.M. CDT. Crouching behind a picture window at the back end of the trailer the three tired but exhilarated crew members arranged themselves, Neil to the president's left, Buzz to the right, and Mike in the middle. Nixon welcomed the men back to Earth on behalf of people all over the world and told them that he had called each of their wives the day before to congratulate them. He also invited them and their wives to a state dinner in Los Angeles. The president closed his remarks by calling the eight days of Apollo 11 "the greatest week in the history of the world since the Creation." It proved to be a controversial statement, especially for many Christians. Neil regarded Nixon's statement as hyperbolic: "It was an exciting time. A lot of times when you are exuberant, you tend to be a little exaggerative."

On her front lawn, Janet thanked all of the people who helped make the flight successful: "We thank you for everything—your prayers, your thoughts, just everything. And if anyone were to ask me how I could describe this flight, I can only say that it was out of this world!" In Wapakoneta, Neil's parents rejoiced.

• • •

As the *Hornet* steamed toward Honolulu, the astronauts could not yet fully relax, as there were still more post-flight medical exams to undergo. Neil had fluid in one of his ears; caused by the stress of reentry, it cleared up the next day. With the doctors interested in how eight days in zero g had affected their bodies, it almost seemed like the mission was still occurring as the men were in isolation and couldn't see day or night. When the tests were over, an impromptu cocktail hour broke out inside the small living room of the mobile quarantine facility; Neil drank scotch. Then came a dinner of grilled steaks and baked potatoes. That night, in soft beds with real pillows, the crew slept hard for nearly nine hours. Their rest was timed to restore a regular sleeping pattern, soon to be disrupted by the loss of six hours traveling east from Hawaii to Houston.

After a hearty breakfast, there was work to do. *Columbia* had been brought on board and its precious rock boxes and other treasures needed to be unloaded. Through a plastic tunnel Neil, Buzz, and Mike walked to their grizzled spacecraft, scarred as it now was from the heat of reentry, and with the help of a recovery engineer, they took the boxes out of the command module and loaded them in a special sterilization unit. A few hours later, the boxes were flown off the aircraft carrier to Houston.

That afternoon witnessed another shipboard ceremony. The captain presented each of them a plaque, inscribed drinking mugs, and caps. Armstrong continued to act as the crew spokesman, as he would in all public events. Someone in the trailer innocently remarked, "And now it begins," a comment that would become the astronauts' refrain in the coming weeks.

For two nights the Apollo 11 crew stayed on the *Hornet,* an experience Neil alone was accustomed to from his days in the navy. He passed some of the time by playing a marathon game of gin rummy with Mike, while Buzz read or played solitaire. They also started autographing pictures earmarked for NASA and White House VIPs.

The scene was wild as they arrived at Pearl Harbor on Saturday morning, July 26. The first time Neil had steamed into Pearl was aboard the *Essex,* as a midshipman, eighteen years earlier. People were cheering, a band was playing, and flags were waving. A broomstick was flying from the *Hornet*'s mast, the symbol of a mission well done. But, as Neil related, "We weren't in a very good position to see all that stuff." Commander of Pacific Forces Admiral John Sidney McCain Jr., father of the future U.S. senator, greeted the crew upon their arrival, as Nixon had on the *Hornet,* through the rear window of their trailer. They stayed at Pearl only long enough to transfer to an airplane for their flight to Houston. Their trailer got lifted onto a flatbed truck and was then driven, at a speed of ten miles per hour, to nearby Hickam Field. Crowds of people lined the streets. Finally reaching Hickam, the MQF was loaded into the cavernous belly of a C-141 Starlifter transport. The long flight to Houston meant just that much more time inside the MQF. According to Neil, "It was pretty much like everything else. Here we were confined to a very small place—but a bigger place than we had been in for quite a while. We had more room. We had hot food. We had cocktail hour. We had lots of things to do. Anytime we had spare time there were lots of things we wanted to write down or talk about."

Arriving at Ellington Air Force Base in Houston around midnight, they got rolled onto another flatbed truck. The off-loading at Pearl and the loading at Hickam had both gone well, but it took three tries before they were loaded onto the truck. At the window with the astronauts, Dr. Bill Carpentier quipped, "They can send men safely to the Moon and back, but they can't get the men off the airplane."

Safely on the back of the truck, they were driven across the tarmac to an awaiting crowd of several thousand people and a host of television cameras. The mayor of Houston addressed the astronauts, as did MSC head Bob Gilruth. "Everybody was assembled to greet us," Armstrong remembered, including the crew's wives and children. The astronauts spoke to their loved ones through special telephone hookups. Neil did not recall what he said, or what they said to him, except "glad to have you back."

Not until 1:30 A.M. did the truck leave Ellington and head slowly down NASA Road 1 to the Manned Spacecraft Center. Regardless of the late hour, people were still clamoring in the street. It was not until around 2:30 that they arrived at the Lunar Receiving Laboratory, where they were to spend the rest of their twenty-one-day quarantine. With its special air-conditioning system, no air was supposed to escape from the LRL without passing through a number of filters and pumps.

The LRL was safe, secure, and quiet. Besides private bedrooms for each crew member, it had a kitchen and a dining area. It also had a large living room and recreation area where, besides a television, recent Hollywood movies were projected on a big screen. The population of the LRL included two cooks, a NASA public relations officer, another doctor who was a lab specialist, and a janitor. It was big enough to accommodate everyone without crowding the astronauts. NASA had even agreed to embed a journalist, John Macleish, who issued a stream of communiqués. From the LRL Neil made his first calls to his family, including his mother.

Mike and Buzz eventually found the time inside the LRL boring, but not Neil; he welcomed the refuge from the turmoil surrounding them. "We really needed that time to be able to do all of the debriefings and talk to all the various systems guys. The subsequent Apollo crews were very interested in this question and that question that had to do with their own mission planning—what they thought they might reasonably do and whether we had ideas on how they might improve their own flights. Mostly, the discussion revolved around what was doable on the surface, because that affected the planning substantially. So that time was very valuable to us personally, as well as to everyone else."

The days in the LRL also gave them plenty of time to review batch after batch of mission photographs that were being developed and printed by a special MSC photo lab. "Those were dribbling in to us a few at a time," Neil recollected. "They would run one roll of film and, as soon as that was ready, they would get copies of them to us. As we went through the pictures themselves, a lot of questions also came up that the other flight crews were interested in having answers to. The photos helped them ask their questions and helped us answer them." Some of the debriefings required the astronauts to write out long pilot reports covering their special responsibilities in the mission; other debriefings were filmed in a room resembling a TV control booth, the astronauts sitting at a table opposite questioners behind a glass wall. The mission was documented down to the minutest detail, resulting in 527 pages of single-spaced transcripts.

On August 5, the LRL chef surprised Neil with a cake on his thirty-ninth birthday.

Near the end of their stay, each astronaut, as federal government employees, was asked to fill out an expense report for their flight to the Moon. Filled out for them to sign, the forms read: "From Houston, Tex., to Cape Kennedy, Fla., to the Moon, to the Pacific Ocean to Hawaii and return to Houston, Tex." Their total reimbursement was for $33.31.

Only one time during the quarantine did any tension simmer between the three astronauts. It came when Aldrin, during the debriefings, elaborated at extensive lengths on the phenomenon of the flashing lights that all three of them had seen during the outbound journey. Buzz sensed Neil's growing irritation, and the discussion ended.

As busy as they were with the debriefings, day after day in seclusion gave Armstrong and his mates plenty of time to think about their future with the space program; one day Deke even directed them to consider whether or not they wanted to return to flight status. Neil thought it was too early to come to any conclusion, though he hoped he would be able to fly again.

The men also considered how all the glamour and publicity would affect their personal and professional lives, and the lives of their families. Just prior to reentry, Jim Lovell had warned the Apollo 11 astronauts, "Backup crew is still standing by. I just want to remind you that the most difficult part of your mission is going to be after recovery."

Armstrong understood Lovell's message, as he thought back on it years later: "We were not naïve, but we could never have guessed what the volume and intensity of public interest would turn out to be. It certainly was going to be more than anything any of us had experienced before in previous activities of flight. And it was."

Their quarantine came to an end at 9:00 on Sunday evening, August 10. By then, even Neil was very ready for it to be over. They had been in physical seclusion for over a month. Outside the LRL, a NASA staff car and driver waited to drive them home individually. The crew of Apollo 11 went their separate ways, but not for long.

Their short trips home that night presaged the astronauts' lot for years to come. The moment each car passed through the NASA gate, a different TV crew pulled behind to follow the famous passengers. Reporters and photographers awaited them in front of their residences.

Neil wanted none of it, certainly not then. As soon as his NASA car pulled into his driveway, he bolted for the front door. Janet was waiting to shut it quickly behind him.

Armstrong's life on the dark side of the Moon had begun.

CHAPTER 26

For All Mankind

In the months following *Columbia*'s return to Earth, Armstrong and his two crewmates would be asked almost endlessly to express themselves about the Moon landing and its meaning for history and the global community. By all accounts, Neil, center stage, performed superbly well. Even today his first wife, Janet, who accompanied him on all the immediate post–Apollo 11 goodwill trips, proudly relates that Neil was "never comfortable speaking . . . but he did it, and he did a great job of it."

Post-quarantine, Neil stayed at home one full day to take refuge from reporters. As a matter of courtesy, the legitimate press had agreed to leave all three of the astronauts alone until Wednesday, though casual onlookers and paparazzi continued to stake out the crew's homes. One carload of photographers pursued Aldrin and his wife, en route to buy him a new suit for what was to be a one-day, coast-to-coast Apollo 11 celebration tour, even after Aldrin diverted into Ellington Air Force Base. Neil spent that Monday indoors, catching up on personal mail, visiting with the family, and watching Janet get herself and the boys ready to join the cross-country trip. The next day he returned to his office at the Manned Spacecraft Center, where huge bags of mail awaited, some fifty thousand letters a week, for several weeks.

That same afternoon the first postlanding press conference was held in the MSC auditorium. Computer program alarms, the fuel situation during the lunar descent, and the other problems involved in the landing dominated the questions, which then turned to Neil's unique experiences. Asked if there was ever a moment on the Moon when he was "just a little spell-bound by what was going on," Neil replied with a smile, "about two and a half hours." Asked about the primary difficulty during the EVA, he offered, "We had the problems of a five-year-old boy in a candy store. There were just too many things to do." Asked what he thought about the imminent three-cities-in-one-day tour to New York, Chicago, and Los Angeles, Neil shook his head slowly and admitted it was "certainly the last thing we're prepared for."

At five A.M. the next morning, Wednesday, August 13, the Armstrong family of four, the Collins family of five, and the Aldrin family of five boarded the transport jet Air Force Two, which President Nixon had sent to Houston for the trip. Mike and Neil spent the flight preparing their speeches (Buzz's unease with extemporaneous speaking had motivated him to begin days in advance). At LaGuardia Airport, Mayor John Lindsay and his wife greeted their honored guests, then flew them by helicopter to a pier near Wall Street in full view of a salute by a flotilla of fireboats. A string of open convertibles waited for them. Into the lead car went the three astronauts, followed by a security car, a car with the wives, another security car, a car with all eight of the astronauts' children, and yet another security car. Buzz remembered, "We were advised not to reach out to shake hands because we could be pulled from cars and couldn't be rescued easily."

Not even the revelry at the end of World War II or the parade for Lindbergh in 1927 matched in size this New York City celebration. A blizzard of ticker tape enveloped their parade as it moved between the skyscrapers through the Financial District, along Broadway and Park Avenue, past a Manhattan-record turnout of an estimated four million.

"I had never seen so many people in my life," Janet exclaimed, remembering "people cheering and waving and dropping confetti that floated down from everywhere out of buildings, from out of the sky."

"They also threw out IBM punch cards," Neil added. "Sometimes they threw a whole stack of punch cards from the eighty-seventh floor of a building, and, when they didn't come apart, it made like a brick. We had a couple of dents in our car from cards that didn't quite open."

At City Hall, the handsome Mayor Lindsay presented them with keys to the city, and all three astronauts made brief remarks. Then onward they went to the United Nations, where they shook hands with Secretary-General U Thant. The crew received a book of commemorative stamps representing all the UN member nations. Neil was the only astronaut to speak.

As wild as the crowds were in New York, they were wilder in Chicago. By the time the parade of open limousines crept its way to the massive new Civic Center, Aldrin recalled, "we were covered with confetti and streamers and perspiring so much that they were glued to us. We were deaf from the shouting, and jaws ached from smiling." Following a public ceremony at Richard J. Daley's City Hall—where the rough-mannered boss mayor directed the astronauts' photo shoot by saying, "Hey you, over here"—the astronauts were surprised to find themselves, before heading to O'Hare International Airport, in Grant Park to address a gathering of some fifteen thousand young people.

"It was exciting to be in these cities as there was electricity in the air from the joy these people were expressing on behalf of the achievement," Janet noted. Neil explained, "That's probably the first time we had seen such large aggregations of people . . . *really* a lot of people. It was just one event after another, big parades, ending up with the Nixon state dinner in Beverly Hills."

Arriving at Los Angeles International Airport, the plane was met by Mayor Sam Yorty, then helicopters took the party to the posh Century Plaza Hotel. The children of the three astronauts would not be attending the black-tie affair, instead partaking of a spread of hamburgers, French fries, and chocolate malts in front of a color TV tuned to the live telecast.

President Nixon, his wife Patricia, and their two grown daughters, Julie and Tricia, hosted the astronauts and their wives in their presidential suite prior to joining dinner guests Mamie Eisenhower, widow of the former president; Esther Goddard, widow of rocket pioneer Robert Goddard; Chief Justice and Mrs. Warren E. Burger; former Vice President and Mrs. Hubert H. Humphrey (among the few Democrats invited); Arizona senator and 1964 Republican presidential nominee Barry Goldwater; and current Vice President Spiro Agnew and his wife. Government notables filled the high-domed and elegantly chandeliered banquet hall: NASA and other space program officials, more Cabinet members than sometimes attended Cabinet meetings, governors of forty-four states (including California governor Ronald Reagan), members of the Joint Chiefs of Staff, diplomatic corps members representing eighty-three nations, and a battery of Congressional leaders. U.S. and international aviation pioneers were represented by Jimmy Doolittle, the man who had headed the NACA when Neil began his government career in 1955, Wernher von Braun, and Willy Messerschmitt. From Hollywood and show business came entertainers Rudy Vallee, Gene Autry, Jimmy Stewart, Bob Hope, Red Skelton, Rosalind Russell, Art Linkletter, and a score of others. Evangelist Reverend Billy Graham was there. Howard Hughes and Charles Lindbergh had been invited, but neither aviator came out of his self-imposed seclusion to attend. Ironically, not a single member of the Kennedy family attended, indebted as was the occasion to the inspiration of former President John Kennedy. On July 18, the day Apollo 11 approached lunar orbit, Massachusetts Senator Edward "Ted" Kennedy, following a party, had plunged off a bridge at Chappaquiddick Island, near Martha's Vineyard, an accident that had killed twenty-eight-year-old campaign worker Mary Jo Kopechne.

While the Kennedys remained in seclusion, peace and antipoverty protestors did not, waging an orderly demonstration outside the hotel where a fleet of black limousines—glistening Cadillacs, Imperials, Continentals, and Rolls-Royces—sat in the parking lot. To the protestors the glory of Apollo 11 was temporary or shallow, or both. The mood of Vietnam-era America remained highly agitated, and these particular taxpayers were not eager to pick up the tab for Nixon's $43,000-plus gala, with its 1,440 guests. The president himself had approved the menu right down to the *clair de lune* dessert, a sphere of dimpled ice cream topped with a tiny American flag.

After the meal, Vice President Agnew, the chairman of the administration's National Aeronautics and Space

Council, presented the three astronauts with the Medal of Freedom, the nation's highest civilian honor, for their participation in "a unique and profoundly important adventure. Their undertaking will be remembered as long as men wonder and dream and search for truth on this planet and among the stars." Flight controller Steve Bales, who, according to his citation that night, "made the decision to proceed with the lunar landing when computers failed just before *Eagle*'s landing on the Sea of Tranquility," earned a Medal of Freedom of his own. As factually misleading as that citation was, Bales's honor was an important symbolic gesture on behalf of the estimated four hundred thousand persons who had contributed to the Apollo program.

When it came Armstrong's turn to address the throng, he was, by all accounts, emotional: "Neil Armstrong choked back tears as he groped for words to tell America how the Apollo 11 astronauts feel about their country and the honor it has given them," opened the UPI wire coverage. And *Time* magazine reported, "Neil Armstrong's words to President Nixon in Los Angeles last week seemed all the more eloquent because they were unstudied, and because for once the usually phlegmatic voice of the first man on the Moon quavered with emotion."

"We were very privileged to leave on the Moon a plaque endorsed by you, Mr. President, saying, 'For all mankind.' Perhaps in the third millennium a wayward stranger will read the plaque at Tranquility Base. We'll let history mark that this was the age in which that became a fact. I was struck this morning in New York by a proudly waved but uncarefully scribbled sign. It said: 'Through you, we touched the Moon.' It was our privilege today to touch America. I suspect that perhaps the most warm, genuine feeling that all of us could receive came through the cheers and shouts and, most of all, the smiles of our fellow Americans. We hope and think that those people shared our belief that this is the beginning of a new era—the beginning of an era when man understands the universe around him, and the beginning of the era when man understands himself."

No one in the audience was prouder of Neil than his own family. "My parents were there as guests," Neil related, "as well my grandmother and sister and brother and their families. I had very little time to see them, but they were there. It was an impressive occasion for everyone."

On Saturday, an estimated 250,000 gathered in Houston (a city of only 1.2 million in 1969)—"Spacetown USA"—to throw ticker tape, confetti, and enough "Moon certificates," fake $100 and $1,000 bills, to cover the streets in two to three feet of litter. The ultimate Texas barbecue was held in the Astrodome for a by-invitation-only crowd of 55,000. Placards in the grandstands read: "You've come a long way, baby. Welcome home," and "We're proud of y'all." Frank Sinatra served as master of ceremonies and entertained with singer Dionne Warwick and comedians Bill ("José Jiménez") Dana and Flip Wilson, all big stars in 1969.

The day before the parade, Neil, Mike, and Buzz had taped NBC's *Meet the Press* for Sunday morning broadcast, a morning they were also to appear live on CBS's *Face the Nation.* In regard to the crew's future as part of the U.S. space program, Neil answered that the next several decades would be even more exciting. "We can do much more in ten years than we would expect. And if we judge that will probably be true, then I think in ten years we will be looking at the planets." Presaging a recommendation early the following month by a presidentially appointed Space Task Group chaired by Vice President Agnew, Armstrong added that he thought the goal of reaching Mars was possible, perhaps with an intermediate landing point. Asked by Howard Benedict of the Associated Press whether man can survive for months on end in space, Neil replied, "I would like to take a trip of [up to two years], and perhaps a considerably larger vehicle would allow us to take the families along." Ending the program, correspondent David Schoumacher of CBS News asked all three astronauts whether—and when—they would return to space. Collins announced that Apollo 11 was his last flight; Aldrin anticipated a future Apollo mission. Neil said, "I am available to serve in any capacity that they feel I can contribute best."

Returning to Houston, Neil was looking for a place to get away for a week's vacation. Colorado governor John Love remembered Sleeping Indian Ranch, owned by Harry Combs, the chief of aircraft distributor Combs Gates Denver, Inc., who had attended the Apollo 11 launch with the governor. Neil, Janet, and the boys stayed at the rough ranch in gorgeous countryside with bear, elk, and deer.

A week of near-total relaxation in Colorado's mountain air bolstered Neil and Janet mentally and physically for the incredibly frenetic schedule to come. It began in Wapakoneta, Armstrong's hometown, on Saturday, September 6, 1969. More than five hundred police officers were brought in. Service stations ran out of gasoline. A local movie theater stayed open all night free of charge to give visitors a place to rest. With Bob Hope serving as the parade's grand marshal, the arriving crowd numbered in excess of ten times that of the town's normal 7,000 population. Neil's Purdue University Marching Band provided music. The small town was wrapped almost entirely in red, white, and blue bunting. Streets along the parade route were renamed for their favorite son—"Lift-Off Lane," "Apollo Drive," "Eagle Boulevard"—in the tradition of "Neil Armstrong Drive," site of the senior Armstrongs' residence. Downtown headquarters, directed by Charles Brading Jr., the son of Neil's boyhood employer, welcomed visitors (and some 350 credentialed journalists) to "Tranquility Base." Heading the Neil Armstrong Homecoming VIP Committee was Fred Fisher, Neil's boyhood friend. Governor James Rhodes announced—though Neil had yet to be consulted—that the state of Ohio would be moving forward with plans to build a Neil Armstrong Museum in Wapakoneta.

Neil took it all in very good humor, content to repeat what he had told them during his visit back in 1966, "I'm proud to stand before you today and consider myself one of you." Then, to the delight of his mostly Ohio audience, he added that though news reports indicated that he and Buzz had not found anything "organic" on the lunar surface, "I think you know better now. There was a Buckeye on the Moon."

From Wapakoneta Neil and Janet flew to Washington, leaving the boys with Neil's parents. On Monday the ninth of September they attended NASA's Apollo 11 Splashdown Party at the Shoreham Hotel, preceded by the formal unveiling at the U.S. Post Department of the commemorative Moon landing stamp, the ten-cent stamp that Neil and Buzz canceled after they got back into *Columbia* on July 22. The following week the Armstrongs returned to Washington, where the Apollo 11 crew was to be honored at a midday joint session of

Congress. Promptly at noon, the astronauts were led by a bipartisan delegation up to seats on the Speaker's rostrum. Following a long and loud standing ovation, Armstrong stepped first to the microphone and addressed the assembly, saying that the whole venture began in those halls with the Space Act of 1958. Neil then introduced Buzz, followed by Mike, to the great chamber. After their remarks, he took the podium and reiterated the fact that they had traveled on a peaceful mission for all of mankind. He concluded with:

"We carried on Apollo 11 two flags of this Union that had flown over the Capital, one over the House of Representatives, one over the Senate. It is our privilege to return them now in these halls which exemplify man's highest purpose—to serve one's fellow man.

"We thank you on behalf of all the men of Apollo, for giving us the privilege of joining you in serving—for all mankind."

With thunderous applause, the U.S. Congress seemed poised to vote strong support for the future of the space program. That was not to be. Immediately following their speeches, there was a photo shoot of the trio. Following that, the wives and families of congressmen awaited the astronauts' narration of Apollo 11. Remembered Aldrin, "No one had previously mentioned this to any of us. My reaction was tempered by my elation of the moment, but both Mike and Neil were justifiably furious." Buzz recalled, "We raised hell" with NASA Headquarters for letting the session run for over two hours.

The next morning's State Department briefing afforded the crew's first details on their impending world tour of a minimum of twenty-three countries in forty-five days. Logistics of travel aboard Air Force Two would be managed by a "support team" of six PR officers from the space agency, a White House representative, two men from the United States Information Agency, two secretaries, a doctor, a baggage man, two full-time security officers, a photographer-cameraman, plus four men from the Voice of America.

The astronauts ranked their stated objective "to demonstrate goodwill to all people in the world and to stress that what we had done was for all mankind," over State Department and NASA agendas "to visit the American embassies anxious to score social coups." "We would take care of Americans in America," they insisted at the briefing. The tour called "Giant Leap" pledged to go "around the world to emphasize the willingness of the United States to share its space knowledge." The trip would kick off from Houston on September 29, then travel to Mexico City, Bogotá, Buenos Aires, Rio de Janeiro, Grand Canary Island, Madrid, Paris, Amsterdam, Brussels, Oslo, Cologne, Berlin, London, Rome, Belgrade, Ankara, Kinshasa (Congo), Tehran, Bombay, Dacca, Bangkok, Darwin (Australia), Sydney, Guam, Seoul, Tokyo, Honolulu, and back to Houston. Neil did not keep a diary of the trip as George Low had for the Latin American tour three years earlier, but he did tape-record a travelogue.

On October 8 in Paris, France, Neil reported, "A representative of the Aero Club of France gave us their gold medal, which had previously only been given to the Wright Brothers and Charles Lindbergh from America and it had been given to no other crews from spaceflights. I responded with some remarks concerning recollection of that welcome by Charles Lindbergh."

Janet, who herself had "a wonderful trip" representing her country, had a particularly distinct memory of the visit to Belgium and the Netherlands, regarding meeting the two kings and two queens in one day: "That was really something. We had lunch with one and dinner with the next one. We were told you were never supposed to turn your back on a king or a queen. Well, Mike Collins got caught in a situation in Belgium where the king was ahead of him and the queen was behind him and Mike was in between and so he had to sidestep up twenty-five or thirty stairs in the palace. He was just so good about it. We all joked about that later." On the way back to the hotel, Collins reportedly said, "I think I broke my goddamn ankle!"

Attendance at a "typical press conference" numbered, as it did in Cologne and Bonn, Germany, on October 12, "a thousand or more people." The next day in Berlin assembled "an extremely large crowd estimated as two hundred thousand to three hundred thousand, but I expect it to be probably closer to a million. We proceeded upstairs to a gigantic reception." On October 14 in London, England, he noted, "We had canceled our two scheduled TV programs, with BBC and an independent network, and did only the press conference on television. Because of colds and laryngitis. All the press reports [led] with our bleary eyes and sore throats." In Bogotá, Dr. Carpentier had prescribed anti-anxiety pills for Buzz. One night in Norway Buzz felt so depressed he had stayed in his room all evening while all the others, including his wife, went out to dinner. Buzz wrote that this was the only night of the trip that he drank too much, but the problem for him was "there was liquor everywhere . . . bottles of scotch or gin in every hotel room, a jug of mimosas on the breakfast tray every morning." In Rome, attending "an elegant party right out of *La Dolce Vita* at Gina Lollobrigida's," without his wife, Buzz did not return to their hotel room until after dawn and was "in the doghouse" for the rest of the day. Later, in Iran, the couple had "one of the more memorable fights" of their marriage; Buzz remembered, "I was informed either begin to stay at home more or plan to move out."

Both Collins and Armstrong knew that something serious was bothering their crewmate. "The trip produced some disturbing symptoms in Buzz," Mike wrote, "causing him to withdraw into stony-faced silence from time to time," resulting in "obvious stress" to Joan. Neil remembered different indications of an onset of depression but "wasn't smart enough to recognize the problem. It bothered me then, and bothers me now, that I wasn't up to the job. I've thought to myself: Had I been more observant or more attentive, I might have noted something that could have helped Buzz's situation, and I failed to do that. It was sometime after the tour that he started having real problems."

Whether any of Buzz's depression was a carryover from how he felt about Neil being deemed the first man on the Moon is unclear, but there is no question that the situation still grated on him—and on his father. Back at the ceremonial unveiling of the ten-cent Apollo 11 stamp at the U.S. Post Office building in Washington just a few weeks earlier, Buzz had been chagrined to see that the stamp, which showed Neil stepping down onto the lunar surface, bore the caption "First Man on the Moon." Aldrin recalled in his autobiography, "Lord knows what prompted the caption under the stamp, but it caused me to feel rather useless and it positively infuriated my father . . . 'Men' would have been more accurate, and I must confess, my feelings were hurt."

Janet Armstrong feels that the fact that Neil made most of the toasts and did such an outstanding job as the crew's spokesman exacerbated Buzz's sense of being aggrieved. "It was hard to follow Neil—he always did so well. Buzz used notes and that bothered Buzz. He was not as comfortable speaking as Neil and Mike were. Neil was not comfortable speaking, either, but he did it, and he did a great job of it," as the tour continued from Tehran, where the astronauts visited the Shah of Iran, to Tokyo, where they were received by Emperor Hirohito.

Thus ended the astronauts' Giant Step forty-five-day world tour. After a fuel stop in Anchorage, Alaska, Air Force Two flew directly to the nation's capital. Shortly before landing at Dulles National, each astronaut received a memo parodying national protocols:

> *Your next stop is Washington, DC, USA. Here are a few helpful reminders. 1. The water is drinkable, although it is not the most popular native drink. 2. You can always expect student demonstrations. 3. Never turn your back on the President. 4. Never be seen with the Vice President. 5. If you leave your shoes outside the door, they will be stolen. 6. It is unsafe to walk on the street after dark. 7. Do not discuss the following sensitive issues with the natives: Vietnam War, Budget, Foreign Aid, Import-Exports. 8. Rate of exchange is .05 cents per one dollar (American).*

On the White House lawn, with the Marine Band playing, President and Mrs. Nixon welcomed them home. That night the astronauts and their wives dined and slept at the White House. "The president was quite nice," Neil remembered. "He was very interested in everything we had to report about the tour, about the various leaders we had met, what their reaction was and what did they say. He had been trying for years to get a meeting with Romanian president Nicolae Ceausescu and after leaving the *Hornet* he was able to get an appointment. President Nixon said something to the effect, 'That meeting alone paid for everything we spent on the space program.' "

During dinner Nixon asked all three men in turn what they wanted to do next in their lives. Collins said he would like to continue doing goodwill work for the State Department, upon which, right at the table, Nixon instantly phoned Secretary of State William Rogers, asking him to get Mike set up. Aldrin told the president he felt he could contribute more by staying in technical work. When Nixon then asked the commander of Apollo 11 if he wouldn't like to serve somewhere as a goodwill ambassador, Neil politely said he, too, would be honored to serve as an ambassador, but he was not sure in what kind of role he could serve best. Nixon told him to think it over and requested a personal reply.

During Giant Step, between 100 million and 150 million people were estimated to have seen the astronauts, and as many as 25,000 of these actually shook hands with them or received autographs. In the immediate aftermath of the trip, Armstrong certainly felt like it had done some good. Speaking to an audience at Ohio's Wittenberg College in November 1969, Neil said, "More can be gained from friendship than from technical knowledge," quite an admission coming from the devoted aeronautical engineer.

• • •

Armstrong next joined Bob Hope's Christmas 1969 USO tour to entertain the U.S. and allied troops in Vietnam, with stops along the way in Germany, Italy, Turkey, Taiwan, and Guam. Actresses Teresa Graves, Romy Schneider, Connie Stevens, Miss World 1969, the "Golddiggers" showgirls, and Les Brown and His Band of Renown completed the cast. Under Hope's tutelage,

Armstrong, decked out in chino pants, a red sport shirt, and a jungle hat, often played the straight man:

Hope: *Your step on the Moon was the second most dangerous of the year.*
Neil: *Who took the most dangerous?*
Hope: *The girl who married Tiny Tim.* [Tiny Tim was a long-haired ukulele player with a very high-pitched voice who had become a pop icon.]

During a show in Vietnam, a GI asked, "Do you think it's possible that one day humans will live on the Moon?" "Yes, I think they will. We will see a manned scientific base being built on the Moon. It'll be a scientific station manned by an international crew, very much like the Antarctic station. But there's a much more important question than whether man will be able to live on the Moon. We have to ask ourselves whether man will be able to live together down here on Earth."

Armstrong had a serious message for the soldiers: "I tried to use the occasion to have the troops in Vietnam consider increasing their education when they got back home. I tried to make the point that in today's world this would be a good time for them to do it—many of them—before they got too many other commitments."

The Christmas 1969 tour was spared "any [enemy] fire or even any explosions in the distance. Some of the places were fairly close to combat zones, but I don't remember any action." At Lai Khe, the troupe performed for the U.S. 1st Infantry Division, which had seen some of the heaviest fighting in the war to date. So battle-weary were the soldiers that, when Hope reiterated his personal assurances from President Nixon of a plan for peace, more than a scattering of boos arose, a startling precedent for Bob Hope, whose USO shows dated back to World War II.

For the first time, Neil faced scandalous publicity. Stories appeared in the gossip columns that he and actress Connie Stevens had become romantically involved, and that after their return, Neil had been spotted in the audience of Stevens's Las Vegas act. The truth was that the singer-actress and the astronaut had done nothing more than play cards to pass time on the tour.

In May 1970, Armstrong traveled to the Soviet Union, as only the second American astronaut to make an official visit. "I was invited to give a paper to the thirteenth annual conference of the International Committee on Space Research." On May 24 he arrived at the Leningrad airport on a flight from Warsaw. A red carpet awaited him but no crowds, as the Soviet government had not released news of Armstrong's arrival. Serving as his hosts were Georgy T. Beregovoy and Konstantin P. Feoktistov, the cosmonauts whose goodwill visit to America came two months after Apollo 11. Neil received a great welcome from a predominantly Russian audience and was mobbed by scientists seeking autographs. After five days in Leningrad, Neil was given permission to visit Moscow. At the Kremlin, he met with Premier Alexei N. Kosygin for an hour. On behalf of President Nixon, Neil presented him with some chips of a Moon rock and a small Soviet flag that had been carried aboard Apollo 11. The next morning Kosygin sent Neil bottles of vodka and cognac. The great Russian aircraft designer Andrey N. Tupolev and his son Adrian "took me to the airfield hangar where they kept their supersonic TU-144—the 'striking' Concorde lookalike. Apparently I was the first Westerner to see the airplane. The Tupolevs gave me a model of the TU-144, which Andrey Tupolev signed. When I got back, I gave that to the Smithsonian." Armstrong met several other Soviet astronauts. In a secluded forest outside of Moscow, he spent the day at the Cosmonaut Training Center, which was part of the space complex of Zvezdyny Gorodok ("Star City"), Russia's version of Houston's Manned Spacecraft Center. His hostess was Valentina Tereshkova, the first woman to fly in space, whom he found charming. Neil toured their training facilities, simulators, and spacecraft mockups, "which struck me as being functional but a bit Victorian in nature." Tereshkova also took him to the office of the late Yuri Gagarin, whose personal effects had been preserved as a shrine to the first human space traveler. Neil's lecture was attended by many of the cosmonauts. Afterward "they brought two ladies up—one was Mrs. Gagarin and the other Mrs. Vladimir Komarov. Because we had left medallions on the lunar surface in their husbands' honor, it was kind of a touching little ceremony." Neil told Soviet media he had been "most emotionally moved" by his meeting with the widows.

"That night the cosmonauts invited me to a dinner. There was much toasting going on. . . . They presented me a very nice shotgun inscribed with my name on the stock—a twelve-gauge double-barrel side-by-side that the U.S. government permitted me to keep.

"After dinner, around midnight, Georgy Beregovoy, my host, invited me to his apartment for coffee. At one point Georgy talked a little bit on the phone, then someone called him and he went over and turned on his television set. It was the launch of Soyuz IX. It wasn't live; it was a tape of the launch that had occurred earlier in the day at Baikonur. And the occupant was

Andrian G. Nikolayev, who was Valentina's husband, as well as Vitaly I. Sevastyanov. So I had spent the whole day with Tereshkova and the whole evening with all the colleagues of the two cosmonauts, and it was never mentioned once that they were having a launch that day. I concluded that Valentina was either awfully good at keeping a secret or she was dreadfully misinformed."

The launch went well, or Neil would never have seen it. Vodka was brought out for toasts. Beregovoy smiled broadly when he told Armstrong, "This launch was in your honor!"

From July 1969 to June 1970, Armstrong traveled the half million miles to the Moon and back followed by nearly 100,000 miles on Earth. He would have been happy to keep flying as an active member of the astronaut corps, but his superiors in Washington had other ideas for their great American hero than risking him undertaking another dangerous space mission.

PART SEVEN

ICON

I think that people should be recognized for their achievements and the value that adds to society's progress. But it can be easily overdone. I think highly of many people and their accomplishments, but I don't believe that that should be paramount over the actual achievements themselves. Celebrity shouldn't supersede the things they've accomplished.

—NEIL A. ARMSTRONG TO AUTHOR,
CINCINNATI, OH, JUNE 2, 2004

CHAPTER 27

Standing Ground

Following the Moon landing, Armstrong recalled, "I never asked the question about returning to spaceflight, but I began to believe that I wouldn't have another chance, although that never was explicitly stated." Both George Low and Bob Gilruth "said they would like me to consider going back to aeronautics and take a deputy associate administrator job in Washington. I was not convinced that that was a good thing for me. Probably because of all the time I had worked at the field centers, I more or less looked down at Washington jobs as not being in the real world."

Private-sector opportunities were plentiful, from business ventures, hotel and restaurant property development, to commercial banking. People suggested he run for political office, as fellow Ohio astronaut John Glenn had done. But Neil wanted to stay in engineering.

"Thinking it over, I concluded that the NASA aeronautics job was something I could do." Janet felt that Neil was not unhappy with the change: "He was a pilot, and he was always happier when he was flying." She worried that he was not a desk job type of person.

Armstrong's principal contribution to NASA aeronautics during his time in Washington was his support for the new technology of fly-by-wire. Until Neil became deputy associate administrator for aeronautics, no one at NASA Headquarters had given the radical concept of flying an airplane electronically (and with only one of its inputs being the pilot's controls) much credence. Neil stunned a team of Flight Research Center engineers when they visited his office in 1970 asking for modest funding to conduct flight research with an airplane installed with an *analog* fly-by-wire system. As NASA historian Michael H. Gorn has written, "To their surprise, Armstrong objected. Why analog technology?" he asked. Rather than a system of human impulses transmitted by mechanical linkages from the cockpit to the control surfaces, Neil proposed employing a more advanced system, one based on counting—on *digital* fly-by-wire (DFBW). The FRC engineers knew of no flight-qualified digital computer. "I just went to the Moon and back on one," said Armstrong. According to Gorn, "The visitors from the Flight Research Center admitted with embarrassment that they had not even thought of it."

Out of this initiative arose NASA's innovative F-8C Crusader DFBW flight test program, undertaken at Dryden Flight Research Center from 1972 to 1976. Proven reliable, DFBW untied the hands of high-speed aircraft designers, coaxing them to venture forward with radical new aerodynamic configurations, including airplanes possessing absolutely no innate stability of their own minus the computerized control system. DFBW stands as another major contribution to aeronautics that needs to be associated with the First Man.

Neil's main frustration was not with his aeronautics job per se, but with the ongoing "requests" from NASA,

Congress, and the White House for "appearances on demand," which Neil came to find "a real burden." But he didn't have a choice. Many an evening was spent on the Washington dinner-party circuit. Recollected Janet, "We were able to meet a lot of Washington people. They enjoyed meeting Neil personally and congratulating him for what Apollo 11 accomplished for our country and the world. Since we were still on the government pay scale [Neil's annual salary was $36,000], there was not much money available. Dottie Blackmun, the wife of Supreme Court Justice Harry Blackmun, was a fine seamstress and operated a dress store in Minnesota before their move to DC. She was a dear friend and helped with [evening] clothes."

Armstrong took every opportunity to fly airplanes, including NASA transports en route to the Ames, Lewis, Langley, or Dryden laboratories. "I was able to maintain my flight currency—not as current as I would have liked, but better than not at all. And being around the research programs involving the field centers gave me the opportunity to accept some invitations to fly other aircraft"—including England's Handley-Page 115, a small aircraft to test the Concorde supersonic transport's highly swept wing shape; and Germany's large Akaflieg Braunschweig SB-8 sailplane, innovative for its use of structural composite materials.

NASA's foremost aeronautical endeavor of the 1960s fell victim to politics when, on March 24, 1971, in one of the most dramatic roll calls in modern U.S. Senate history, fifty-one senators voted to deny further funding for the American supersonic transport program. Although Armstrong's office did not have any responsibility for the SST, Neil was in favor of continuing the program.

The demise of the American SST had no bearing on

Neil's decision, in August 1971, to resign from NASA for a teaching post at the University of Cincinnati. "I had always told people it was my intention to go back to the university. That was not a new thought for me. I didn't want to leave NASA precipitously, though it was never my intention to be in that bureaucracy job that long. I had met the president of the University of Cincinnati on several occasions. His name was Walter C. Langsam, a historian of early-twentieth-century Europe. Walter had talked to me and written me a couple of nice little notes saying how much he would like me to come to his university. Walter said, 'If you come, we will make you a full professor and you can do whatever you want.' I decided to accept the invitation. I had had a lot of university invitations by this point but most of them—far and away the majority of them—were invitations to be considered for university presidencies. I just wanted to be a professor." Curiously, NASA offered little resistance to Armstrong's departure. Neil's time in civilian government service totaled sixteen and a half years.

Some people in and around NASA thought that Neil must be "batty" for landing in a place like Cincinnati. Friends and associates recalled Neil's interest over the years in writing an engineering textbook. Most suspected a geographic pull, which Neil flatly denied: "Returning to Ohio wasn't a consideration. I thought Cincinnati's was a pretty good department and it was small, about a dozen people" and unlikely to protest Neil's immediate full-tenure status without the customary year of probation. Department head Dr. Tom Davis was a fairly well-known specialist in the burgeoning field of computational fluid dynamics (CFD). The program offered a Ph.D., though Neil only had a master's recently awarded at the University of Southern California after more than ten years of off-and-on graduate work. Armstrong's title was University Professor of Aerospace Engineering. Students called him "Professor Armstrong" or "Dr. Armstrong," even though the only doctorates that he had at this point were honorary (eventually he wound up with 19 honorary doctorates). The faculty, many of whom wound up being good friends, called him Neil.

Neil could have gotten away with a light teaching load, but that wasn't what he wanted. He taught some core courses, and taught three quarters a year with summers off. "I was usually there every day. I did travel some, but tried to make sure it didn't interfere with my normal full schedule of responsibilities."

At the end of Neil's first day of class, the hall was packed with reporters. Chaos broke out, and Neil slammed the door shut and refused to come out. In 1974 Italian actress Gina Lollobrigida appeared unannounced at the door to his classroom. "She came to town ostensibly to take pictures for a book she was doing, but it turned out that it wasn't for a book at all but rather for a magazine article [*Ladies' Home Journal,* August 1974]. I really liked Gina from my visits with her in Mexico and Italy [during the 1969 Giant Leap world tour], but I was really disappointed in her not being truthful about her objective."

Armstrong personally invented two courses for the department. The first was aircraft design. The second was experimental flight mechanics, and both were graduate courses.

Students were surprised to find their celebrity professor such an excellent teacher. Though a serious lecturer and a demanding grader, near the end of an academic quarter he was known to relate some of his favorite flying stories.

Armstrong ultimately failed to navigate the Byzantine labyrinth of university politics. "I really couldn't work the system. I had determined not to take any work from NASA; I wouldn't make proposals to them because I thought it might be viewed as taking advantage of my past association, which I wouldn't do. In retrospect, I was probably wrong about that. I probably should have been active, because I would have known exactly where to go to get some satisfying research projects done. It would have been easier in terms of funding sources had I taken that route."

Two major changes at the University of Cincinnati ultimately led, in 1980, to Armstrong's leaving the school. "It was burdened with lots of new rules," related Armstrong of UC's shift from independent municipal university to state-school status. "In order to escape being bound to the rules of the faculty collective bargaining group, it was required that I be less than full-time. So a strategy some of us tried was going to half-time teaching and half-time in a research institute." In July 1975, the university approved Armstrong and three prominent researchers—George Rieveschl, a UC chemist famous for inventing Benadryl, the first antihistamine; Edward A. Patrick, an electrical engineering professor; and Dr. Henry Heimlich, Cincinnati's famous inventor of the Heimlich maneuver, who practiced medicine at the local Jewish hospital—coming together as the Institute of Engineering and Medicine.

"Establishing the institute was not something that had been high on my priority list. It was just kind of a necessary evil. Once getting into the work, though, I did find some of it very interesting and tried to actively participate in it." Neil added, "Yet the university's rules were still so cumbersome that I just went completely to half-time. Really it was half-time in name only—what it really amounted to was half pay,"* and ultimately "a conflict of the instructions between my basic job as it had been offered to me by President Langsam and the new rules." Beyond that, "I could not expect the volume of requests coming my way to subside—some of which were good opportunities with good people and quality institutions. I realized that, in my situation, I couldn't remain in that kind of job. On the other hand, taking board of directors' positions provided a livelihood without obliging myself to spend all of my time with any one of them."

For Armstrong, his last years at the university were not especially stressful, "just irritating." In the autumn of 1979 he wrote a short note of resignation, effective the first of the year.

In January 1979 Neil, after turning down any number of lucrative promotional offers, agreed to become a national spokesperson for the Chrysler Corporation. His first TV commercial for the American car manufacturer came during the telecast of that month's Super Bowl XIII, which Neil attended in Miami in the company of Chrysler execs. More TV spots aired the next day as did splashy print ads in fifty U.S. newspaper markets, showing Neil endorsing Chrysler's new five-year, fifty-thousand-mile protection plan. At the 185-acre Warren County farm northwest of Lebanon, Ohio, that the Armstrongs had purchased after moving from Bethesda, a small fleet of Chrysler automobiles—a New Yorker, Fifth Edition, Cordoba, W200 four-wheel-drive pickup truck, two different front-wheel-drive Omnis,

* Armstrong's tax form for 1979 showed that he earned an income of $18,196 from the University of Cincinnati. From his own personal service corporation that year, he earned $168,000. Beyond that, he earned about $50,000 in fees for serving on boards of directors for various companies.

and a Plymouth Horizon—had been seen parked for days at a time. According to Janet, Neil had told Chrysler, "'I need to try your product first.'"

The press asked questions: Why is Armstrong starting to do advertising now, after all this time? And why Chrysler, of all companies? Neil later explained: "In the Chrysler case, they were under severe attack and in financial difficulty, but they had been perhaps the preeminent engineering leader in automotive products in the United States, just very impressive. I was concerned about them and when their head of marketing approached me to take a role that was not just as a public spokesman but also as someone to be involved in their technical decision-making process, I became attracted to that. I visited Detroit, where I talked to Chrysler head Lee Iacocca and other leading company executives. I had a look at the projects they were working on. I got to know some of their people and concluded that it was something I should try. It wasn't an easy decision, because I hadn't done anything like it before. Yet I decided to try, on the basis of a three-year agreement. I loved the engineering aspects of the job, but I didn't think I was very competent in the role as a spokesman. I tried my best, but it wasn't something I was good at. I was always struggling to do it properly."

In the coming months Armstrong forged professional relationships with General Time Corp. (a subsidiary of Tally Industries) and the Bankers Association of America. He made promotional commitments on a case-by-case basis, in General Time's case by conceiving his involvement not as a commercial endorsement of the company's "Quartzmatic wristwatch" but rather a technological breakthrough. "The Quartz watch company had built the timer in the lunar module, so that was the connection there—the technology was good. As it turned out, the product quality was not as good as I thought it should be. As for the American Bankers Association, it was not a commercial organization, but rather did an institutional kind of advertising. We made a couple of ads, but it just didn't come together." Armstrong's trial run as a public spokesman for select American products proved to be temporary, but corporate concerns became his primary focus for the rest of his professional life.

Simultaneous with leaving UC, Neil entered into a business partnership with his brother, Dean, and their second cousin Richard Teichgraber, owner of oil industry supplier International Petroleum Services of El Dorado, Kansas. Dean, formerly the head of a General Motors' Delco Remy transmission plant in Anderson, Indiana, became the IPS president; Neil became an IPS partner and the chairman of Cardwell International Ltd., a new subsidiary that made portable drilling rigs, half of them for overseas sale. Neil and his brother stayed involved with IPS/Cardwell for two years, at which time they sold their interests in the company. Dean later bought a Kansas bank.

By 1982, Neil had several different corporate involvements: "I think some people invited me on their boards precisely because I *didn't* have a business background, but I did have a technical background. So I accepted quite a few different board jobs. I turned down a lot more than I accepted."

The very first board on which Armstrong had agreed to serve, in 1972, had been with Gates Learjet, then headed by Harry Combs. Chairing its technical committee and type-rated in the Learjet, Neil flew most of the new and experimental developments in the company's line of business jets. In February 1979, he took off in a new Learjet from First Flight airstrip near Kill Devil Hill in North Carolina and climbed over the Atlantic Ocean to an altitude of fifty-one thousand feet in a little over twelve minutes, setting new altitude and climb records for business jets.

In the spring of 1973 Neil joined the board of Cincinnati Gas & Electric, an engineering company for power generation.

Armstrong linked his connection with Cincinnati-based Taft Broadcasting to Taft's dynamic CEO and president Charles S. Mechem Jr., who was "one of the seven or eight Cincinnati people I invited as my guest to Gene Cernan's Apollo 17 flight in December 1972." Mechem gives a very clear impression of the strengths Armstrong brought to the corporate boardroom: "Typically you ask somebody to go on the board and they say, 'Terrific. When is the first meeting?' Well, it wasn't that way with Neil. After probing as to why I wanted him and what he could bring to the board that didn't have anything to do with his being the first man on the Moon," Armstrong came on board.

Armstrong joined United Airlines in January 1978, and in 1980 he joined Cleveland's Eaton Corporation, as well as their AIL Systems subsidiary, which made electronic warfare equipment. In 2000, AIL merged with the EDO Corporation, which Neil chaired until his retirement from corporate life in 2002.

In March 1989, three years after the explosion of Space Shuttle *Challenger,* Armstrong joined Thiokol, who had made the Shuttle's solid-rocket boosters (SRBs). With Neil's help, Thiokol managed not only to survive but also grow, in the expanded form of

Cordant Technologies, into a manufacturer of solid-rocket motors, jet aircraft engine components, and high-performance fastening systems worth some $2.5 billion, with manufacturing facilities throughout the United States, Europe, and Asia. In 2000, Cordant was acquired in a cash deal by Alcoa, Inc., at which time the Thiokol board on which Neil had been serving for eleven years dissolved.

Reluctant to assess the value of any of the corporate contributions he made over the past thirty years, Armstrong only said, "I felt that in most cases I understood the issues and usually then had a view on what was the proper position on that issue. I felt comfortable in the boardroom."

For the first time in his life, Armstrong also made a good deal of money. Besides handsome compensation for his activities as a director, he was also receiving significant stock options and investing his money wisely. By the time he and Janet divorced in 1994, the couple was worth well over $2 million.

Though he never made a show of his philanthropy, Neil was regularly involved in promoting charitable causes, particularly in and around Ohio. In 1973, he headed the state's Easter Seal campaign. From 1978 to 1985, Armstrong was on the board of directors for the Countryside YMCA in Lebanon, Ohio. From 1976 to 1985, he served on the board of the Cincinnati Museum of Natural History, for the last five years as its chairman. From 1988 to 1991, he belonged to the President's Executive Council at the University of Cincinnati. Right up to his death in 2012, he actively participated in the Commonwealth Club and Commercial Club of Cincinnati, having presided over both. In 1992–93, he sat on the Ohio Commission on Public Service. In 1982, he narrated the "Lincoln Portrait" with the Cincinnati Pops Orchestra.

According to Cincinnati Museum of Natural History director Devere Burt: "His name gave us instant credibility. Anywhere you went looking for money, you simply had to present the letterhead, 'Board Chairman Neil A. Armstrong.' "

For his college alma mater, Neil was perhaps the most active. He served on the board of governors for the Purdue University Foundation from 1979 to 1982, on the school's Engineering Visiting Committee from 1990 to 1995, and from 1990 to 1994 he cochaired with Gene Cernan the university's biggest-ever capital fund-raiser, Vision 21. Its goal set at a whopping $250 million, the campaign raised $85 million more, setting an American public university fund-raising record.

Dr. Stephen Beering, Purdue's president from 1983 to 2000, recalled Armstrong's essential contributions to Vision 21: "Neil was really the PR piece of it. He might say to an alumni group, 'You know, my landing on the Moon was really facilitated by my Purdue experiences—it goes back to my very first semester when I had a physics professor who had written our textbook and for the first Friday recitation I anticipated that I would need to regurgitate the assigned chapter. Instead, the professor said, "I'm curious what you *thought* about this material." At that moment I realized what Purdue was about: it was about teaching problem-solving, critical thinking, analyzing situations, and coming to conclusions that were in detail and original. When I was flying the LM down onto the Moon, that's exactly what I had to do—take my training but then solve problems, analyze situations, and find a practical solution for myself. Without Purdue, I couldn't have done it.'

"And whenever he was on campus, you could see in his eyes how much he enjoyed it. His pure joy showed just standing there with his arm around some band member at a football game. He was thrilled like a kid when he was asked to be the one to bang the big Boilermaker drum. 'I've never done that! I'd like to do it.' And he marched with the baritones, which he played back when he was in the band. Not for a moment did he act like a celebrity."

Armstrong was also involved in a few benevolent causes at the national level. From 1975 to 1977, he cochaired, with Jimmy Doolittle, the Charles A. Lindbergh Memorial Fund, which by the fiftieth anniversary of Lindbergh's historic flight, in May 1977, raised over $5 million for an endowment fund supporting young scientists, explorers, and conservationists. In 1977–78, Neil accepted an appointment to Jimmy Carter's President's Commission on White House Fellowships. In 1979, he served as the on-air host for *The Voyage of Charles Darwin,* a seven-part documentary broadcast on PBS. The National Honorary Council's USS *Constitution* Museum Association counted him as a member from 1996 to 2000.

Some have said that Neil did not have a political bone in his body: "I don't think I would agree in the sense that I have beliefs, I participate in the process, and I vote my conscience. But what is true is that I am not in any way drawn to the political world." He turned down opportunities to chair the Nixon reelection campaign in Ohio in 1972 and to run as a Republican against Democratic U.S. Senator John Glenn in 1980. In terms of American political traditions, Armstrong always identified most strongly with the moderate roots of Jeffersonian Republicanism. "I tend to be more in favor of the states retaining their powers unless it's something only the federal government can do and it's in everyone's interest. I'm not persuaded that either of our current political parties is very right on the education issue. But it's not politic to express those views to anyone today. So I don't."

CHAPTER 28

To Engineer Is Human

"I am, and ever will be, a white-socks, pocket-protector nerdy engineer. And I take substantial pride in the accomplishments of my profession." So Armstrong declared in his February 2000 address to the National Press Club honoring the Top 20 Engineering Achievements of the Twentieth Century as determined by the National Academy of Engineering, an organization he had been elected to in 1978. Neil went on to say: "Science is about what is; engineering is about what can be."

Spaceflight ranked only twelfth on the NAE list. In regard to pure engineering achievement, however, Armstrong regarded human spaceflight as one of the greatest achievements of the century, if not *the* greatest.

Armstrong never lost touch with the U.S. space program. Back in April 1970, just as he was transferring from the astronaut corps to the aeronautics office, the Apollo 13 accident occurred. Halfway to the Moon, an oxygen tank exploded in the service module, causing another tank to leak as well. Commander Jim Lovell ordered his crew—Fred Haise and Jack Swigert—into the LM, where the three astronauts rationed the module's limited supply of oxygen and electricity long enough to slingshot back around the Moon and return home safely. The Apollo program could only continue when and if NASA discovered the cause of the accident.

NASA asked Armstrong to serve on its internal investigation board under Dr. Edgar M. Cortright, the director of NASA's Langley Research Center. Neil aided F. B. Smith, NASA's assistant administrator for university affairs, with production of a detailed and accurate chronology of pertinent events from a review of telemetry records, air-to-ground communications transcripts, and crew and control center observations, as well as the flight plan and crew checklists. After nearly two months of investigation, Cortright's Apollo 13 review board released its report on June 15, 1970. Typical of so many technological accidents, what happened to the spacecraft "was not the result of a chance malfunction in a statistical sense but, rather, it was the result of an unusual combination of mistakes coupled with a somewhat deficient and unforgiving system," a complicated label for what Chris Kraft has called "a stupid and preventable accident." Tank manufacturer Beech Aircraft Co. was supposed to have replaced a 28-volt thermostat switch that heated up the liquid oxygen with a 65-volt switch, but had failed to do it. The Apollo Program Office was not diligent in cross-checking its own orders, so it overlooked the omission.

One of the most questionable conclusions of Cortright's panel was its recommendation that the service module's entire tank needed to be redesigned—at a cost of $40 million. A number of Apollo managers thought the costly change was unnecessary since Apollo 13's problem pertained not to the tank but the thermostat. In the following weeks, Kraft and Cortright fought it out all the way through NASA Headquarters. "After taking up the new job," Armstrong recalled, "I was released from active involvement in the Apollo 13 investigation," or he might have taken some active role in supporting Kraft's position.

Naturally, the public valued Armstrong's thoughts on U.S. space exploration, present and future, and he was often quoted on the topic. In an age when environmental concerns were growing more urgent, Armstrong spoke thoughtfully about the subject, usually from the perspective of his lunar exploration: "To stand on the surface of the Moon and look at the Earth high overhead leaves an impression not easily forgotten. Although our blue planet is very beautiful, it is very remote and apparently very small. You might suspect in such a situation, the observer might dismiss the Earth as relatively unimportant. However, exactly the opposite conclusion has been reached by each of the individuals who has had the opportunity to share that view. We have all been struck by the similarity to an oasis or island. More importantly, it is the only island that we know is a suitable home for man. The very success of the human species over eons of time now threatens our extinction. It is the drive that made for success that must now be curbed, redirected or released by expansion into a new world ecology. If we can find people skillful enough to reach the Moon, we sure can find people to solve our environmental problems."

Armstrong's typical reserve earned him the moniker "The Lunar Lindbergh" from some disgruntled members of the press. "Armstrong Has No Comment for Last Shot" one frustrated reporter complained in the buildup to Apollo 17 in December 1972, the last flight in the manned lunar program. Armstrong's special secretary at the university, Ruta Bankovikis, stated: "Mr. Armstrong does not wish to speak to reporters. He does not give exclusives. He does not give out interviews. It would be indiscreet of me to tell you where he is staying at Cape Kennedy while he watches the Moonshot."

Armstrong's resolute unwillingness to play any direct public role other than that of his own choosing proved especially frustrating to proponents of the U.S. space program, including some of his fellow astronauts. Jim Lovell has said: "Sometimes I chastise Neil for being too Lindbergh-like. I tell him, 'Neil, Charles Lindbergh flew across the Atlantic on private funds and had a private group build his airplane and everything else, so he had all the right to be as reclusive as he wanted. But you went to the Moon on public funds. The public taxpayers paid for your trip and gave you all that opportunity and fame, and there is a certain amount of return that is due them.' And Neil's answer to that was, 'I'd be harassed all the time if I weren't reclusive.' And he's probably right."

Yet, Armstrong did regularly speak out. "I've given a large number of press conferences. When I've visited other countries, I've usually given them. Every Apollo anniversary we hold press conferences. I feel no obligation to have a press conference just for the purpose of creating feature material which is not newsworthy; it's just human interest. I don't feel that that's required and, consequently, I try to avoid those kinds of situations.

"I've had some bad experiences with individual interviews where the journalist wasn't honest about what he was after. Once they report something erroneously, there is not much you can do that's effective in correcting it. So, a long time ago, I concluded that I just would not do individual interviews with journalists. They would be restricted to press conferences, because when there is a number of journalists present who are all hearing the same thing, they are much less inclined to tell it differently than they heard it."

Snapshots of Armstrong frequently appeared in the social section of Cincinnati newspapers on those occasions that he attended charity balls and other civic functions. Appearing semiregularly were feature stories and personality profiles written specifically about him or about him as part of the select group of men who had walked on the Moon, though he only rarely agreed to be interviewed for the piece.

In November 1978, at his Lebanon farm, Neil severed the ring finger of his left hand when his wedding ring caught on a door while jumping off the back of a truck. The injury and successful emergency microsurgery by a special team at the Jewish Hospital in Louisville, Kentucky (Neil was airlifted there), inspired a fresh spate of headlines. Postsurgery, the finger became fully functional except the uppermost joint.

One notable activity that Neil kept out of the press was his trip to the North Pole, which he made in April 1985 under the direction of the professional expedition leader and adventurer, California's Michael Chalmer Dunn, and in the company of the world-famous climber of Mount Everest, Sir Edmund Hilary, Hilary's son Peter, and Pat Morrow, the first Canadian to reach Everest's summit. "I found the trip to the North Pole tremendously interesting," Armstrong recalled, "predominately because it was so different from everything we normally see in our usual life. It's so very different up there. It was well worth the troubles of the trip."

A month before he journeyed to the North Pole with the Hilary expedition, Neil had become a member of a fourteen-member commission named by President Ronald Reagan to "devise an aggressive civilian space agenda to carry America into the twenty-first century." Chaired by former NASA Administrator Dr. Thomas O. Paine, Neil's colleagues in the task included U.N.

Ambassador Jeane J. Kirkpatrick, astronaut Dr. Katherine Sullivan, and space futurist Dr. Gerard K. O'Neill. According to Neil, "We worked off and on for several months, collected a lot of information from all kinds of different sources, had meetings and presentations, and then tried to develop a long-range plan for the nation's future in space."

However, the commission's recommendations were largely ignored due to the tragic events of January 28, 1986, when the Space Shuttle *Challenger* disintegrated, killing commander Dick Scobee and pilot Mike Smith along with three mission specialists: flight test engineer Ellison Onizuka, the first American of Asian descent to fly in space; physicist Ron McNair, the second black American in space; and electrical engineer Judy Resnik, the second American woman in space. Dying with them were two payload specialists, Gregory Jarvis, a designer of satellites, and Christa McAuliffe, a high school social studies teacher from Concord, New Hampshire, who had been selected from a list of over eleven thousand applicants to be the first teacher in space. With the deaths of the *Challenger* 7, as they came to be known, representing as they did a microcosm of American society, the U.S. space program entered a deep and prolonged period of crisis and depression.

At Reagan's request, Armstrong joined the Presidential Commission on the Space Shuttle *Challenger* Accident. Reagan and former secretary of state William P. Rogers, who had agreed to chair the commission, both wanted Neil to serve as its vice chair. "The morning after the accident, I received information that the White House was trying to get in touch with me. I called the switchboard and after talking to one of the president's staff, I was put on the line with Mr. Reagan. It is very difficult to turn down a president. Our job was to get the report to the president in four months—a hundred and twenty days—from the time he gave us the job."

The swearing in of the thirteen panel members took place in Washington on February 6. Going in, Armstrong was privately concerned that the *Challenger* investigation would be conducted by an outside body and not by NASA, as with the Apollo 1 fire or the Apollo 13 accident. "As it happened, the hardcore investigators were out there doing their work anyway, and they weren't so much encumbered with having to deal with public hearings and other affairs that the commissioners were stuck with. So perhaps in the long run, being public didn't affect the investigation's timetable very much."

Rogers's rationale for running the commission in a very public way was persuasive even to Armstrong. "At the start Bill talked to all of the commissioners about what his expectations were and some of the things he thought were very important. He thought it was very important, for example, for the commissioners to be aware of how public opinion was being expressed through the media. So he encouraged everyone to read the *Washington Post* and the *New York Times* every morning, stuff that I certainly wouldn't have thought of or encouraged. He understood that side of the equation.

"He was of a firm opinion—and I certainly agreed with this—that there ought to be one investigation, and that we had to find ways to placate the other constituencies out there that would like to be doing our job—or at least would like to be catching some of the limelight from it. So Bill was busy early on going over and talking with committee chairmen in the House and the Senate.

"So our compromise was that we, the commission, would report to Congress periodically. We would go over up to the Hill and testify to the progress of the investigation and what some of its difficult points were, some of the items we were making progress on, and what our outlooks were at that point. Then the legislators would get a certain amount of media coverage on what they were thinking and doing in Congress, but without it really affecting what was going on in the investigation.

"We had a lot more public hearings than I had ever been exposed to in any of the accident investigations that I had participated in before. So that was a new wrinkle for me. The fact that the investigation panel was a public entity had pluses and minuses. The pluses were that it did give us an opportunity to give a status report to the public at large as to what was happening, but it was also an opportunity for some people to play to the cameras."

As vice chair of the commission, Armstrong sat ex officio on all subcommittees. "I probably spent the most time on the accident itself, because my feeling was that, if we didn't get the accident pinned down precisely, then all the rest was for naught. So I wanted to be the closest to that." Each committee chair decided which matters they would look into, and each set up their own schedule of hearings and presentations as well as site visits where they examined hardware to understand how things worked. "From the Justice Department, we borrowed a system for keeping track of all the data and documents and filing them properly so that we could retrieve anything that we wanted at any time. So everything was both written down and computerized, which was good since the investigation generated almost 6,300 documents totaling more than 122,000 pages, as well as almost 12,000 pages of investigative and 2,800 pages of hearings transcript," Neil added. Armstrong did some of his own private investigating, seeking information or insights from some of the private contacts he had developed over thirty years in NASA and the aerospace industry. "On occasion I would talk to people privately. We had no such prohibitions from our chairman, so I didn't hesitate to do that.

"I think we pretty much would have made the same conclusions without the public hearings. As for whether the investigation would have concluded any faster without them, because proof of our hypothesis [that 'the cause of the *Challenger* accident was the failure of the pressure seal in the aft field joint of the right solid rocket motor . . . due to a faulty design unacceptably sensitive to a number of factors' including cold temperature] only came when we finally got that last piece of debris off the bottom of the ocean, the end result couldn't really be hurried up."

Armstrong was pleased with the commission's final conclusions and recommendations. "I think the conclusions and findings were right on, and I think our descriptions of how the accident occurred were very close to precisely correct. There have been only a few contrary opinions or hypotheses, but none of them have stood the test of time." Neil played the key role in laying out the basis for the manner of thinking that went into the commission's final report. "I made the case to fellow commissioners that the effectiveness of our recommendations was going to be inversely proportional to their number. The fewer, the better. Second, let's make sure that we don't tell NASA to do something it can't do." The panel came up with sixty or so recommendations, which were then reduced to nine.

As for Richard Feynman's famous Minority Report to the Space Shuttle Challenger Inquiry, contrary to stories that the commission tried hard to suppress its publication (because it was allegedly "anti-NASA"), Armstrong was okay with the colorful physicist expressing his unique take on the subject and attaching it to the commission's final report as an appendix, as long as Chairman Rogers was. Armstrong knew the truth of what Richard Feynman wrote at the end of his Minority Report, because he had been living and breathing it ever since he first took flight in an airplane forty years earlier: "For a successful technology, reality must take precedence over public relations, for nature cannot be fooled."

On Saturday morning, February 1, 2003, a morning phone call from a friend drove Armstrong to the television in his study. Another Space Shuttle was lost. Just minutes before its scheduled landing at the Cape, STS-107, *Columbia,* had come to pieces over Texas, high in the Earth's atmosphere, following a sixteen-day mission.

As soon as Neil heard reports that debris was being found, "I knew at that point the vehicle was lost. There was no chance." Another tragic loss of a crew of astronauts: Rick Husband, the commander; Willie McCool, the pilot; five mission specialists: Kalpana Chawla, Laurel Clark, Mike Anderson, David Brown, and Ilan Ramon, payload specialist.

The height at which the *Columbia* had been ripped apart was a little over thirty-nine miles. Neil could not help but ponder the irony that the Shuttle had broken up at almost exactly the same height as his highest flight in the X-15—207,500 feet. This time around, the investigation would be handled very differently—more internally within NASA. The Bush White House telephoned to ask if Neil and Carol, Neil's second wife, would attend the memorial service for the *Columbia* crew, scheduled for Monday, February 3, at the Johnson Space Center in Houston, which they quickly agreed to do. Neil did provide a statement to the press: "The *Columbia* disaster saddened everyone and reminded us that there is no progress without risk. Our charge is to maximize the former while minimizing the latter. So long as humans have an independent, creative, and curious mind, we will continue to challenge the frontiers."

In January 2004, President George W. Bush announced a "new vision" for the U.S. space program. The president proposed a commitment to a long-term human and robotic program to explore the solar system, starting with a return to the Moon that, in the view of the White House, "will ultimately enable future exploration of Mars and other destinations." Two months later, Armstrong, in Houston to accept the Rotary National Award for Space Achievement, lent his support to the Bush plan. There would be many critics of the Bush plan both inside and outside the space community; Neil's philosophy was to favor anything that moved the technology forward.

Neil Armstrong never considered himself an *explorer* per se: "What I attended to was the progressive development of flight machinery. My exploration came totally as a by-product of that. I flew to the Moon not so much to go there, but as part of developing the systems that would allow it to happen."

CHAPTER 29

Dark Side of the Moon

Not surprisingly, one of Armstrong's boyhood heroes was Charles A. Lindbergh. Neil first met Lindbergh along with his wife, Anne, at the launch of Apollo 8. "I was given the job of helping with touring him around and taking him and showing him the facilities. The night before the launch, I took him out to look at the Saturn V; it was all illuminated with the xenon lights. As Frank Borman's backup, I couldn't spend more than just a little time with him."

Privately, Neil had the chance to talk with Lindbergh "several times" following Apollo 11. "We both went to the Society of Experimental Test Pilots meeting in Los Angeles in late September 1969. He was being inducted as an honorary fellow, and we were seated next to each other at the banquet." The two fliers also corresponded, as Neil would later do with Anne Morrow before and after Neil came to cochair the Lindbergh Memorial Fund. Lindbergh posed to Neil the rhetorical question: "I wonder if you felt on the Moon's surface as I did after landing at Paris in 1927—that I would like to have had more chance to look around."

At the SETP banquet in September 1969, Lindbergh had offered Neil one piece of advice: "He told me never to sign autographs. Unfortunately, I didn't take his advice for thirty years, and I probably should have."

The fan mail came in like a tsunami. For several months following the return of Apollo 11, Neil was getting ten thousand pieces of mail per day. Calculated at five days a week for just the first six months following the Moon landing, that came to three hundred thousand items. The mail came in the form of letters, cards, telegrams, presents, and more. During the same stretch of time, Neil, Mike, and Buzz made their forty-five-day trip world tour to twenty-three countries; Neil made a three-week Bob Hope USO tour; and he also made a ten-day trip to the Soviet Union. Waiting for him in Houston was nearly a third of a million cards and letters that he was now supposed to answer, with thousands more to come.

NASA did its best to help while he was still employed by the government, assigning four clerks to help him deal with his mountain of mail. But NASA could not keep up, as one of its public affairs officers in Houston explained to a complainant whose gift he had mailed to Neil did not receive a prompt reply. "We regret very much that you have been caused any unhappiness by what seemed to you a lack of proper handling of your gift to Mr. Armstrong. At the time of the lunar landing, the Astronaut Office was not prepared for the veritable avalanche of mail and gifts that flooded the office from all over the world. . . . The sheer volume posed logistic, administrative, and clerical problems. Correspondence had to be answered; gifts had to be stored safely and recorded and acknowledged—all in addition to the normal workload of the office and without additional personnel. . . . The astronauts themselves, when time permitted, helped in writing the cards of acknowledgment; and it is quite possible that the handwritten card you received was written by Mr. Armstrong himself. I am taking the liberty of sending your letter to Mr. Armstrong, because if he knew how you felt, he would want to write to you directly and thank you in a more appropriate manner for your gift. He is that kind of person."

During the eight years he was at the University of Cincinnati, most of Armstrong's fan mail came through the campus post office, his only well-known address. When he left the university, where the regular assignment of two secretaries was to help answer his mail, Neil soon realized that handling his mail on his own was an impossible burden. In February 1980, he rented a small office in Lebanon, Ohio, and hired an administrative aide. Vivian White worked full-time for him for about ten years; after that, she "cut back" to four and a half days a week.

"For the first twelve to fifteen years, he would sign anything he was asked to sign, except a first-day cover. Then about 1993, he realized that his autographs were being sold over the Internet. Many of the signatures, he found, were forgeries. So he just quit signing. Still we get letters saying, 'I know Mr. Armstrong doesn't sign anymore, but would you ask him to make an exception for me?' "

After 1993, form letters under Vivian's signature went out in answer to 99 percent of the requests. In the few instances that Armstrong accepted the invitation, he composed and signed a personal letter. If he chose to answer someone's technical question, according to White, he would "write out his answer, I'd type it up and then put underneath it, 'Mr. Armstrong asked me to give you the following information,' and I'd sign it. We never answered personal questions. They were just too much an invasion of privacy." In Vivian's filing system, they would go into "File 11," the waste basket.

On the outbound journey to the Sea of Tranquility from aboard *Columbia,* Neil had made sure to pass along a "hello to all my fellow Scouts at Farragut State Park in Idaho having a National Jamboree there this week; Apollo 11 would like to send them best wishes." For several years thereafter he took the time to write letters congratulating boys who had achieved the ultimate rank of Eagle Scout. Once his address was posted on the Internet and he was deluged with requests (some 950 letters asking for congratulatory letters to new Eagle Scouts in the first five months of 2003 alone), he could no longer write personal letters to Scouts.

Armstrong's belated decision to follow Charles Lindbergh's advice provoked disappointment and even antagonism, mostly from profiteer, or more commonly, hobbyist "collectors" of autographs and space memorabilia. Without question, Armstrong's signature remains by far the most popularly sought-after astronaut autograph. Items with his signature can quickly sell today at auctions or on the Internet for $10,000 and more. Neil Armstrong forgeries far outnumber the authentic examples, with one estimate placing the fake Armstrong signatures as high as 90 percent of the eBay catalog.

On the third anniversary of Apollo 11 in July 1972, the Neil Armstrong Air and Space Museum opened in Wapakoneta. The pride of Ohio governor James Rhodes, the facility began with half a million dollars earmarked by the state legislature even before the Apollo 11 mission was over. Its exterior designed to resemble a rising full Moon, the museum's grand opening featured an appearance by the twenty-six-year-old presidential daughter Tricia Nixon, who said, "Because of what you, Neil, have done, the heavens have become a part of our world." Before a crowd of five thousand, Tricia then presented the museum with one of Apollo 11's Moon rocks: "It is a rock which symbolizes mankind's ability for great achievement to build a better America and a better world."

Armstrong put on a relatively happy face for the crowd that day, many of them old friends and neighbors, but he was not at all happy with how the entire museum project had come together: "I should have been asked. The policy I followed from the start had been that I neither encouraged nor prohibited the use of my name on public buildings, but I did not approve their use on any commercial or other nonpublic facility. If the organizing committee had asked me, I'm sure I would have said okay, because it was in the town where my parents lived. Nevertheless, I would have been happier had they not used my name or, if they used my name, they would have used a different approach for the museum. I did try to support them in any way that I could by presenting them with such materials as I had available, either gifting or loaning items. From the outset I was uncomfortable because that museum was built as the 'Neil Armstrong Museum.' A number of people came to believe that it was my personal property and a business undertaking of mine. The Ohio Historical Society in Columbus was actually going to be overseeing the museum, and I told its director that I felt uncomfortable. I asked him as well as another member of the planning board if there was anything that could be done about the public image issue and to respond to me about what they thought. They said they would, but they did not."

Armstrong's relationship with the museum leadership remained strained throughout the forty years until his death in 2012. In the mid-1990s, for example, came the issue of a picture postcard of Neil as an astronaut on sale in the museum gift shop. The image came from an official NASA photograph, taken when he was a federal government employee. For him it was a question of ownership. The rights to the picture belonged to the people, the same visitors, Neil believed, who "think I own the place." The seal of the Ohio Historical Society was displayed inside the main door, but according to Neil, "it's so low profile that most people don't notice it." Eventually, Armstrong relented on the matter of the picture, granting then museum director John Zwez "my permission on a limited-time basis."

As for the namesake Wapakoneta airport, "Again, they just didn't ask. It's a public airport so, had they asked, I probably would have said sure, okay. The problem is that there were businesses on that airport that took the name of the airport, like the 'Neil Armstrong Electronics Shop.' "

In the 1990s, Armstrong had a run-in with Hallmark, the greeting card company.

"The Hallmark case was simple," Neil related. "They put out a Christmas tree ornament. It had a little spaceman inside it. It also had a recording that played my voice, and it had my name on the box." Hallmark advertised the product by saying, "The Moon glows as the famous words spoken by Neil Armstrong when he stepped out on the Moon and into history." Unfortunately, Hallmark's people had not received or even asked his permission. Nor did the popular card company follow NASA's established procedures for such matters. So, in 1994, Neil sued Hallmark. Wendy Armstrong, the wife of his son Mark, served as his attorney. At the end of 1995, the two parties settled out of court: "Hallmark Cards announced today that it had settled a lawsuit with Apollo 11 astronaut Neil Armstrong over the use of his likeness in a Christmas ornament last year. Armstrong had claimed that his name and likeness was used without his permission on the ornament, which celebrated the twenty-fifth anniversary of the Apollo 11 landing. The size of the settlement was undisclosed, but said by one source to be substantial. Armstrong plans to donate the settlement from the Kansas City–based company, minus legal fees, to Purdue University, his alma mater." Purdue later confirmed that it received the money.

Neil felt that "NASA hadn't been very careful about the matter, either. Up to then, it had been pretty careless in the treatment of individual rights. Now, I get letters that correctly state what NASA's position is about getting my approval, and before that I never did. I get many such requests, some of which I've granted [some without charge and some for a fee] and some of which I haven't.

"In many cases where they are either nonprofit or government public-service announcements, I will approve them. At first, I wasn't very careful about keeping records of this and would just say, 'Yes, that's all right.' Then, after being exposed to the legal world, I recognized that you have to have all kinds of files of proof."

An even more loathsome legal matter concerned the sale of some of Neil's hair. In early 2005, the Lebanon, Ohio, barbershop that Neil had patronized for more than twenty years sold some of its famous client's locks for $3,000 to a Connecticut man who, according to Guinness World Records, had amassed the largest collection of hair from "historical celebrities." In a private conversation in the back of the shop, Neil asked his barber to either return the hair or donate the $3,000 to a charity of Armstrong's choosing. When neither result followed, Neil's attorney sent the barber a two-page letter, one that referenced an Ohio law protecting the names of its celebrities. Instead of settling the matter quietly, the barber sent the letter to local media. The strange story attracted international attention.

Armstrong also found himself innocently immersed in religious controversy, none of it of his own making. Many religious groups wished to connect the narratives of their belief system to the exploration of space; some critics of the Apollo program even asserted that walking on the heavenly body that was the Moon was a "godless" act. Buzz Aldrin was rumored to have been a Freemason, and there was even a rumor that Neil Armstrong converted to Islam after hearing a voice singing in Arabic while walking on the Sea of Tranquility. Only later, after returning to Earth, did Armstrong realize that what he heard on the lunar surface was the *adhan*, the Muslim call to prayer. Neil then allegedly converted to Islam, moved to Lebanon (the country in the Middle East, not Lebanon, Ohio), and subsequently visited several Muslim holy places, including the Turkish masjid where Malcolm X once prayed.

The story of Armstrong's conversion grew so far and wide by the early 1980s that, not only Armstrong himself, but also an official body of the United States government, found it necessary to respond. In March 1983, the U.S. State Department sent the following message to all embassies and consulates in the Islamic world denying that idea:

1. *Former astronaut Neil Armstrong, now in private life, has been the subject of press reports in Egypt, Malaysia and Indonesia (and perhaps elsewhere) alleging his conversion to Islam during his landing on the Moon in 1969. As a result of such reports, Armstrong has received communications from individuals and religious organizations, and a feeler from at least one government, about his possible participation in Islamic activities.*
2. *While stressing his strong desire not to offend anyone or show disrespect for any religion, Armstrong has advised department that reports of his conversion to Islam are inaccurate.*
3. *If post receive queries on this matter, Armstrong requests that they politely but firmly inform querying party that he has not converted to Islam and has no current plans or desire to travel overseas to participate in Islamic religious activities.*

Whatever help the State Department might have been in clarifying Armstrong's views, it wasn't enough. Requests for him to appear in Muslim countries and at Islamic events became so frequent in the mid-1980s that Neil felt compelled to act. "We were getting such a barrage of information, just inundated with questions about this, predominately from the Islamic world but also from the non-Muslim world, the latter of which was saying, 'This can't be true, can it?' Finally we decided that we needed to have something official that journalists could refer to. We again used the State Department, this time to assist in setting up." Eventually Neil set up a telephone press conference to Cairo, Egypt, where a substantial number of journalists from the Middle East could attend and "be told that there was no truth to this persistent rumor and be able to ask me questions and get my response." "Just how much that helped is impossible to know, but it certainly didn't completely stem the questions." Some clung to the notion that the U.S. government didn't want their great American hero to be known as a Muslim, and thus was somehow forcing him publicly to deny his faith.

Subsequently the story even got embellished to include the assertion that Apollo 11 discovered that the Earth emitted radiation (which it does) and that the source of the radiation came from the Kaaba in Mecca, proving that Mecca is "the center of the world." In the last years of Neil's life, Vivian White tried hard to set the record straight with a form letter that states, "The reports of his conversion to Islam and of hearing the voice of the *adhan* on the Moon and elsewhere are all untrue." Yet even today, an Internet search of "Neil Armstrong + Islam" results in 573,000 hits.

Armstrong understood why such projections were made onto him. "I have found that many organizations claim me as a member, for which I am not a member, and a lot of different families—Armstrong families and others—make connections, many of which don't exist. So many people identify with the success of Apollo. The claim about my becoming a Muslim is just an extreme version of people inevitably telling me they know somebody whom I might know."

Armstrong, because he was so hard to know, turned out to be myth personified, an enigma prime to be filled with meaning.

Back in the 1970s, *Chariots of the Gods?* (1969) author Erich von Däniken had tried to turn Armstrong into a collaborator on his sensational (and best-selling) theory of "ancient astronauts," extraterrestrial beings who had visited Earth in the remote past and left various archaeological traces of their civilization-building activities. In August 1976, Armstrong had accompanied a Scottish regiment, Black Watch and the Royal Highland Fusiliers, on a scientific expedition into the vast Cueva de los Tayos ("Caves of the Oil Birds") in a remote part of Ecuador first discovered by the Argentinian Juan Moricz. At the time, Neil was unaware that in *The Gold of the Gods*, Däniken's 1972 follow-up to *Chariots of the Gods?*, the controversial Swiss author had described his own exploration of the Cueva de los Tayos, in which he claimed to have found considerable archaeological evidence of an extraterrestrial presence, including that certain doorways in the cave were too square to have been made naturally. "But it was the conclusion of our expedition group," declared Neil, "that they were natural formations."

Newspaper reports of the Los Tayos expedition and Armstrong's role in it made it clear that Däniken's claims about the caves were false. In a two-page letter written to Neil from his home in Zürich, Switzerland, on February 18, 1977, Däniken told the world's most famous astronaut that Armstrong's "expedition cannot possibly have been to my cave." Däniken urged Armstrong "to participate in a cave expedition which I am presently planning" whereby "relics from an extraterrestrial civilization—will be inspected." Armstrong responded politely: "Because of my Scottish ancestry, and the fact that the U.K. side of this project was largely Scottish, I was invited to act as honorary chairman of the expedition, and I accepted. . . . I had not read your books and did not know of any connection that you might have had with the caves. I made no statements regarding any hypotheses you may have put forth. . . . I appreciate your kind invitation to join you in your forthcoming expedition, but am unable to accept."

• • •

What of "Mr. Gorsky"?

Just before reentering the LM after Apollo 11's EVA, Armstrong supposedly made the enigmatic remark, "Good luck, Mr. Gorsky." Some reporters at Mission Control attributed the remark as referencing a rival Soviet cosmonaut. However, there was no Gorsky in the Russian space program. Over the years many people questioned Armstrong as to what his statement about Mr. Gorsky meant, but Armstrong always just smiled. The story resumed in 1995 during an address in Tampa, Florida, when Armstrong finally responded to a reporter's question about the story. Mr. Gorsky had finally died, so Neil felt he could answer the question. When he was a kid, he was playing baseball with a friend in the backyard. His friend hit a fly ball that landed in front of his neighbor's bedroom window. His neighbors were Mr. and Mrs. Gorsky. As he leaned down to pick up the ball, young Armstrong heard Mrs. Gorsky shouting at Mr. Gorsky, "Oral sex! You want oral sex?! You'll get oral sex when the kid next door walks on the Moon!"

As a story, "Mr. Gorsky" always gets a laugh, which was what comedian Buddy Hackett was counting on when he first delivered the joke (which Hackett apparently invented) on NBC's *Tonight Show* sometime around 1990. In spite of the ease with which the story can be debunked, and in spite of various attempts on the Internet (a search for "Armstrong" and "Gorsky" generates 558,000 hits) to expose it for the urban legend that it has become, the story is funny enough that countless people continue to read it and pass it along, no matter its origin. "There is absolutely no truth to it. I even heard Hackett tell the story at a charity golf outing."

• • •

Even during the time of Apollo 11, some believed that the Moon landings never really took place—that they were a fraud foisted upon the world for political reasons by the U.S. government. The Flat Earth Society maintained an active membership. But the idea of a Moon hoax picked up greatly in 1977 because of *Capricorn One*, a Hollywood conspiracy fantasy, not about the Moon landing, but about the first manned mission to Mars. In the tale, NASA attempted to cover for a highly defective spacecraft by forcing its astronauts before cameras in a desert film studio to act out the journey and trick the world into believing they made the trip. Though a mediocre movie, *Capricorn One*'s notion of a government conspiracy never fell out of favor with a small number of skeptics.

Inevitably, there were people who not only chose to believe in some version of the lunar conspiracy theory, but who saw a way to profit from it. In 1999, Fox TV broadcast a "documentary" entitled *Conspiracy Theory: Did We Land on the Moon?* The program was based largely on a low-budget commercial video produced by a self-proclaimed "investigative reporter" from Nashville, Tennessee. Called *A Funny Thing Happened on the Way to the Moon*, it speculated that the Moon landings were an ingenious ploy of the U.S. government to win the Cold War and stimulate the collapse of Soviet communism by forcing the Kremlin into investing massive sums of money on its own lunar program, thereby ruining the Russian economy and provoking the internal downfall of the government.

No matter that every piece of "evidence" raised by the sensationalistic program was parroting the same uninformed arguments about Apollo that had been around for over two decades—i.e., that the American flag planted by Apollo 11 appears to be waving in a place where there can be no wind; that there are no stars in any of the photographs taken on the lunar surface; that the photographs taken by the Apollo astronauts are simply "too good" to be true; that the 200-degree-plus Moon surface temperatures would have baked the camera film; that the force of the LM's descent engine should have created a crater under the module; that no one can travel safely through the "killer radiation" of the Van Allen Belts; and more. Some members of the TV viewing audience succumbed to the trickery, others to its darker legacy.

When Armstrong answered the conspiracy theories in writing, he usually did so through his secretary, Vivian White, with his statement but her signature. Neil's explanation was direct and logical, as an engineer would handle it: "The flights are undisputed in the scientific and technical worlds. All of the reputable scientific societies affirm the flights and their results. The crews were observed to enter their spacecraft in Florida and observed to be recovered in the Pacific Ocean. The flights were tracked by radars in a number of countries throughout their flight to the Moon and return. The crew sent television pictures of the voyage including flying over the lunar landscape and on the surface, pictures of lunar scenes previously unknown and now confirmed. The crews returned samples from the lunar surface including some minerals never found on Earth." Vivian would add that "Mr. Armstrong believes that the only thing more difficult to achieve than the lunar flights would be to successfully fake them."

"People love conspiracy theories," Neil told this author. "They are very attracted to them. As I recall, after Franklin D. Roosevelt died, there were people saying that he was still alive someplace. And, of course, 'Elvis lives!' There is always going to be that fringe element on every subject, and I put this in that category. It doesn't bother me. It will all pass in time. Generally, it's almost unnoticeable except for the peaks that occur when somebody writes a book or puts out an article in a magazine or shows something on television." Sadly, the time for belief in this conspiracy has not yet passed. A poll published by a British national newspaper in 2016 reported that "52% of Brits don't believe it really happened."

Over the years Armstrong experienced his share of "crazies"—amid his private papers (now in the possession of the Purdue University Archives) rest several folders of cards and letters that Neil labeled "Quacks." Most of these individuals were harmless, but some were downright bothersome, even scary. Occasionally, Neil and his family even needed to call in police to assess the potential threat.

The greatest and most frequent nuisance came from the man who made the video "A Funny Thing Happened on the Way to the Moon." The pest showed up with a video-camera-carrying assistant on several occasions, including at the annual meeting of EDO Corporation stockholders in New York City in 2001. EDO president James Smith recalled the scene: "This guy shows up with a Bible and shouts out, 'Neil Armstrong, will you swear on this Bible that you went to the Moon?' Well, the audience immediately started booing the intruder very loud, but he went right on, 'Everybody else in the world knows you didn't, so why don't you just admit it?!' It quickly turned into a kind of pushy-shovy thing, so I and a few other men got the guy out of there.

Subsequent to that, we never had a meeting where we didn't hire special security."

"Had I the opportunity to run that episode over in my life," Armstrong commented, "I wouldn't have allowed my company people to usher me out of the room. I would have just talked to the crowd and said, 'This person believes that the United States government has committed fraud on all of you, and simultaneously he wants to exercise his right protected by the U.S. government to state his opinions freely to you.' "

A few months after the EDO meeting, on September 9, 2002, the same man with Bible in hand confronted Buzz Aldrin outside of a Beverly Hills hotel. A resident of the Los Angeles area, Buzz had arrived at the hotel thinking he was to be interviewed by a Japanese educational television network. At first Aldrin, his stepdaughter in tow, tried to answer the man's questions, then did his best to get away from him. But the insistent independent filmmaker dogged him out of the hotel and kept directing his assistant to keep the camera running, while shouting at Buzz, "You are a coward and a liar." Harassed to the point of complete exasperation, the seventy-two-year-old Aldrin, all 160 pounds of him, decked the thirty-seven-year-old 250-pounder with a quick left hook to the jaw. The man from Nashville filed a police report but, after watching the accuser's own tape of the incident, the L.A. County District Attorney rather forcefully declined to file charges. As the self-proclaimed "victim" later told reporters, "If I walked on the Moon and some guy said swear on a Bible, I'd swear on a stack of Bibles."

Even before the EDO and Aldrin incidents, the same individual entered uninvited into the Armstrongs' suburban Cincinnati home. Neil's second wife, Carol, related what happened: "Neil was at the office. This guy knocked at the door and there was a big dog with him, and he had a package. I opened the outside door while leaving the screen door shut, and the man said, 'Is Neil here?' I said, 'No, he's not. May I help you?' He opened the screen door and just walked in, bringing along his dog. He said, 'I want him to sign this,' and I said, 'Neil doesn't sign things anymore.' 'He'll sign this,' he uttered, and then he left. It sort of hit me three minutes later. All of a sudden I felt shaky." In the following weeks, the interloper started putting letters and other things in the Armstrongs' mailbox. Some of the materials had religious overtones and most were about the Moon landing being faked. The local police department responded, "It's probably nothing, but why don't you just bring the tapes and letters and we'll take a look at them," until a call to the ABC TV station in Nashville revealed that he had never worked there, but instead was an independent filmmaker who had operated a business called ABC Video. A few weeks later, Carol received a phone call from her neighbor: "Carol, there's this car parked out here and it's been out here for a long time." When the neighbor went out to investigate, she saw a lot of camera equipment in the backseat. The siege continued for three days, culminating in a car chase involving the Armstrongs, the intruder, and the police.

A final indication of the extraordinary iconography associated with Armstrong and his trip to the Moon came five years after his death, in the summer of 2017, when the sale of a small, empty cloth bag stained with a few particles of lunar dust and used by Neil on his EVA became "the most valuable space artifact to ever sell at auction." On Thursday, July 20, 2017, on the forty-eighth anniversary of Apollo 11's Moon landing, world-famous Sotheby's auction house sold the "lunar sample return" bag (12 by 8.5 inches) for $1.8 million (speculation had been that it might sell for as much as

$4 million) as part of its first auction dedicated entirely to the relics of the U.S. space program. (The sale took place at Sotheby's New York City gallery as well as on eBay.)

Sotheby's description of the item (Lot 102) read: "This [Apollo 11 lunar sample return decontamination bag] is indeed the rarest and most important space exploration artifact to ever be offered. A true first of firsts: an item used to protect the first lunar sample, collected by the first man on the moon, during the first lunar landing." In August 2015, the small bag with its zippered pouch, which was used as the "outer decontamination bag" to protect roughly five hundred grams of Moon dust and twelve 12 rock fragments (the "contingency sample") that Neil had collected soon after stepping onto the Sea of Tranquility, had been mistakenly sold in an online auction run by Gaston & Sheehan Auctioneers for the U.S. Marshals Service, which back in 2003 had found and taken the bag into custody while executing a search warrant inside the garage of Max Ary, the director of the Cosmosphere space museum in Hutchinson, Kansas. "How the sample bag ended up at Max Ary's home is not exactly known," reported a Space.com article in August 2016 while covering the news of two lawsuits then in litigation over who rightfully owned the bag. (While maintaining his innocence, Ary served two years in prison and forfeited the bag—which Ary claimed was part of his own legal personal space collection—to help pay $132,274 in restitution. Today, Ary is the director of the Stafford Air & Space Museum in Oklahoma.) The person who bought the bag in the August 2015 auction was an Illinois lawyer, Nancy Lee Carlson, who won the bag with a bid of just $995. Wanting to know exactly what she had, Carlson contacted Ryan Zeigler, the Apollo sample curator at

NASA's Johnson Space Center, whose testing not only identified the dust in the bag as authentic, but also as originating from Apollo 11, which hitherto had not been known. Realizing the historic importance of the contingency sample bag from humankind's first steps onto the Moon, NASA confiscated the bag and placed it under lock and key at Johnson Space Center—that was, until a federal court ruled that it was to be returned to what it deemed was its rightful owner, Nancy Lee Carlson, who ultimately arranged for it to be auctioned at Sotheby's.

Clearly, the high price of celebrity—even for a bag—pressed a heavy burden upon the unique historic legacy of the First Man on the Moon.

CHAPTER 30

Into the Heartland

The few puffy clouds over the ski slopes at Snowmass were a meek harbinger of the major blizzard sweeping toward Aspen's four snowcapped summits that February day in 1991. Neil, age sixty, rode to the top of the intermediate ski run known as Upper Hal's Hollow with Doris Solacoff, whose husband, Kotcho, was Armstrong's boyhood friend from Upper Sandusky. Neil's brother, Dean, recently divorced, completed the ski quartet, who had just finished lunch. Neil had eaten a big bowl of chili with plenty of onions.

Neil remained so quiet throughout the ascent that Dorie, a registered nurse, took notice. A few hundred feet into her run, she observed Neil skiing down ever so slowly. "I don't feel too well," he said. Noticing that his face was ashen, Dorie insisted on going for help. "No, just wait a second." Neil hesitated, knowing what sort of fuss would be made over him. "I feel real weak. I think I'm going to sit down and rest here for a minute."

Dorie raced to contact the ski patrol. "I have a friend that I believe is having a heart attack, and I'll tell you right where you need to come."

Down at the bottom of Upper Hal's Hollow, Kotcho and Dean had started to worry. Finally, Dorie approached, shouting, "Neil has had a heart attack, and the ski patrol is bringing him down in the rescue toboggan!"

The doctor on duty at the lodge infirmary confirmed a heart attack and administered atropine to stabilize a cardiac arrhythmia through an IV line. An ambulance transported him to Aspen Valley Hospital, where he was placed in the intensive care unit. There, Armstrong experienced repeated episodes of bradycardia, or abnormal slowing of heartbeat.

Armstrong's heart rate soon stabilized enough for a transfer to Denver, but the blizzard kept him in Aspen for three days. Practiced at protecting celebrities, the little resort hospital kept word of his heart attack secret.

Kotcho, an Ohio physician, helped arranged for a transport by medivac from Colorado to a hospital in Cincinnati. There, a team of heart specialists carried out a catheterization that linked the attack to a tiny aberrant blood vessel. The rest of his coronary arteries were clear of blockages; his heart tissue sustained only the slightest amount of permanent damage.

Released the next day with no major restrictions, Armstrong took the heart specialist at his word and flew to a business meeting. Six months later, he passed his flight physical and was put back on full flight status.

In the coming years, he would make many more visits to the Colorado ski slopes, once or twice with Kotcho, Dorie, and Dean.

The day Neil had his heart attack, he was in the process of separation from Janet. What role stress played in Armstrong's illness is unknown, but difficulties in his personal life had mounted during the previous year. His father, Stephen, died on February 3, 1990. His mother,

Viola, passed away barely three months later. Both age 83, his parents had been married for 60 years. Shortly before his mother's death, Janet had left Neil, citing years of emotional distance.

With Neil's departure from NASA in 1971, Janet Armstrong had hoped for a new beginning in suburban Cincinnati. "My husband's job was there, so that was where we went. He wanted to realize a more quiet life," having "spent all those years in the program with little time for himself."

Lebanon was a rural bedroom community for Cincinnati and Dayton. "I had never lived in a small town. We went into the ice cream parlor and just kind of cased the place. It seemed like a safe community and a good place to raise the children."

The nineteenth-century farmhouse had to be gutted. "Neil did not like debt and wouldn't take out another loan, so it took seven years as we paid cash for the work to be done. It got so that the builder could answer the telephone if I wasn't there and go pick the kids up at school! He just became part of the family! It was difficult on the kids and it was difficult on me.

"It was easier for Mark than it was for Rick," but both boys were teased for being Neil Armstrong's son. According to Rick, "It was rough, but I learned to ignore it." In Rick's view, Mark had a little easier time of it: "He was much more of the social butterfly." Rick recalled farm life as "an isolation that, I think, was driven a lot by what Dad was experiencing, and it had a trickle-down effect on the rest of us." Janet (and undoubtedly Neil as well) did not know that the boys were having it so rough: "It took me a couple of years before I caught on to that, because the boys wouldn't say anything to me," Janet said later.

Neil did some chores around the three-hundred-plus-acre farm, if not as many as Janet would have liked. "We started by carrying between seventy and ninety head of cattle. We grew corn, soybeans, hay, and wheat." Asked whether she actually enjoyed doing the farm work, Janet replied, "It was something that had to be done. It was really difficult to shovel poop during the day and go out to a dinner party at night."

In 1981, a year after Neil resigned from UC, Neil and Janet became empty-nesters when Mark went to Stanford University (Rick had since graduated from Wittenberg College in Ohio). "I don't think it affected Neil at all, but it certainly affected me. I felt that this was a time when we could really do things together." As it turned out, however, Neil, with all of his new corporate board responsibilities, wasn't home any more than he had been before. "The kids have gone, Neil is gone, our dog, Wendy, had been stolen. We had no security system. I was stuck out there in the country. Finally, I got tired of all this, and in 1987, I started a travel agency. I sold that agency in 1993."

Janet's frustrations with Neil rose as her dissatisfaction with her own life increased. She tried in vain to help him get better organized. "He had so many requests for speeches and so many this and so many that—he didn't know where to start. He had to make decisions—and decision-making seemed to be especially difficult for him at that time.

"The man needed help. I couldn't help him. He really didn't want me helping him. He didn't want to get angry with me, I suppose, or he didn't want me to get angry at him. That was probably smart on his part. Vivian White [his administrative assistant] used to get just beside herself. She just learned to go with the flow."

Janet also tried to plan vacations for the two of them, but Neil couldn't commit—his schedule was always too busy. "I could not continue to live like that. He'd look at all sides of everything, and sometimes he'd discuss them—and I'd say, 'Just do it!' But he couldn't, or just didn't.

"In November 1987, I asked him to go skiing, but he couldn't work it into his schedule for another year." Finally, in late 1988, they made it out to the slopes at Park City, Utah, where Janet persuaded him it would be fun to have a vacation home. "He had free travel, the boys could come out, and we could have a place that was so convenient, and everybody liked to ski." In early 1989, they bought a brand-new chalet-style home on the outskirts of Park City, one of the sites for the 2002 Winter Olympics. It could have been a turning point in their marriage if the couple had chosen to approach it that way, which neither did. "The fact was, it took a whole year to get on his schedule to go away for a weekend! In a sense, I resented it. It really put the handwriting on the wall."

A few months after purchasing the vacation home, Neil came back from a business trip to find a note from Janet on the kitchen table of their Lebanon farmhouse. The note said that she was leaving him.

"We had family. We had grandchildren. It was a long hard decision for me. It wasn't an easy thing to do—I cried for three years before I left." Janet had prolonged her decision because "the children were still there, the nest wasn't empty, there were still things going on. I always had hoped our life together would improve with time.

"I realized the personality. I just couldn't live with the personality anymore."

Neil took it hard. "Can't you do something about it, Neil?" his friend Harry Combs asked. "No, I just can't," Neil answered. "Jan has just given up on us. She doesn't want to live that kind of life." Said Combs, "He was in the deepest depression that I've ever seen. It was awful. He would just sit there and glare at the table—not even move. I would ask him, 'Is there any improvement?' and he would say, 'The children are supportive, but I have no sign of ever getting her back.' There were two or three years of this stuff."

Dean confirms that Neil became very depressed: "He begged her for a long time to come back."

The separation was tragically bookended by the deaths of Neil's parents; first Stephen, then Viola. Their last few years of life had been sad and problematic. Stephen had suffered a series of minor strokes, and thought they did not have enough money to live on. The children moved their parents into a duplex in Bisbee, Arizona, where June and her husband, Jack Hoffman, lived. Viola adapted well, but Stephen hated the desert. In the summer of 1989, Neil moved them to a retirement community in Sidney, Ohio, just south of Wapakoneta.

Stephen lived unhappily in an unassisted private apartment at the nursing home for six months, and made life even more difficult for Viola. Neil was with him on February 3, 1990, when he succumbed to another series of strokes. "Dad sat straight up in bed, looked at us, and laid down and died," Neil remembered. A few days before, Stephen had motioned his wife over, whispering, "I love you."

After grieving for her husband, Viola was ready to go on living. A previous diagnosis of pancreatic cancer turned out to be a heart problem. Unfortunately, her health was more fragile than anyone suspected. On Monday, May 21, 1990, back in Ohio, she died suddenly. A few days earlier, she surprised her daughter by saying, "I am not sure there really is a God. But I am very happy that I believed." That next winter following the death of his parents and his separation from Janet, Neil suffered his heart attack. His cardiac health recouped quickly, but it would take longer to cure the heartache.

Out of ashes, if a person is lucky, a brand-new life can rise. For Neil, rejuvenation—and a type of personal redemption—began when he met Carol Held Knight.

Born in 1945, Carol was a recent widow. Her husband, 49-year-old Ralph Knight, had been killed in a small plane crash in Florida in 1989. Carol was left to raise her two teenage children, Molly and Andrew, and also run the family business, a small Cincinnati construction company.

The meeting between Neil and Carol in the summer of 1992 was surreptitiously arranged by mutual friends, Paul and Sally Christiansen, at a pre-golf tournament breakfast at their club in suburban Cincinnati. Out of embarrassment at sitting next to the famous astronaut, Carol said little, then left early to tend to her ill mother. Neil escorted her out to her car.

"A couple of weeks later, my son, Andy, and I were out in the backyard. I could hear the phone ringing. There was a very quiet voice on the other end, 'Hello.' And I said, 'Who is this?' And this quiet voice said, 'Neil.' And I said, 'Neil who?' And he said, 'Neil Armstrong.' And I said, 'Oh, what do you want?' 'What are you doing?' 'Well, actually, my son and I are trying to cut down a dead cherry tree.'

"Neil came to life and said, 'Oh, I can do that.' 'Well, you know where I live,' I answered, 'across the street from Paul and Sally.' 'Well, I'll be right over.' Thirty-five minutes later, there's a pickup truck in the driveway. Andy answered the door, and Neil's standing there with a chainsaw in his hand. Andy comes back in the kitchen and he says, 'Do you know who's at the door?' I said, 'Oh, I forgot to tell you.' "

Carol and Neil were married after Neil and Janet's divorce became final in 1994. There were two wedding ceremonies. Planning the family gathering, Carol said, " 'How does that look, Neil, June eighteenth?' He opened up his date book and said with a very serious expression, 'I have a golf tournament.' Then he looked up at me very sheepishly and said, 'But I could change it.' "

Because the state of California required a blood test for a marriage license plus a waiting period of five days, Carol and Neil first married in Ohio. The mayor of Carol's village (also a friend) presided on June 12, 1994. The Christiansens stood as their witnesses. Their California wedding took place at San Ysidro Ranch, near Calabasas Canyon in the Los Angeles area. With them that day were only the couple's four adult children, plus Mark's wife, Wendy, and their two children.

The new Mr. and Mrs. Neil A. Armstrong decided to build a brand-new house on the same property where Carol's old house was standing. The one-story English-country-style home was finished in 1997. "We talked about whether we would like to live anyplace else. But all our friends were here, and we had come to the stage in life where that network was really priceless."

Did Carol give much thought to what it might mean to be Mrs. Neil Armstrong? "I'm sure the attention is so much less than it was thirty years ago. We have noticed most of that when we travel out of the country. But he's not recognized that much anymore. I definitely run interference. I will politely explain, 'Neil doesn't sign autographs anymore.' We try to give them something instead: 'How about a picture?' You have to respect their feelings, too.

"There have been a few times when I've been actually scared, maybe twice in the U.S. and a few times in other countries. I remember coming into an overseas airport around two in the morning. I didn't think we'd be able to get to the car, just people all over! We needed help from half a dozen policemen just to get in the car.

"Once we came back from London and we had just gotten home after the flight and just taken our suitcases in the bedroom, when the doorbell rang. I went to the door and opened it and this woman said in a British accent, 'I'm from the *London Times* and I missed you in Britain. I wanted an interview. Could I have one now?' And I just looked at her and said, 'You must be kidding.'

"Neil and I are a good balance, so we have a good partnership."

Those who knew Armstrong during his second marriage agreed that Carol did a lot to make Neil very happy.

Today, Janet Armstrong resides close to her two sons and six grandchildren in suburban Cincinnati, after spending twenty-five years on her own in Utah. After Rick graduated from Wittenberg College in 1979 with a major in biology, he trained dolphins and sea lions for a company in Gulfport, Mississippi, then went on to Hawaii, after which he began doing dolphin shows at Ohio's Kings Island. Today, Rick and his wife are divorced, their three children still living in northern Cincinnati suburbs. He plays the guitar semiprofessionally and tours the world to watch performances of his favorite music group, Marillion. Mark majored in physics at Stanford, where he also played on the golf team and helped set up the university's first student computer lab. He went to work with Symantec in Santa Monica, then he joined his former college roommate's startup, WebTV, which was eventually bought by Microsoft. Mark stayed with Microsoft in Silicon Valley until 2004, when he moved his wife and three children to the Cincinnati area. It was through Mark's interest in Apple's original Macintosh that Neil first became enthusiastic about computers.

In doing several hours of interviews for the original edition of *First Man* published in 2005, it was clear that Janet still struggled to understand him:

"Everyone gives Neil the greatest credit for not trying to take advantage of his fame, not like other astronauts have done."

"Yes, but look what it's done to him inside. He feels guilty that he got all the acclaim for an effort of tens of thousands of people. Someone like Jim Lovell was a different personality completely! He would just walk on and not let it bother him. Neil would let it bother him. He always was afraid of making a social mistake, and he has no reason to feel that way, for he was always a well-mannered gentleman.

"He's certainly led an interesting life. But he took it too seriously to heart.

"He didn't like being singled out, or to feel that people were still wanting to touch him or get his autograph. Yet he wouldn't quit signing autographs for twenty years because probably, in the bottom of his heart, he didn't think most people were trying to make money selling them."

"Are you saying that if he had gone out in the public more times over the years, that the interest in him would have dwindled—that he's made himself into a type of target?"

"I agree."

In the last years of his life, Neil Armstrong seemed to be a very happy man—perhaps happier than at any other time in his life. Although he technically "retired" in the spring of 2002, he remained as busy as ever; traveling around the world, giving speeches, attending events, visiting children and grandchildren, reading books, writing essays, playing golf. He attended meetings of the American Philosophical Society, and frequently participated in annual sessions of the Academy of the Kingdom of Morocco, in which he was a member since King Hassan II established it in 1980.

As for his personal flying, he seized the occasional opportunity to take control of an interesting aircraft. In 1989, when he became chairman of AIL Systems, Inc., he was invited to fly the B-1 bomber. In 1991, he flew the B-1 again for a television series called *First Flights*. For that series, he also flew a number of other aircraft types, including the Harrier, helicopters, gliders, and an old Lockheed Constellation.

In the late 1990s, Neil sold his Cessna 310, but kept his pilot's license current for those occasions when he would be offered the chance to fly a special aircraft. In 2001, in association with his directorship of RMI Titanium Co., he flew an Airbus 320 at Airbus's headquarters in Toulouse, France. "I have been blessed to have many exciting events to be involved in and remember," Neil wrote a friend. "This week I was in Toulouse flying the flight test Airbus 320 over the Pyrenees. It was not overly exciting, but it certainly was fun." In the summer of 2004, he flew the new Eurocopter and AStar helicopters, and an assortment of light aircraft. In 2011, just a year before his death, he even accepted an offer from Qantas Airlines, while visiting Australia, to fly its Airbus A380 simulator. The A380 is a double-deck, widebody, four-engine jet airliner that is the world's largest passenger airliner. As often as he could right up to his death, he still went aloft in sailplanes, a relaxing sporting activity that he had enjoyed since the early 1960s. "He was always a natural at that," Janet recalled. "He could actually hear the thermals. It was a wonderful relief for him to be up there flying by himself."

In 2002, Neil agreed to an authorized biography, resulting in the publication of the original edition of *First Man: The Life of Neil A. Armstrong*. Many people wondered why he had finally accepted a proposal for his life story to be written, after having turned down offers from some of the most prominent authors in America, including James Michener, Herman Wouk, and Stephen E. Ambrose. No more explicit answer came from Neil or his family other than to say, "It was time." His single compliment to the author was, "You wrote exactly the book you told me you would write."

Initially Armstrong, then seventy-five years old, agreed to do three interviews related to the book, always making it clear that he was the subject of the book but in no way its author. Several media sources sought Armstrong's interview. In the end, he agreed to sit down with the CBS newsmagazine *60 Minutes,* and to make that his only interview. The story aired on November 1, 2005, to a record number of viewers, the evening before *First Man* became available in bookstores. CBS billed the interview as "the first television profile the Moonwalker has ever agreed to do." The reaction to the piece was universally positive. One of Neil's good Cincinnati friends, John G. Smale, the chief executive officer of Procter & Gamble, who later became chairman of General Motors, sent him a handwritten note: "The *60 Minutes* interview was simply great. You came across to a national audience exactly as you are." The host for the segment was Ed Bradley; joining him for part of the interview, held at Cape Canaveral, was Walter Cronkite, the legendary news anchor who had helmed all the televised U.S. manned space launches for CBS from Mercury through Apollo.

Neil's succinct, quick, witty, and thoughtful responses to their questions showed a personal side of the astronaut that surprised viewers, many of whom were not alive when the *Eagle* landed in July 1969: "I knew we [the Apollo program] would have a limited life. But I must say it was a bit shorter than my expectation. I fully expected that by the end of the [twentieth] century, we would have achieved substantially more than we actually did. When we lost the competition [factor in our race with the Soviets], we lost the public will to continue." Neil also explained to Ed Bradley why he did not warrant the acclaim he was afforded by the milestone-making Apollo 11 flight, or by his "one small step . . . one giant leap" onto the lunar surface. "I just don't deserve it [the attention for being the first man on the Moon]. I wasn't chosen to be first. I was just chosen to command that flight. Circumstance put me in that particular role. That wasn't planned by anyone." He went on to explain one of the disappointing parts of gaining astronaut celebrity status: "Friends and colleagues, all of a sudden, looked at us [and] treated us slightly differently than they had months or years before, when we were working together. I never quite understood that." He also commented on the impact that being an Apollo astronaut and the resulting fame from being first on the Moon had on his personal life and family: "The one thing I regret was that my work required an enormous amount of my time and a lot of travel, and I didn't get to spend the time I would have liked with my family as they grew up." As part of the episode, CBS arranged for Neil to pilot a sailplane, with a TV camera on board, out of a small airfield outside Orlando. The piece ended with Bradley asking Neil whether, with NASA then considering putting humans back on the Moon by 2018, such treks were something that Neil would consider now that he was seventy-five years old. "I don't think I'm going to get the chance," he responded, adding with a smile, "But I don't want to say that I'm not available."

In 2010, Armstrong did get a chance to make himself available for two trips to the Middle East as part of the "Legends of Aerospace" tours, organized by Morale Entertainment in association with Armed Forces Entertainment. The purpose of the tours was to "lift the spirits of our brave men and women in uniform." Joining him was quite a crew, led by Jim Lovell and Gene Cernan, the commanders of Apollo 13 and Apollo 17. Accompanying them on the Legends trip in March (lasting ten days) was Steve Ritchie, the only USAF fighter pilot "ace" since the Korean War, and Robert J. Gilliland, the chief test pilot who was first to fly the SR-71 Blackbird. Serving as event moderator in front of the troops on both trips was David Hartman, a passionate aerospace enthusiast who for many years had hosted ABC's *Good Morning America*. The itinerary was extensive, with the entourage flying a total of 17,500 air miles; stopping in six countries (Germany, Turkey, Kuwait, Saudi Arabia, Qatar, and Oman); visiting six military bases and their hospitals; helicoptering out to the USS *Dwight D. Eisenhower*, an aircraft carrier at war; and interacting with fifteen-thousand-plus troops. The October trip (totaling seven days) started in Bahrain, traveled out to the USS *Harry S. Truman* at the head of a task force of eleven U.S. warships in the Persian Gulf, and finished at Joint Base Balad, an Iraqi air force base located forty miles north of Baghdad. After the astronauts visited his base, Air Force Senior Master Sergeant Bradley Behling, of the 386th Expeditionary Force Support Squadron, said, "This visit was a dream come true. Tours like this show everyone over here fighting that we aren't forgotten. When Neil Armstrong comes across the globe to say thank you personally for the things you are doing, I don't see how anyone can feel anything but inspired to continue our mission until it's done."

During the March 2010 trip, Armstrong, Lovell, and Cernan spent a good deal of time together discussing the space policy of the Obama administration. All three of the former astronauts were unhappy with President Obama's decision to cancel "Constellation," the human spaceflight program that had been developed by NASA from 2005 to 2009, and whose goals were completion of the International Space Station and a return to the Moon no later than 2020, with a crewed flight to the planet Mars as the ultimate goal. Estimated in 2004 to cost some $230 billion over twenty years, Constellation was to involve the creation of an Ares launch vehicle, an Orion crew exploration vehicle, and an Altair lunar lander. Over the years, Neil had chosen not to speak out publicly in opposition to any political decisions about the U.S. program, but the cancellation of Constellation really bothered him. With Jim and Gene egging him on, Neil agreed to be part of a panel that would testify before Congress.

Before the U.S. Senate Committee on Commerce, Science, and Transportation on May 12, 2010, Neil joined his two fellow Apollo commanders in speaking out against the cancellation of Constellation. It was not politics, and certainly not partisanship, that motivated Armstrong to go public with his concerns. If Lovell and Cernan, along with former NASA administrator Mike Griffin, had not so strongly urged Neil to share his thoughts about NASA's errors in direction, Neil likely would not have done it. But once he made up his mind to do it, he gave it everything he had and made it the best, clearest statement of national concern that he could compose. "I believe that, so far," he told the U.S. House Committee on Science, Space, and Technology on May 26, "our national investment in space exploration, and our sharing of the knowledge gained with the rest of the world, has been made wisely and has served us very well. America is respected for the contributions it has made in learning to sail upon this new ocean. If the leadership we have acquired through our investment is allowed simply to fade away, other nations will surely step in where we have faltered. I do not believe this would be in our best interest."

In private, Neil held much blunter views of the political environment affecting American space policy. Writing to a friend who was a retired USAF colonel and NASA employee in August 2010, he wrote: "The President could not be expected to know much about our world [that is, the aerospace community] and it is just a small ball compared to the many big balls he has in the air right now: Afghanistan, Health Care, Stimulus, Oil Spills, etc. So I hoped he would be guided by good advisors. I have concluded that he did not have any. The cabal that did advise him had their own agenda. They hoped they could get it through by doing an end-run around Congress by eliminating the normal review process of the proposed NASA Budget Submission during the period from Thanksgiving to February. I have yet to find a Senator, Representative, Program

Director, Senior Air Force official, or National Academy official who had any idea what the NASA plan proposal would be. So when the President announced the plan on 1 February [2010], he made many congressmen mad—on both sides of the aisle. I am certain their reaction surprised him. The cabal made some quick changes which were announced at his 3/15 speech at JSC [Johnson Space Center]. It was clear to many that they were poorly crafted. So I was disturbed by the process the President used. Most of the initiatives he did announce were very vulnerable to elimination or modification by future congresses or administrations. That's always true to some extent, but lack of a backbone to the plan suggested human space flight could just slowly disappear over a few years. I am still concerned but remain hopeful. There is still work to be done." In 2011, Neil once again testified before Congress, this time to the House Committee on Science, Space, and Technology. In addition, he sent a number of letters to the chairpersons and other members of that committee, reinforcing his views.

The following year, his public testimony backfired somewhat. On March 25, 2012, five months before his death, Armstrong grew upset about a *60 Minutes* story hosted by CBS's Scott Pelley, featuring SpaceX's Elon Musk and the progress that was being made to "commercialize space." Over a video of Neil's testifying, Pelley asserted that "There are American heroes who don't like the idea. Neil Armstrong and Gene Cernan have both testified against commercial space flight in the way that you are developing it . . . that the Obama Administration's drive to commercialize space could compromise safety and eventually cost the taxpayers. . . . I wonder what you think of it." A visibly emotional Musk, tears welling in his eyes, replied: "I was very sad to see that. Those guys are heroes of mine, so it's really tough. I wish they would come and visit [and] see the hard work that we're doing here. I think that it would change their mind."

Armstrong always took the truth very seriously, and demanded, in this case, that it be made literal. He wrote to *60 Minutes*, correcting the record: "My wonder was where you obtained that information. . . . I combed [all my testimony to Congress and] I found nothing in [it] to confirm your assertions. . . . I generally refer to the viewpoints of others rather than my own. It is impossible for me to reconcile your assertions with my testimony. It is true that the Committees have expressed skepticism with regard to NASA's plans and programs in the Commercial Cargo and Commercial Crew areas and hearings have been held specifically on that subject. Testimony certainly exists that you could have used to support your claim. So I found it very surprising that you chose to create a 'anti-position' for me. . . . I ask that you explain to me how you came to the false assertions that you clearly stated to your television audience." Attached to his letter to CBS were excerpts from his testimony to the Senate committee, which Neil called "the only comments included in my testimony that are related to the so-called 'commercial space' endeavors."

On behalf of CBS News, Scott Pelley wrote back to Armstrong, but not until June 12. Apparently, Neil's letter had been "mishandled on our end," causing the ten-week delay in responding. Pelley apologized, but explained why *60 Minutes* had felt justified in coming to the conclusion it did about Neil's position on the commercialization of space: "By way of explanation, we took note of your congressional testimony when you raised concerns about the Obama administration program. Part of that testimony included:

> *"I am very concerned that the new plan, as I understand it, will prohibit us from having human access to low Earth orbit on our own rockets and spacecraft until the private aerospace industry is able to qualify their hardware under development as rated for human occupancy. I support the encouragement of the newcomers toward their goal of lower-cost access to space. But having cut my teeth in rockets more than 50 years ago, I am not confident. The most experienced rocket engineers with whom I have spoken believe that it will require many years and substantial investment to reach the necessary level of safety and reliability."*

Anyone hearing or reading Neil's statement could reasonably infer, as Pelley did, that Neil was not supportive of handing over the design and operation of human spaceflight to commercial business, not in the short term and possibly not for many years. Still, Pelley admitted to Neil that there was some lack of precision in what aired: "We should have made it explicit in our story that, while you were 'not confident' that the newcomers could achieve safety and cost goals in the near term, you did want to encourage them. We also should have spelled out more clearly that your concerns were directed toward the 'newcomers' in general and not SpaceX in particular." Pelley invited Neil to write a statement that "we would post prominently in its entirety," but Neil chose instead to accept the issuance of a corrected statement written by Pelley on behalf of *60 Minutes*.

Much more than testifying before Congress, in the last years of his life, Armstrong enjoyed his marriage to Carol; their quiet home and circumscribed social life in suburban Cincinnati; and their vacation home in the Rocky Mountain ski resort of Telluride, Colorado. He also enjoyed getting closer to Rick and Mark—often on the golf course, which the whole family loved, with near-annual trips to Scotland and Ireland with one or both of his sons. In addition, he became better acquainted with his two stepchildren (Carol's two grown children), Andy (Knight) and Molly (Knight Van Wagenen), and caring for their "Brady Bunch" of eleven grandchildren. Neil also still traveled a great deal, sometimes to faraway destinations, often with Carol, but sometimes alone. In July 2007, they visited Israel, with a tour of Masada, and a stop at the Yad Vashem World Holocaust Remembrance Center. Neil gave public talks in Haifa and Tel Aviv, and did a Q&A with fifty children at the Haifa Science Museum. In 2008, they cruised to Scandinavia with a group from Purdue; and in 2009, into the South Atlantic to the Falklands and Antarctica, a twenty-six-day journey with a National Geographic Explorer expedition. Almost annually, during the late summer Neil would attend the "Ranch Meeting" of the Conquistadores del Cielo, the highly private fraternity of top airline executives and accomplished flyers whose meeting places were held at undisclosed, "top secret" locations around the U.S. These meetings were only for relaxation and such sporty recreations as "fast draw," fly casting, horseshoes, knife throwing, skeet, and shuffleboard.

Armstrong's final international destination was Australia in August 2011. Many people found the basis for Neil's traveling Down Under to be rather unusual: he agreed not only to give a talk at the 125th anniversary meeting of the Certified Practicing Accountants of Australia, but also to participate in a rare, tape-recorded one-on-one personal interview with Alex Malley, the

CPAA's chief executive, who during a business trip to Ohio the previous year had persuaded Neil to do the unlikely event. "I knew something a lot of people didn't know about Neil Armstrong—his dad was an auditor," Malley told his membership and the interested Australian press. Neil agreed to appear before the assembly of Australian accountants to honor his father, Malley declared.*

But the event turned out to be yet another betrayal of Neil's trust. The fifty-minute Malley interview was posted on the CPAA website in four parts, with access supposedly restricted to CPAA members. But the interviews quickly went viral. An Australian friend, Len Halprin, wrote to Neil a few weeks later: "I have to tell you that the series of video interviews you gave to Alex Malley from CPA Australia last year has exploded over the media here in the last 48 hours, and it appears to be spreading like wildfire. Media outlets from all over Australia and the world are chasing Alex for interviews as to how he got this world exclusive with you. One local station here in Melbourne has been talking about the interview incessantly for the last hour and chased Alex down in Vietnam to interview him. I hope that the media respect your privacy, and that you are not inconvenienced in any way." To which Neil replied, "Yes,

---

* While in Sydney, Armstrong also met with some university students and business leaders, and enjoyed a cruise around Sydney Harbor in a 1903 steamboat piloted by Captain Richard Champion de Crespigny, the Qantas pilot who saved his Airbus A380 from disaster after an uncontained engine failure in November 2010—the first of its kind for the big passenger airliner—that required an extraordinary emergency landing at Singapore Changi Airport. Captain de Crespigny, during this visit, also gave Neil a tour of the Airbus A380 simulator, leading to a long discussion between the two great pilots on the differences between Europe "fly-by-wire" technology and U.S. fly-by-wire.

the release of those interviews came as quite a surprise as they had been produced for the CPAA internal use only. I am getting comments from around the world." Neil was not happy, and had his lawyers write to Malley charging breach of agreement. Australian media later reported that Malley—besides bragging that his Armstrong interview reached one billion people—offered it for sale to the Nine Network's *60 Minutes* as well as to the Australian Broadcasting Corporation. CPA Australia spokesmen insisted that the organization had complied fully with the arrangement, but by 2017, opposition to Malley's leadership of the Australian CPA, part of it due to the Armstrong scandal, led to termination of his contract. Meanwhile, Malley had tried to continue hawking Neil Armstrong's name.

Armstrong's last public appearance before his death was his keynote address at the Lowell Observatory in Flagstaff, Arizona, before some 730 guests. The "First Light" Gala celebrated the opening of the new Discovery Channel Telescope (DCT) after a decade's worth of construction. The highlight of his talk was his blow-by-blow commentary on what he saw as he descended from lunar orbit on his way down to the Sea of Tranquility on July 20, 1969. (He first gave this presentation during his Australia CPA talk in August 2011.) The sensational imagery came from cameras on board NASA's Lunar Reconnaissance Orbiter, or LRO, which had begun imaging the six different Apollo landing sites in July 2009. Over the years since the first Moon landing, countless people had asked Neil to recall the details of the landing, almost always in vain. Now with this advanced side-by-side view—on the left side of his screen showing the original 1969 movie film that Apollo 11 had taken from the window of the lunar module, on the right the high-definition animation compiled from the LRO camera acquisition—Neil relished this new opportunity to explain how the historic landing happened:

> The actual powered descent of the lunar module took twelve minutes and thirty-two seconds, and this is just the final three minutes, the part that is really interesting as you get close to the surface of the Moon. . . . On the left this took place forty-two years ago, and the pictures on the right took place in the last two years. You'll hear the crewmen talking. You might hear my copilot giving altitude descent rates, and you'll hear people in the background talking from Mission Control on Earth. We've been descending about two thousand meters; we're down now to below a thousand meters in altitude. My computer tells me that it's taking us to a landing just on the right side of that big crater up in the upper left-hand corner. The slopes are steep and the rocks look very large, the size of automobiles, certainly not a place that I want to land. So I took over manually from the computer—the autopilot—and flew it like a helicopter on out to the west to try to find a smooth, more level landing spot. The computer is complaining now and then; you will hear it give alarms—1202s and 1201s—which is telling us the computer is a little bit concerned about its operation. But everything looks good, and the people in Mission Control tell us we can continue. Now we are about one hundred meters, looking down at this surface at this thirty-meter crater that is about eight meters deep. It looks like a real geological treasure; I want to go back and take a look at that if I ever get the chance while I am on my own, on foot. We're looking for a smooth spot beyond that

> crater, and I see a smooth spot right there at the top of the screen. It looks like that is a good spot to be, and I am running low on fuel; I have less than two minutes of fuel. We're getting down to below seventy meters now . . . fifty meters, still looking good. . . . In the left side, you will see in the old movie that the rocket engine is starting to kick up some dust off the surface. We get a thirty-second fuel warning; we need to get it down here on the ground pretty soon before we run out. The picture on the left is more accurate, but there is more dust. There you see the shadow of my landing leg coming on the surface, on the blowing dust. We're very close to the surface right now.

When Neil's presentation ended, with the famous recorded voices of Buzz ("Twenty feet, down a half. Drifting forward just a little bit. Good. Okay. Contact light. . . . Shutdown.") and Neil ("Okay. Engine stop. . . . Houston, Tranquility Base here. The *Eagle* has landed."), the audience in Flagstaff rose to its feet in thunderous applause. The applause may never have ended if they had realized that the great astronaut would die only a few weeks later.

On Saturday, August 25, 2012, Neil Armstrong died in a suburban Cincinnati hospital from complications following a quadruple coronary bypass surgery that had been performed nineteen days earlier, on August 6. Just the day before, August 5, he had celebrated his eighty-second birthday. Shortly after his death, his family released the following statement:

> We are heartbroken to share the news that Neil Armstrong has passed away following complications

> resulting from cardiovascular procedures. Neil was our loving husband, father, grandfather, brother, and friend. Neil Armstrong was also a reluctant American hero who always believed he was just doing his job. He served his nation proudly, as a navy fighter pilot, test pilot, and astronaut. He also found success back home in his native Ohio in business and academia, and became a community leader in Cincinnati. He remained an advocate of aviation and exploration throughout his life and never lost his boyhood wonder of these pursuits. As much as Neil cherished his privacy, he always appreciated the expressions of good will from people around the world and from all walks of life. While we mourn the loss of a very good man, we also celebrate his remarkable life and hope that it serves as an example to young people around the world to work hard to make their dreams come true, to be willing to explore and push the limits, and to selflessly serve a cause greater than themselves. For those who may ask what they can do to honor Neil, we have a simple request. Honor his example of service, accomplishment, and modesty, and the next time you walk outside on a clear night and see the moon smiling down at you, think of Neil Armstrong and give him a wink.

The news of his death stunned the entire world, made headlines on the front page of virtually every newspaper on the planet, and brought outpourings of deeply felt sentiments about the greatness of Neil Armstrong; not just as an astronaut, test pilot, naval aviator, and engineer, but as a highly honorable man. NASA administrator and former astronaut Charles Bolden commented, "As long as there are history books, Neil Armstrong will be included in them." President Barack Obama said, "Neil Armstrong was a hero not just of his time, but of all time." The British astronomer Sir Patrick Moore said, "As the first man on the Moon, he broke all records. He was a man who had all the courage in the world." Harvard University astrophysicist Dr. Neil deGrasse Tyson commented, "No other act of human exploration ever laid a plaque saying, "We came in peace for all mankind." Buzz Aldrin said that he was "deeply saddened by the passing. I know I am joined by millions of others in mourning the passing of a true American hero and the best pilot I ever knew. I had truly hoped that on July 20, 2019, Neil, Mike and I would be standing together to commemorate the 50th Anniversary of our Moon landing. Regrettably, this is not to be." Michael Collins said of Neil: "He was the best, and I will miss him terribly."

Neil could have died on so many occasions during his extraordinary life in air and space—in combat over North Korea; when test-flying highly dangerous unproven airplanes; in fiery launches atop powerful rockets known to blow up occasionally; while spinning dizzily and nearly blacking out in an out-of-control spacecraft following rendezvous and docking in Earth Orbit; managing to eject from a Lunar Landing Training Vehicle a fraction of a second before the perverse machine exploded; flying long and running out of fuel in *Eagle* on the way down to a rocky and cratered landing site on the surface of the Moon. That a man who had escaped death and major injury so many times during his extraordinary flying career was to die in a hospital bed from surgical complications seems inappropriate and unfair.

What exactly happened with Neil's medical condition leading to his death on August 25, 2012, may never be known outside the Armstrong family and the medical staff that attended to him. Few of the actual facts leading up to his death are publicly known. Here is what little is known: (1) Neil went into the hospital on August 6, the day after his eighty-second birthday. As the author of this book knows from an email that Neil sent him at 3:53 P.M. on August 11, he had been suffering from "an apparent reflux problem," was told by his cardiologist to come immediately to the hospital, and that the doctor did "a nuclear stress test leading to an angiogram, leading to a quad bypass." In sum, Neil's failing a cardiac stress test on August 6 prompted an urgent four-vessel coronary bypass graft surgery, done the morning of August 7. (2) The hospital to which Neil went was Fairfield Mercy Hospital, a 293-bed facility located in a northern suburb of Cincinnati in Butler County, Ohio. (3) Again, as Neil's email to the author on August 11 stated: "Recovery is going well but golf will be on the back burner for a while. Hope to be kicked out of the hospital in a day or so." In other words, Neil expected to be released for recovery at home on August 12 or 13. (4) Neil never made it home; he died at Fairfield Mercy on August 25, two weeks after his hopeful email to the author. (5) All the family would say was that he "died from complications that resulted from cardiovascular surgery."

Clearly something bad happened in those two weeks. Heart surgery of any kind is truly major surgery—so much can go wrong, especially when the patient is eighty-two years old. But one might think that Neil's positive message to the author five days after the bypass surgery, that he expected to go home in a day or two, indicated that Neil had done well surviving the major postsurgery dangers. Not just something bad, but something unexpected, must have led to Neil's death. Maybe someday the world will know what that was. Whether the world deserves to know, and whether Neil would want us to know, are two questions that this biographer has struggled with mightily over the past few years. As it stands, history must respect the family's right to privacy, for whatever reasons they are keeping it private.

A private funeral for family and close friends was held on Friday, August 31, at the Camargo Golf Club in Indian Hill, the Cincinnati suburb in which the Armstrongs had lived since their marriage in 1994, and the club at which they were longtime members. An estimated two hundred people attended, including Neil's relatives and close friends. There was a navy ceremonial guard and a bagpiper, along with tight security to keep out the press and the uninvited. Mike and Buzz were there, as was John Glenn and Jim Lovell, along with several other astronauts, space program officials, and aerospace notables, past and present. Giving eulogies were Ohio congressman Rob Portman, a family friend, and Charles Mechem, Neil's longtime friend and former head of Taft Broadcasting. Both Rick and Mark Armstrong presented short talks about their dad, sharing personal anecdotes—and some of their dad's favorite jokes—that lifted the spirits of the grieving. Carol's son, Andrew Knight, read from Corinthians I, and her granddaughter Piper Van Wagenen read the Twenty-third Psalm. Metropolitan Opera's Jennifer Johnson Cano, a mezzo-soprano, sang the pop standard "September Song," a favorite of Neil's and a metaphor comparing a year to a person's life span from birth to death. At the end of the ceremony, everyone walked out onto the ninth fairway to witness a flyover of F-18 fighter jets peeling away in the "missing man" formation.

So loved and admired was Neil Armstrong that many Americans, led by Ohio congressman Bill Johnson, called for President Obama to grant him a state funeral, a highly formal event steeped in tradition and usually only held for former presidents. A state funeral was in fact offered to the Armstrong family, but Carol declined the offer. (President Obama directed all American flags to be brought to half-staff throughout the nation until sunset on Monday, August 27, as well as "all United States embassies, legations, consular offices, and other facilities abroad, including all military facilities and naval vessels and stations.") A large public memorial service was held on Wednesday, September 13, at Washington National Cathedral, in the lovely northwest quadrant of the nation's capital. A magnificent Gothic structure, the cathedral was an especially appropriate place to hold the Armstrong service, as its "Space Window" depicts the Apollo 11 mission and holds a sliver of Moon rock amid its stained-glass panes. Before an overflowing crowd, Mike Collins led the mourners in a prayer. Eulogizing Neil was his good friend and Purdue mate Gene Cernan, the Apollo 17 mission commander and last man to walk on the Moon; and Charles Bolden, the NASA Administrator. Also speaking at the service were John H. Dalton, former U.S. Secretary of the Navy; and John W. Snow, a fellow Buckeye and former CEO of CSX Corporation who had served as U.S. Secretary of the Treasury under President George W. Bush. One of Neil's favorite contemporary singers, jazz contralto Diana Krall, sang "Fly Me to the Moon." Although Neil was a deist and not a doctrinally religious man, the Reverend Gina Gilland Campbell read a passage from the Book of Matthew, and the Right Reverend Mariann Edgar Budde delivered a homily.

The following day, September 14, Armstrong's cremated remains were scattered in the Atlantic Ocean off Jacksonville, Florida, during a burial-at-sea ceremony aboard the USS *Philippine Sea* based at Naval Station Mayport. Neil's wife Carol; son Rick; son Mark, his wife, Wendy, and daughter Kali; Neil's sister, June, and her husband, Jack Hoffman; Neil's brother, Dean, and his wife Kathryn; stepdaughter Molly Van Wagenen and her husband, Brodie; and stepson Andrew Knight and his wife, Cristina, were on board for the service. A U.S. Navy firing squad fired volleys in Neil's honor, followed by a playing of taps. Thus, Neil remained a navy man to the end—many people reasonably thought that was the reason he chose to be buried at sea. Or perhaps it was because Neil was always so humble and private, and did not want the fuss and attention a traditional gravesite would surely bring. As Secretary of the Navy Raymond E. Mabus asserted in Neil's honor on the day of the burial, "Neil Armstrong never wanted to be a living memorial, and yet to generations the world over, his epic courage and quiet humility stands as the best of all examples."

In the last years of his life, Neil Armstrong had received many prestigious national and international awards, always accepting them humbly and modestly as only the rarest of men with "The Real Right Stuff" ever could. He had been given many awards over the years since Apollo 11 (Presidential Medal of Freedom with Distinction, 1969; Dr. Robert H. Goddard Memorial Trophy, with Mike Collins and Buzz Aldrin, from the National Space Club, 1970; Sylvanus Thayer Award, from the U.S. Military Academy, 1970; Congressional Space Medal of Honor, 1978; National Aviation Hall of Fame, 1979; U.S. Astronaut Hall of Fame, 1993; Langley Gold Medal, from the Smithsonian Institution, in 1999), but those he received in the twilight of his life for career achievement were truly special.

In 2006, Armstrong received NASA's Ambassador of Exploration Award, in a ceremony held at the Cincinnati Museum Center at Union Terminal. The award was a beautifully crystal-encased lunar sample, part of the 842 pounds of Moon rocks and soil returned during the six lunar expeditions from 1969 to 1972. Such an award was presented that year to the thirty-eight astronauts and other key individuals (or surviving family members) who had participated in the Mercury, Gemini, and Apollo programs. Former senator John Glenn spoke at the ceremony, saying, "I don't envy many people. But for Neil, I make a big exception." Armstrong, who always wanted to offer his audiences something educational, went beyond a thank-you to share what he called "a thin slice of natural history." Standing next to the piece of Moon rock he had just been given, Neil outlined the Moon's geologic development by reference to its parent rock, which Neil dubbed "Bok." Neil offered, "I was the strange creature that kidnapped Bok," then referred to the lunar sample that was part of his award as "a chip off the old Bok." The inscription on the award described Neil's Moon rock as "a symbol of the unity of human endeavor and mankind's hope for a future of peace and harmony."

On Monday, July 20, 2009, on the occasion of the fortieth anniversary of the first Moon landing, Armstrong, Collins, and Aldrin were guests of President Obama at the White House. Obama hailed the three men as "genuine American heroes" and declared that "the touchstone for excellence in exploration and discovery is always going to be represented by the men of Apollo 11." Neil always felt it was a great honor to receive a handshake from the president of the United States, as he felt also in this case, but in the coming months, as we have seen, Neil and other notable space advocates would grow increasingly critical of the space policy of the Obama Administration.

The preceding evening, Neil and his mates together gave the annual John H. Glenn Lecture in Space History at the National Air and Space Museum, the museum's premier annual event for exploring the role of space, science, and technology in modern American life. The day before that, on Saturday evening, the NASM staff also coordinated a large celebration at the museum for NASA headquarters, which had decided to have one big event to celebrate all of the Apollo Program anniversaries, rather than figuring out how to commemorate each of the individual missions separately. For the two weekend events, twenty different Mercury, Gemini, and Apollo astronauts attended the weekend in person or, if deceased, as represented by close family members. In addition, the entire crew of STS-125, which had just performed the final Hubble Space Telescope mission in May 2009, attended as guests.

Dr. Margaret Weitekamp was in her first year coordinating the Glenn Lecture for NASM's Space History Department. It proved to be a true baptism by fire, not just with handling the logistics of three speakers (Neil, Buzz, and Mike, with three very different personalities), but also assisting in planning the evening, from the opening reception through the talk itself to the speakers' return to the hotel. Security for Neil was always a special concern, and the night of the NASA gala proved no exception. As Weitekamp recalled: "At some point during NASA's reception and program, the run of show called for moving Mr. Armstrong from a ready room near the museum's theater to a stage that had been constructed on the far side of the immense central Milestones of Flight Hall. But the audience was already packed into rows and rows of chairs, and it was standing room only."

As Weitekamp wallked with Neil, "well-wishers leaned over me to greet him from all sides, reaching in to tap his shoulder, get his attention, or try to shake his hand." The NASA curator became "increasingly aware of how narrow a path had been cleared and how many taped-down wires crisscrossed the floor to power the audiovisuals and speakers. There were only three of us to offer him some protection." She "took his arm and guided him carefully through the crowd." But that short walk gave her "a brief glimpse into what it must have been like to be the subject of so much well-intentioned but overwhelming attention."

Neil was initially reluctant to speak about the Apollo 11 landing; indeed, when asked to offer the Glenn Lecture as a solo speaker in 2006, he had chosen to talk about his engineering work on the X-15. In the end, in giving the fortieth anniversary lecture, he had "a little fun with his professorial style of speaking." Neil introduced his talk as being called "Goddard, Governance, and Geophysics," a title that was "so academic-sounding that the audience actually chuckled aloud." Neil paused, smiled, held up one finger, and said, "Part one, Goddard." And the audience realized that he was serious. He had crafted a fine talk about the research background that supported the Apollo 11 lunar landing. The audience, according to Weitekamp, was "utterly silent in their seats listening." Prior to Neil's formal speech, Mike Collins had given brief, casual, witty, and self-disparaging remarks that charmed the audience in NASM's IMAX theatre. Buzz spoke second, relying on teleprompters as a U.S. president would and proclaiming his vision of America's future in space through a long series of elaborate PowerPoint slides. Nothing could have better illustrated the tremendous differences in personality between the three members of the Apollo 11 crew than how they handled their remarks to the Smithsonian audience that evening.

In 2010, Neil received what he considered to be one of the greatest honors of his career when he was inducted into the Naval Aviation Hall of Honor, in Pensacola, Florida, where he had trained to become a naval aviator sixty years earlier. Located at the National Naval Aviation Museum, the award recognized individuals "who by their actions or achievements made outstanding contributions to Naval Aviation."

In 2011, he was given the Congressional Gold Medal, bestowed by the U.S. Congress to persons "who have performed an achievement that has an impact on American history and culture that is likely to be recognized as a major achievement in the recipient's field long after the achievement." At the same ceremony, held in the Capitol Rotunda, Mike Collins, Buzz Aldrin, and John Glenn also received the gold medals, the first recipient of which had been George Washington in 1776.

In 2013, Neil posthumously received the General James E. Hill Lifetime Space Achievement, the highest honor given by the Space Foundation, a Colorado-based nonprofit organization whose mission is "to advance space-related endeavors to inspire, enable and propel humanity."

A large number and wide variety of things have been named in Armstrong's honor. Across the United States, more than a dozen elementary, middle, and high schools have been named in his honor, and many places around the world have streets, buildings, schools, and other places named for him. In 1969, folk singer John Stewart recorded "Armstrong," a tribute to Neil and his first steps on the Moon. In October 2004, Purdue

University, Neil's alma mater, announced that its new engineering building would be named the "Neil Armstrong Hall of Engineering." At a cost of $53.2 million, the building was dedicated on October 27, 2007, during a ceremony at which Neil was joined by a dozen other Purdue astronauts, including Gene Cernan, John Blaha, Roy Bridges, Mark Brown, Richard Covey, Guy Gardner, Gregory Harbaugh, Gary Payton, Mark Polansky, Jerry Ross, Loren Shriver, and Charles Walker.

Appropriately, a lunar crater was named after Neil, established by the International Astronomical Union several years before his death. The "Armstrong" crater sits thirty-one miles to the northeast of the Apollo 11 landing site in the southern part of the Sea of Tranquility. Collins and Aldrin also have craters named after them—in fact, the three form a neat little row located a short distance due north of the bright crater Moltke. All three are rather small, Neil's being the largest at 2.9 miles across, followed by Buzz's at 2.1 miles, and Mike's at 1.5 miles. Astronomy books state that, with a six-inch telescope and steady air, one should be able to pick out all three at high magnification starting about the time the Moon is six days old, or right before the first quarter phase. Neil's is the easiest to see.

There is also an asteroid named after Armstrong: "6469 Armstrong," a stony Flora asteroid from the inner regions of the asteroid belt, approximately three kilometers in diameter. The asteroid was discovered by Czech astronomer Antonín Mrkos at Klet' Observatory in August 1982.

In September 2012, a few weeks after Neil's death, the U.S. Navy announced that its first "Armstrong-class vessel" would be named the R/V *Neil Armstrong*. Christened on March 28, 2014, the research vessel was launched on March 29, 2014, passed sea trials on

August 7, 2015, and was delivered to the navy on September 23, 2015. Neil's namesake is a highly advanced research platform capable of supporting a wide range of oceanographic research activities. Working out of Woods Hole Oceanographic Institution in Massachusetts, the R/V *Neil Armstrong* supports oceanographic research in tropical and temperate oceans around the world, while specifically serving the academic community's ongoing need for a general-purpose ship based on the East Coast of the United States. Neil's ship has already been playing a pivotal role studying the role that the ecosystems of the North Atlantic and Arctic Ocean play in Earth's changing climate.

Armstrong absolutely never campaigned to get anything named for him—just the opposite was true in many cases. Such was certainly true in the case of what in 2014 became NASA Armstrong Research Center, the former NASA Dryden Research Center (and NACA/NASA Flight Research) where Neil had worked as a research pilot from 1956 to 1962. Neil greatly respected the life and achievements of Dr. Hugh L. Dryden, a pioneering aeronautical research scientist who had become the NACA's Director of Research in 1946 and NASA's first Deputy Administrator after the space agency's establishment in 1958, and he did not care to have Dryden's name removed from the history-making desert facility. But Southern California's congressional delegation was convinced that the government aerospace flight research center needed a "makeover" of sorts, and that the name Neil Armstrong could help significantly in that. Moreover, Armstrong's achievements as a test pilot and astronaut merited the name change. In January 2014, President Obama signed congressional resolution HR 667, renaming the flight facility the "Neil A. Armstrong Flight Research Center." The new law still paid homage to Dryden by naming the area surrounding the center the "Hugh L. Dryden Aeronautical Test Range." This had been at least the third time since 2007 that Congress had tried to rename the facility for Neil Armstrong. If Neil had still been living, no doubt he would have spent his entire talk at the renaming ceremony detailing the illustrious career of Dr. Dryden.

No wonder that in surveys conducted by the Space Foundation, Neil has perennially come out ranked as the number-one most popular space hero; or that, in 2013, *Flying* magazine ranked him number one on its list of the "51 Greatest Heroes of Aviation."

A final legacy for Neil Armstrong is his Apollo 11 space suit. NASA transferred the space suit to the National Air and Space Museum in 1971, five years before the facility on the National Mall in Washington, D.C., opened to the public. When NASA contracted for the suit, explained Kevin Dupzyk, a writer for *Popular Mechanics* in October 2015, "it was concerned with only one thing: getting the astronauts to the moon and back." The suit designers—the International Latex Corporation in Dover, Delaware, a division of the company that manufactured Playtex bras and girdles—"didn't care about museum exhibits, so they chose to use a mix of natural and synthetic rubbers with a six-month life span." In fact, ILC Dover could not make the suits too early, or they would degrade before the mission even started.

In the four and a half decades following Apollo 11, the condition of Neil's suit deteriorated greatly. Its rubber became brittle. Its aluminum—primarily the red and blue buttons and arm disconnects—became badly pitted with corrosion. Something needed to be done to save it for posterity—specifically, for display in "Destination Moon," a new permanent exhibit that NASM planned to open in conjunction with the Apollo 11 fiftieth anniversary in 2019. Instead of raising the $500,000 needed for this effort in any traditional manner, museum leadership decided to try a Kickstarter campaign—its first ever—in which it accepted donations from the public via the Internet. Called "Reboot the Suit," the Smithsonian met its goal in just five days and, by the end of the campaign a month later, had received a total of $719,779 from more than 9,400 backers. With this freely given money, the Smithsonian experts immediately began to preserve the suit properly, down to the particles of lunar dust that clung to its surface, for the whole world to enjoy viewing during the golden anniversary of the first Moon landing.

Emily Perry was five years old when she met the First Man on the Moon. It was the summer of 2001, and the former commander of Apollo 11 was seventy-one years old. Nearly forty years had gone by since his own darling little girl, Karen, two years and ten months old, died in January 1962 from brain cancer. Emily was the granddaughter of one of his best friends, Kotcho Solacoff. They had been boys together in Upper Sandusky during the early 1940s. As they each moved into their golden years, the two friends spent many good times together: attending college football games, gliding down ski slopes, and playing golf. No person apart from family members knew Neil as well as did Kotcho.

Emily unknowingly encountered the First Man one day while he was visiting her grandfather and grandmother at the home of the Solacoffs' daughter Kathy and her husband, Chris Perry. The girl was the youngest of the Perrys' three children and a real firecracker. Neil took to Emily quickly, and Emily to him. Soon she had Neil by the hand, leading him on an expedition through her house. "I want to show you a secret, but don't tell anyone. This is a secret no one knows about." Reaching the attic, Emily said to Neil, "Peek over the mattress and look down there." There it was—a great big dead bug. "But don't tell anyone," she whispered. "Oh, I won't," he whispered back.

Next, Emily led him into her bedroom. "This is my clock, and this is my lamp, and this is my mirror, and these are some of my books. This book is on Winnie the Pooh, and this one is about Sleeping Beauty, and this is Cinderella. And, oh, here is a book about Neil Armstrong. He was the first man on the Moon." Then she stopped, hesitated for a moment, looked at the nice older man who had come to visit her in her house, and said, "Oh! Your name is Neil Armstrong, too, isn't it? Would you like me to read you his book?" Neil gave her a generous smile, then sat down on the edge of her bed. "I would very much like to hear you read a book, Emily. But it doesn't have to be the one on Neil Armstrong; it could be the Winnie book or the Cinderella book or Sleeping Beauty. Any of them, I would like." "No, I'd like to read you the Neil Armstrong book, because that's your name. It's not a very long book, and it's very exciting. You'll see."

Crawling into his lap and straightening her skirt, the child opened the book and began to read. She was clearly proud to be the one who got to tell this gentleman, such a good friend of her grandpa's, the story of the First Man on the Moon.

Acknowledgments

Historians may also voyage from the Earth to the Moon. My own epic journey began sixteen years ago, in June 2002, when Neil A. Armstrong signed a formal agreement enabling me to work as his official biographer. Actually the trip began in October 1999 when I first wrote Neil about my ambition to write his life story. A long thirty-three months later, after numerous letters and emails passed back and forth between us (and a critical face-to-face private meeting—our first—in September 2001), Neil gave me his thumbs-up. That approval brought unprecedented access not only to Neil and his personal papers but also to his family, friends, and colleagues—many of whom, in deference to Neil, had resisted speaking openly about him before.

So, first and foremost, I wish to acknowledge Neil Armstrong himself. Without his full and generous support, this book could never have been written.

I am also indebted to Neil for the integrity with which he wanted the project carried out. He wanted the book to be an independent, scholarly biography. Although he took the opportunity to read and comment on every draft chapter, he did so only to guarantee that the book was as factual and technically correct as possible. Not once did he try to change or even influence my analysis or interpretation.

It should be clear, then, that Neil was not in any way a coauthor of this book. In fact, I am quite sure he did not like the book's title. He would never think to call himself the "First Man," insisting as he always did that Buzz Aldrin landed on the Moon at the very same instant he did. Also, it was not to Armstrong's liking that "First Man" sounded so biblical, so epic, so iconic; he never expressed his life or legacy in those terms. But once Neil had decided to trust my effort, he was not about to interfere with my purpose. The result, I believe, is an exceptionally rare type of book: an authorized biography more candid, honest, and unvarnished than most unauthorized biographies.

Face-to-face, Neil gave me one and only one compliment about the book. As I left his home in suburban Cincinnati upon finishing our review of the manuscript in 2004, Neil shook my hand and said, "Jim, you wrote exactly the book you told me you were going to write." For anyone who truly knew Neil Armstrong, it was the greatest compliment I could have ever gotten from Neil, considering how many people over the years since the Moon landing in 1969 had tried to hoodwink and manipulate Neil, telling him one thing and trying to do another. Nothing about writing the book makes me more proud than remembering that compliment.

Just as it took some 400,000 Americans in government, industry, and universities to carry out the Apollo program, this book could not have been produced without the help of a score of people. A complete list of people interviewed for the book has appeared in the bibliography of the two previous editions of this book (2005, 2012); to every one of them, I have expressed my sincere thanks. I will never meet a finer group of individuals. Meeting them and hearing what they had to say about Neil and about their own lives and careers made me think how lucky Neil was to have had them for colleagues and friends. Conducting the oral history took me to eighteen states and the District of Columbia.

Once again, for this new 2018 edition, I owe special thanks to Neil's immediate family: to his sons, Rick Armstrong and Mark Armstrong; his brother, Dean Armstrong; and especially his sister, June Armstrong Hoffman. Over the years June has provided me with many extremely informative and deeply personal insights into Neil and the history of her family, and shared with me all of her mother's photo albums and personal papers. This "Viola material" proved invaluable by significantly deepening my understanding of the family dynamics from which the young Neil emerged. For sharing their mother with me, and thus the world, I have June to thank. Jayne Hoffman and Jodi Hoffman, June's daughters, were also helpful in sorting out the many intricacies and riddles of the Armstrong family genealogy.

From the start I was committed to hearing firsthand from Neil's first wife, Janet Shearon Armstrong. It was impossible to tell Neil's story without telling Janet's. I was interested not only in what Janet had to say about her former husband of thirty-eight years; I was interested in Janet herself. During the Apollo years, Janet, as the wife of an astronaut and then as the wife of the first man on the Moon, became a public figure in her own right. In that context, it was critically important to examine her own experiences as a woman, wife, mother, and role model. As hard as it was for her to do, Janet eventually agreed to a series of interviews. What she contributed, in my view, is a priceless addition to this book.

I also owe a great debt to Neil's second wife, Carol Held Knight Armstrong, not just for the interviews she has granted me over the years and for the generous and caring hospitality she showed me every time I visited the Armstrong home, but also for her friendship. Carol's daughter, Molly Knight–Van Wagenen, has also helped me answer questions about the family and about the later years of Neil's life. Also, I will never forget the delightful sight of Molly's daughter, Piper, when two years old, sitting contentedly on Grandpa Neil's lap. Piper is now a beautiful young lady of sixteen.

A host of historians, librarians, archivists, curators, and other research professionals at various institutions helped enormously with my research. I thanked all of them in the previous two editions; they know who they are, so I will not repeat their names. But rest assured I am thinking of them again for what they did to help me with this book.

For their continued efforts to help in my preparation of this new edition, I must again thank the following:

Without the *Apollo Lunar Surface Journal* assembled over the course of many years by its editor Eric P. Jones, my understanding of what happened on the Sea of Tranquility during the Apollo 11 mission would have been much less informed and precise. I want to thank Eric for his great support of my project and particularly for his keen reading of my draft chapters relevant to the first landing. He saved me from making several major errors. Those that remain are my own. I also wish to thank England's David Woods and Scotland's Ken

McTaggert, two of the editors of the *Apollo Flight Journal*, for their generous assistance.

The founder and editor of the Web site collectSPACE.com, Robert Pearlman, has for many years provided me with a number of important insights into space history and the popular fascination with astronauts and space memorabilia. My incessant email questions and text messages to him about such details as the contents of the astronauts' PPKs could not have been answered more promptly or completely.

I also want to thank Houston's Roger Weiss for all the helpful information he has so graciously provided over many years, as well as his friendship.

The officers and gentlemen of Fighter Squadron 51 deserve special mention for what they contributed to this book. As a group, not even the Apollo astronauts that I interviewed were more impressive to me.

Without the support I received from my academic home, Auburn University, I could never have produced this book in a timely fashion. All of my colleagues are due my thanks for indulging my passion for my subject and my long absences from their company. In particular, I wish to acknowledge the support of my fellow faculty in our Technology and Civilization program, especially Dr. Guy V. Beckwith, Dr. Monique Laney, Dr. David Lucsko, Dr. Alan Meyer, and Dr. William F. Trimble.

Back in the early 2000s, I had an especially thoughtful and talented coterie of doctoral students who never let me give up on the idea of writing the Armstrong biography. They kept cheering me on even when I had given up most hope that the project would work out. Each went on to complete successful doctorates in aerospace history.

I also want to thank the countless undergraduate students I have enjoyed teaching in our freshman Technology and Civilization survey and in my courses in aerospace history and the history of science and technology. All of them are especially dear to me now that I have retired from teaching after thirty-one years at Auburn University.

The editors at Simon & Schuster, Denise Roy on the earlier editions of *First Man* and Emily Graff on this latest one, have done many wonderful things for this book.

I found a kindred spirit and an angel in the intellectually radiant and spiritually magical Laurie Fox of the Linda Chester Literary Agency. Every minute of my work on this book would have been worthwhile even if its only result had been Laurie's friendship. I also want to thank the majestic Linda Chester herself for all her support over the years, as well as my film agent, Justin Manask.

My immediate family has "lived" my Neil Armstrong saga almost as much as I have. Many times at dinner, as I sat silent or dazed, my mind still spinning with that day's thoughts about Armstrong's life, my wife Peggy, daughter Jennifer, and son Nathan would have to reel me in and bring me back down to Earth. But I never felt anything but their loving support for what I was doing. Since the original edition of this book came out in 2005, both of my children have married and their spouses, Cole Gray and Jessica Phillips Hansen, have been wondrous additions to the family. I dedicated the first two editions of this book to my children, Jennifer, now an art historian and entrepreneur in the business world of portraiture, and Nathaniel, a psychiatrist, both of whom live in Birmingham, Alabama. So it is now to my three blessed grandchildren—and to their children's children's children—that I dedicate this 2018 edition of the book.

Finally, I thank you, the reader, for investing in such a big book and, hopefully, reading it from first page to last. For you, for posterity, and for Neil, I have given it my absolute best.

—James R. Hansen
Auburn, AL
March 2018

Bibliography

Primary Sources

NONARCHIVED PRIVATE PAPERS

Papers of Viola Engel Armstrong and Armstrong Family. Hereford, AZ (Property of June Armstrong Hoffman).

Personal Diary of Ensign Glen Howard "Rick" Rickelton, U.S. Navy, Written During V-51 Combat Flight Training & Korean War Service Aboard CV-9 USS *Essex,* Rickelton Family Papers, Elk Grove, CA, and Seattle, WA.

Personal Diary of Robert Kaps, USS *Essex* (CV-9), Carrier Air Group Five, June 28, 1951 to March 25, 1952.

ARCHIVAL COLLECTIONS

Since 2015, the personal papers of Neil A. Armstrong have been preserved in the Purdue University Archives on the campus of Purdue University in West Lafayette, IN, Armstrong's alma mater. Similarly, all the research materials originally collected by James R. Hansen for the publication of First Man, *including the voice recordings of all Hansen's tape-recorded interviews for the book, are also housed in the Purdue Archives.*

Archives of Aerospace Exploration. University Libraries, Virginia Polytechnic Institute and State University. Blacksburg, VA.

Auglaize County Public Library. Wapakoneta, OH. Neil A. Armstrong Newspaper Files.

Emil Buehler Naval Aviation Library. National Museum of Naval Aviation. Pensacola, FL.

John Glenn Archives. The Ohio State University Archives. Columbus, OH.

NASA Dryden Flight Research Center. Historical Archives. Edwards, CA.

NASA Headquarters History Office. Washington, DC.

National Personnel Records Center. Military Personnel Records. St. Louis, MO.

Naval Historical Center. Department of the Navy, Washington Navy Yard. Washington, DC.

Neil A. Armstrong Museum. Newspaper files. Wapakoneta, OH.

Nixon Presidential Materials. National Archives at College Park. College Park, MD.

Ohio Historical Society. Columbus, OH.

Records of NASA Dryden Flight Research Center. National Archives and Records Administration—Pacific Region. Laguna Nigel, CA.

Records of NASA Glenn Research Center. National Archives and Records Administration—Midwest Region. Chicago, IL.

Records of NASA Headquarters. National Archives and Records Administration—East Region. College Park, MD. Record Group 255.

Records of NASA Johnson Space Center. National Archives and Records Administration—Southwest Region. Fort Worth, TX. Record Group 255.

Records of NASA Johnson Space Center. Library and Archives of the University of Houston–Clear Lake. Clear Lake, TX.

Records of NASA Kennedy Space Center. National Archives and Records Administration—Southeast Region. Atlanta, GA. Record Group 255.

Records of NASA Langley Research Center. National Archives and Records Administration—Atlantic Region. Philadelphia, PA. Record Group 255.

Rensselaer Polytechnic Institute University Archives, Troy, NY. George M. Low Papers.

Time-Life Archives. Time-Life Building. New York City.

University of Cincinnati Archives. Cincinnati, OH.
Wyandot County Public Library. Newspaper files. Upper Sandusky, OH.

DOCUMENTS

Works of Neil A. Armstrong, Published and Unpublished

"Future Range and Flight Test Area Needs for Hypersonic and Orbital Vehicles," *Proceedings of Professional Pilots Symposium on Air Space Safety,* 1958. Also appeared in *Society of Experimental Test Pilots* 3 (Winter 1959).

"Flight and Analog Studies of Landing Techniques Pertinent to the X-15 Airplane," *Research-Airplane-Committee Report on Conference on the Progress of the X-15 Project,* NACA-CONF-30-Jul-58, July 30, 1958. Coauthors: Thomas W. Finch, Gene J. Matranga, Joseph A. Walker.

"Test Pilot Views on Space Ventures," *Proceedings of ASME Aviation Conference,* Mar. 1959.

"Approach and Landing Investigation at Lift-Drag Ratios of 2 to 4 Utilizing a Straight-Wing Fighter Airplane," *NASA TM X-31,* Aug. 1959. Coauthor: Gene J. Matranga.

"Utilization of the Pilot in the Launch and Injection of a Multistage Orbital Vehicle," *IAS Paper 60-16,* 1960. Coauthors: E. C. Holleman and W. H. Andrew.

"X-15 Operations: Electronics and the Pilot," *Astronautics* 5 (May 1960): 42–3, 76–8.

"Development of X-15 Self-Adaptive Flight Control System," *Research-Airplane-Committee Report on Conference on the Progress of the X-15 Project,* 1961. Coauthors: R. P. Johannes and T. C. Hays.

"Flight-Simulated Off-the-Pad Escape and Landing Maneuvers for a Vertically Launched Hypersonic Glider," *NASA TM X-637,* Mar. 1962. Coauthors: G. J. Matranga and William H. Dana.

"The X-15 Flight Program," *Proceedings of the Second National Conference on the Peaceful Uses of Space,* Seattle, WA, May 8–10, 1962. Coauthors: Joseph A. Walker, Forrest S. Petersen, Robert M. White.

"A Review of In-Flight Simulation Pertinent to Piloted Space Vehicles," *AGARD Report 403,* 21st Flight Mechanics Panel Meeting, Paris, France, July 9–11, 1962. Coauthor: Euclid C. Holleman.

"Pilot Utilization During Boost," Inter-Center Technical Conference on Control Guidance and Navigation Research for Manned Lunar Missions, Ames Research Center, Moffett Field, CA, July 24–25, 1962. Coauthor: Euclid C. Holleman.

"X-15 Hydraulic Systems Performance," *Hydraulics and Pneumatics,* Dec. 1962.

"Gemini Manned Flight Programs," *Proceedings of the Society of Experimental Test Pilots,* 8th Symposium, 1964.

"Controlled Reentry," *Gemini Summary Conference,* Houston, Texas, Feb. 1967. Multiple coauthors.

"Safety in Manned Spaceflight Preparation: A Crewman's Viewpoint," *AIAA,* 4th Annual Meeting, Oct. 1967.

"Apollo Flight Crew Training in Lunar Landing Simulators," *AIAA Paper 68–254,* 1968. Coauthor: S. H. Nassiff.

"Lunar Landing Strategy," *Proceedings of the Society of Experimental Test Pilots,* 13th Symposium, 1969.

"The Blue Planet," World Wildlife Fund, London, England, Nov. 1970.

"Lunar Surface Exploration," COSPAR, Leningrad, USSR, 1970, and Akademie-Verlag, Berlin, 1971.

"Change in the Space Age," The Mountbatten Lecture, University of Edinburgh, Mar. 1971.

"Out of This World," *Saturday Review/World,* Aug. 24, 1974.

"Apollo Double Diaphragm Pump for Use in Artificial Heart-Lung Systems," *AAMI National* Meeting, Mar. 1975. Coauthors: H. J. Heimlich, E. A. Patrick, G. R. Rieveschl.

"Intra-Lung Oxygenation for Chronic Lung Disease," Benedum Foundation, 1976. Coauthors: H. J. Heimlich, E. A. Patrick, G. R. Rieveschl.

"What America Means to Me," *The Reader's Digest,* June 1976, pp. 75–76.

"A Citizen Looks at National Defense," *National Defense,* Sept.–Oct. 1978.

"The Learjet Longhorn, First Jet with Winglets," *Proceedings of the Society of Experimental Test Pilots,* 22nd Symposium, 1978. Coauthor: P. J. Reynolds.

Commencement Address, University of Cincinnati, June 13, 1982.

"New Knowledge of the Earth from Space Exploration," Academy of the Kingdom of Morocco, Casablanca, 1984. Coauthor: P. J. Lowman.

Wingless on Luna. 25th Wings Club General Harold R. Harris "Sight" Lecture, presented at Inter-Continental Hotel, New York City, May 20, 1988. New York: Wings Club, 1988.

"Research Values in Contemporary Society," Academy of the Kingdom of Morocco, Casablanca, 1989.

"Reflections by Neil Armstrong: We Joined Hands to Meet Challenge of Apollo Mission," *Cincinnati Enquirer,* July 20, 1989.

"The Ozone Layer Controversy," Academy of the Kingdom of Morocco, Casablanca, 1993. Coauthor: Mark S. Armstrong.

"Engineering Aspects of a Lunar Landing," The Lester D. Gardner Lecture, Massachusetts Institute of Technology, May 3, 1994. Coauthor: Robert C. Seamans.

"Pressure Vessel Considerations in Aerospace Operations," National Board of Boiler and Pressure Vessel Inspectors, Anchorage, AK, 1995.

"Observations on Genetic Engineering," Academy of the Kingdom of Morocco, Rabat, 1997. Coauthor: Carol Knight Armstrong.

OTHER PRIMARY DOCUMENTS

Bennett, Floyd V. *Mission Planning for Lunar Module Descent and Ascent.* Washington, DC: NASA Technical Note MSC-04919, Oct. 1971.

CBS Television Network, *10:56:20 P.M., 7/20/69.* New York: Columbia Broadcasting System, 1970.

Godwin, Robert, ed. *X-15: The NASA Mission Reports.* Burlington, Ontario: Apogee Books, 2000.

———. *Dyna-Soar: Hypersonic Strategic Weapons System.* Burlington, Ontario: Apogee Books, 2003.

———. *Apollo 11: The NASA Mission Reports.* 3 vols. Burlington, Ontario: Apogee Books, 1999–2002.

Jones, Eric P., *Apollo Lunar Surface Journal.*

Low, George M. *Latin American Tour with Astronauts Armstrong and Gordon, Oct. 7–31, 1966.* NASA Manned Spacecraft Center: Unpublished mss., Nov. 16, 1966.

NASA Lyndon B. Johnson Space Center. *Biomedical Results of Apollo.* Washington, DC: NASA SP-368, 1975.

NASA Manned Spacecraft Center. *Apollo 11 Onboard Voice Transcription, Recorded on the Command Module Onboard Recorder Data Storage Equipment.* Houston: Manned Spacecraft Center, Aug. 1969.

NASA Manned Spacecraft Center. *Apollo 11 Preliminary Science Report.* Washington, DC: NASA SP-214, 1969.

NASA Manned Spacecraft Center. *Apollo 11 Spacecraft Commentary, July 16–24, 1969.*

NASA Manned Spacecraft Center. *Apollo 11 Technical Air-to-Ground Voice Transcription.* Prepared for Data Logistics Office Test Division, Apollo Spacecraft Program Office, July 1969.

National Commission on Space. *Pioneering the Space Frontier: The Report of the National Commission on Space.* Toronto and New York: Bantam Books, May 1986.

"Neil Armstrong's Comments on Behalf of the Apollo 11 Crew," Langley Medal Awards Ceremony, July 20, 1999, National Air and Space Museum, Washington, DC.

"Remarks by Neil A. Armstrong upon Receipt of National Space Trophy," 2004 Rotary National Award for Space Achievement, Houston, TX.

"Statement by Neil Armstrong at the White House," NASA Release, July 20, 1994.

U.S. News and World Report. *U.S. on the Moon: What It Means To Us.* Washington, DC: U.S. News and World Report, 1969.

NEWSPAPERS AND PERIODICALS

Akron Beacon Journal
Baltimore Evening Sun

Boston Globe
Chicago Tribune
Christian Science Monitor
Cincinnati Enquirer
Cincinnati Post
Cleveland Plain Dealer
Cleveland Press
Columbus (Ohio) *Citizen-Journal*
Columbus (Ohio) *Dispatch*
Dayton Daily News
Florida Today
Houston Chronicle
Lebanon (Ohio) *Western Star*
Life
Lima (Ohio) *Citizen*
Lima (Ohio) *News*
Los Angeles Times
NASA X-Press (NASA Dryden)
National Observer
Newsweek
New York *Daily News*
New York Times
Seattle Daily Times
Space News Roundup (NASA Manned Spacecraft Center/JSC)
St. Marys (Ohio) *Evening Leader*
Time
Toledo Blade
Wall Street Journal
Wapakoneta Daily News
Washington Post

Interviews

CONDUCTED BY AUTHOR

Aicholtz, John, June 5, 2003, Cincinnati.
Aldrin, Buzz, Mar. 17, 2003, Albuquerque, NM.

Anders, Valerie, Apr. 8, 2003, San Diego; July 17, 2004, Dayton, OH.

Anders, William A., Apr. 8, 2003, San Diego.

Armstrong, Carol Knight, June 2004, Cincinnati.

Armstrong, Dean, Nov. 14, 2002, Bonita Springs, FL.

Armstrong, Janet Shearon, Sept. 10–11, 2004, Park City, UT.

Armstrong, Neil A., Cincinnati,

- Aug. 13, 2002.
- Nov. 26, 2002.
- June 2–4, 2003.
- Sept. 18–19 and 22, 2003.
- June 2–3, 2004.

Armstrong, Rick, Sept. 22, 2003, Cincinnati.

Armstrong Hoffman, June,

- Aug. 14, 2002, Wapakoneta, OH.
- Apr. 4–5, 2003, Hereford, AZ.
- June 7, 2003, Wapakoneta, OH.

Baker, Steve, June 5, 2003, Cincinnati.

Barnicki, Roger J., Dec. 11, 2002, Lancaster, CA.

Barr, Doris, Aug. 15, 2002, Cincinnati (telephone).

Bean, Alan, Feb. 7, 2003, Houston.

Beering, Stephen, May 30, 2003, Carmel, IN.

Bennett, Floyd V., Feb. 8, 2003, Houston.

Blackford, John "Bud," July 25, 2003, Concord, NH.

Borman, Frank, Apr. 15, 2003, Las Cruces, NM.

Borman, Susan, Apr. 15, 2003, Las Cruces, NM (telephone).

Brading, Charles, Jr., Aug. 17, 2003, Wapakoneta, OH.

Burrus, David, June 5, 2003, Cincinnati.

Burt, Devere, June 3, 2003, Cincinnati.

Butchart, Stanley P., Dec. 15, 2002, Lancaster, CA.

Cargnino, Larry, Nov. 29, 2002, West Lafayette, IN.

Carpentier, Dr. William, Feb. 8, 2003, Seabrook, TX.

Cernan, Eugene A., Feb. 10, 2003, Houston.

Collins, Michael, Mar. 25, 2003, Marco Island, FL.

Combs, Harry, Oct. 7, 2003, Orlando, FL.

Crossfield, A. Scott, July 17, 2004, Dayton, OH.

Dana, William H., Dec. 9, 2003, Edwards, CA.

Day, Richard E., Dec. 11, 2003, Palmdale, CA.

Frame, Arthur, Aug. 15 and 17, 2002, Wapakoneta, OH.

Franklin, George C., Feb. 5, 2003, Houston.
Friedlander, Charles D., Apr. 8–9, 2003, San Diego.
Glenn, John H., Sept. 23, 2003, Columbus, OH.
Gordon, Linda, Apr. 12, 2003, Prescott, AZ.
Gordon, Richard F., Jr., Apr. 12, 2003, Prescott, AZ.
Gott, Herschel, June 20, 2003, Los Altos, CA.
Gustafson, Bob, Aug. 15, 2002, Wapakoneta, OH.
Heimlich, Dr. Henry, June 5, 2003, Cincinnati.
Hollemon, Charles, Nov. 21, 2002, West Lafayette, IN.
Keating, William, June 5, 2003, Cincinnati.
Keiber, Ned, Aug. 15, 2002, Wapakoneta, OH.
Kinne, Tim, June 5, 2003, Cincinnati.
Kleinknecht, Kenneth S., June 27, 2003, Littleton, CO.
Knight, William "Pete," Dec. 15, 2002, Palmdale, CA.
Knudegaard, Vincent Aubrey, Sept. 11, 2002, Auburn, AL.
Kraft, Chris, Feb. 7, 2003, Houston.
Kranz, Eugene, Feb. 8, 2003, Friendswood, TX.
Kutyna, Donald J., Mar. 20, 2004, Colorado Springs.
Love, Betty, Jan. 30, 2003, Edwards, CA (Conducted by Christian Gelzer).
Lovell, James A., July 17, 2004, Dayton, OH.
Lovell, Marilyn, July 17, 2004, Dayton, OH.
Lunney, Glynn, Feb. 6, 2003, Houston.
Mackey, William A., Sept. 21, 2002, Tuscaloosa, AL.
Mallick, Donald L., Dec. 12, 2002, Lancaster, CA.
Matranga, Gene J., Dec. 12, 2002, Lancaster, CA.
McDivitt, James A., Apr. 7, 2003, Tucson, AZ.
McTigue, John G., Dec. 9, 2002, Lancaster, CA.
Mechem, Charles S., Jr., June 25, 2003, Jackson Hole, WY.
Meyer, Russ, Oct. 7, 2003, Orlando, FL.
North, Warren J., Apr. 11, 2003, Phoenix.
Palmer, George, Nov. 21, 2002, West Lafayette, IN.
Peterson, Bruce A., Dec. 9, 2002, Lancaster, CA.
Preston, G. Merritt, Mar. 27, 2003, Melbourne, FL.
Rogers, James, June 5, 2003, Cincinnati.
Schiesser, Emil, Feb. 4, 2003, Seabrook, TX.
Schirra, Walter M., Jr., Apr. 8, 2003, San Diego.
Schmitt, Harrison H. "Jack," Mar. 16, 2003, Albuquerque.
Schuler, Dudley, Aug. 15, 2002, Wapakoneta, OH.

Schwan, Harold C., Oct. 17, 2002, Chesterfield, MO.
Scott, David R., Feb. 1, 2003, Atlanta.
Shaw-Kuffner, Alma Lou, Aug. 15, 2002, Wapakoneta, OH.
Smith, James M., July 17, 2003, New York City.
Solacoff, Doris, June 1, 2003, Upper Sandusky, OH.
Solacoff, K. K. "Kotcho," June 1, 2003, Upper Sandusky, OH.
Spitzen, Ralph, June 5, 2003, Cincinnati.
Stear, Mark, June 5, 2003, Cincinnati.
Townley, Diana, June 5, 2003, Cincinnati.
Townley, Gary, June 5, 2003, Cincinnati.
Walker-Wiesmann, Grace, Dec. 14, 2002, Reedley, CA.
Waltman, Gene L., Dec. 19, 2002, Edwards, CA.
White, Robert M., Nov. 12, 2002, Sun City Center, FL.
White, Vivian, May 29, 2003, Lebanon, OH.
Wilson, James R., June 23, 2003, Park City, UT.
Windler, Milton L., Feb. 4, 2003, Friendswood, TX.
Zwez, John, Aug. 14, 2002, Wapakoneta, OH.

CONDUCTED BY OTHER RESEARCHERS

Albrecht, William P., Feb. 16, 2001, Edwards, CA (Curtis Peebles).
Aldrin, Edwin E., July 7, 1970, Houston, TX (Robert B. Merrifield).
Algranti, Joseph S., Mar. 15, 1968, Houston, TX (Robert B. Merrifield); Aug. 10, 1998, Chapel Hill, NC (Erik Carlson).
Anders, William A., Oct. 8, 1997, Houston (Paul Rollins).
Arabian, Donald D., Feb. 3, 2000, Cape Canaveral (Kevin M. Rusnak).
Armitage, Peter J., Aug. 20, 2001, Houston (Kevin M. Rusnak).
Armstrong, Janet Shearon, Mar. 12, 1969, Houston *(Life)*.
Armstrong, Neil A.,

- Sept. 1, 1964, Houston *(Life)*.
- Summer 1965, Houston *(Life)*.
- Apr. 6, 1967, Houston (Barton C. Hacker).
- Feb. 23, 1969, Houston (*Life*).
- Mar. 2, 1969, Houston (*Life*).
- Mar. 12, 1969, Houston (*Life*).
- Aug. 7, 1969, Houston (*Life*).
- Sept. 23, 1971, Washington, DC (Robert Sherrod).

• Oct. 11, 1988, Cincinnati (Andrew Chaikin).
• Mar. 6, 1989 (Neil McAleer: phone interview).
• Sept. 19, 2001, Houston (Stephen E. Ambrose and Douglas Brinkley).

Armstrong, Viola, May 9, 1969, Wapakoneta, OH (Dora Jane Hamblin).

Bean, Allan, Apr. 10, 1984, Houston (W. David Compton).

Bond, Aleck C., Sept. 3, 1998, Houston (Summer Chick Bergen).

Borman, Frank, Apr. 13, 1999, Las Cruces, NM (Catherine Harwood).

Bostick, Jerry C., Feb. 23, 2000, Marble Falls, TX (Carol Butler).

Butchart, Stanley P., Sept. 15, 1997, Lancaster, CA (Curt Asher).

Carlton, Robert L., Mar. 29, Apr. 10, and Apr. 19, 2001, Houston (Kevin M. Rusnak).

Carpenter, M. Scott, Mar. 30, 1998, Houston (Michelle Kelly).

Catterson, A. Duane, Feb. 17, 2000, Houston (Carol Butler).

Cernan, Eugene A., Apr. 6, 1984, Houston (W. David Compton).

Charlesworth, Clifford E., Dec. 13, 1966, Houston (Vorzimmer).

Chilton, Robert G., Mar. 30, 1970, Houston (Loyd Swenson).

Collins, Michael, Oct. 8, 1997, Houston (Michelle Kelly).

Crossfield, A. Scott, Feb. 3, 1998, Lancaster, CA (Peter Merlin).

Dana, William H., Nov. 14, 1997, Edwards, CA (Peter Merlin); Mar. 9, 1999, Edwards, CA (Michael Gorn).

Day, Richard E., May 1, 1997, Edwards, CA (J. D. Hunley).

Donlan, Charles J., Apr. 27, 1998, Washington, DC (Jim Slade).

Drake, Hubert M., Nov. 15, 1966, Edwards, CA (Jim Krier and J. D. Hunley); Apr. 16, 1997, Edwards, CA (J. D. Hunley).

Duke, Charles M., Mar. 12, 1999, Houston (Doug Ward).

Fendell, Edward I., Oct. 19, 2000, Houston (Kevin M. Rusnak).

Franklin, George C., Oct. 3, 2001, Houston (Kevin M. Rusnak).

Fulton, Fitzhugh L., Jr., Aug. 7, 1997, Edwards, CA (J. D. Hunley).

Gordon, Richard F., Jr., Oct. 17, 1997, Houston (Michelle Kelly).

Griffin, Gerald D., Mar. 12, 1999, Houston (Doug Ward).

Grimm, Dean F., Aug. 17, 2000, Parker, CO (Carol Butler).

Haines, Charles R., Nov. 7, 2000, Houston (Kevin M. Rusnak).

Haise, Fred, Jr., Mar. 23, 1999, Houston (Doug Ward).

Hodge, John D., Apr. 18, 1999, Great Falls, VA (Rebecca Wright).

Honeycutt, Jay F., Mar. 22, 2000, Houston (Rebecca Wright).

Hutchinson, Neil B., June 5, 2000, Houston (Kevin M. Rusnak).

Kelly, Thomas J., Sept. 19, 2000, Cutchogue, NY (Kevin M. Rusnak).

Kleinknecht, Kenneth S., Sept. 10, 1998, Littleton, CO (Carol Butler); July 25, 2000, Houston (Carol Butler).

Kranz, Eugene F., Jan. 8, 1999, Houston (Rebecca Wright); Apr. 28, 1999, Houston (Roy Neal).

Love, Betty, Apr. 10, 1997, Edwards, CA (Michael Gorn); May 6, 2002, Palmdale, CA (Rebecca Wright).

Lovell, James A., Jr., May 25, 1999, Houston (Ron Stone).

Low, George M., Jan. 9, 1969, Houston (Robert B. Merrifield).

Lunney, Glynn S., Jan. 28, Feb. 8, and Apr. 26, 1999, Houston (Carol Butler).

Mattingly, Thomas K., II, Nov. 6, 2001, Costa Mesa, CA (Rebecca Wright).

McDivitt, James A., June 29, 1999, Elk Lake, MI (Doug Ward).

Maxson, Jerre, May 9, 1969, Wapakoneta, OH (Dora Jane Hamblin).

Mitchell, Edgar D., Sept. 3, 1997, Houston (Sheree Scarborough).

North, Warren J., Mar. 14, 1968, Houston (Robert B. Merrifield); Sept. 30, 1998, Houston (Summer Chick Bergen).

O'Hara, Delores B. "Dee," Apr. 23, 2002, Mountain View, CA (Rebecca Wright).

Preston, G. Merritt, Feb. 1, 2000, Indian Harbor Beach, FL (Carol Butler).

Saltzman, Edwin J., Dec. 3, 1997, Edwards, CA (J. D. Hunley).

Schirra, Walter M., Jr., Dec. 1, 1998, San Diego (Roy Neal).

Schmitt, Harrison H. "Jack," May 30, 1984, Houston (W. D. Compton); July 14, 1999, Houston (Carol Butler).

Schweickart, Russell L., Oct. 19, 1999, Houston (Rebecca Wright).

Seamans, Robert C., Nov. 20, 1998, Beverly, MA (Michelle Kelly); June 22, 1999, Cambridge, MA (Carol Butler).

Shea, Joseph F., May 16, 1971, Weston, MA (Robert Sherrod).

Sherman, Howard, Feb. 11, 1970, Bethpage, NY (Ivan Ertel).

Slayton, Donald K., Oct. 17, 1967, Houston (Robert B. Merrifield); Oct. 15, 1984, Houston (W. D. Compton).

Stafford, Thomas, Oct. 15, 1997, Houston (William Vantine).

Thompson, Milton O., Sept. 22, 1983, Edwards, CA (Larry Evans).

EMAIL AND LETTER CORRESPONDENCE WITH AUTHOR

Aldrin, Buzz, Los Angeles, CA.
Armstrong, Janet Shearon, Park City, UT.
Armstrong, Neil A., Cincinnati.
Armstrong-Hoffman, June, Hereford, AZ.
Baker, Steve, Cincinnati.
Beauchamp, Ernest M., Corona Del Mar, CA.
Bowers, William "Bill."
Brandli, Hank, Melbourne, FL.
Burke, Mel, Edwards, CA.
Burrus, David, Cincinnati.
Campbell, Nick, Denver, CO.
Clingan, Bruce, Troy, OH.
Day, Richard E., Palmdale, CA.
Esslinger, Michael, Monterey, CA.
Friedlander, Charles D., San Diego.
Gardner, Donald A., Clinton, IN.
Gates, Charles, Denver.
Gott, Herschel, Los Altos, CA.
Graham, Herb A.
Hamed, Awatef, Cincinnati.
Hayward, Tom.
Hoffman, Jayne, River Falls, WI.
Honneger, Barbara, Monterey, CA.
Hromas, Leslie A., Rolling Hills, CA.
Huston, Ronald, Cincinnati.
Jones, Eric P., Australia.
Karnoski, Peter, Las Vegas, NV.
Kinne, Tim, Cincinnati.
Klingan, Bruce E., Bellevue, WA.
Koppa, Rodger J., College Station, TX.
Kraft, Chris, Houston.
Kramer, Ken, Houston and San Diego.
Kranz, Eugene F., Houston.
Kutyna, Donald J., Colorado Springs.
Mackey, William A., Tuscaloosa, AL.

Mechem, Charles S., Jr., Loveland, OH.
Pearlman, Robert, Houston.
Perich, Pete, Warren, OH.
Petersen, Richard H., La Jolla, CA.
Petrone, Rocco, Palos Verdes Peninsula, CA.
Rickelton, Glen, Elk Grove, CA.
Rickelton, Ted, Seattle, WA.
Russell, George E. "Ernie," Cashion, OK.
Schwan, Harold C., Chesterfield, MO.
Scott, David R., London, England.
Slater, Gary L., Cincinnati.
Spanagel, Herman A., Satellite Beach, FL.
Spitzen, Ralph E., Columbus, OH.
Stear, Mark, Cincinnati.
Stephenson, David S., King of Prussia, PA.
Thompson, Tom, Rancho Palos Verdes, CA.
Walker-Wiesmann, Grace, Reeedley, CA.
White, Vivian, Lebanon, OH.

Secondary Sources

BOOKS

Aldrin, Buzz, and Malcolm McConnell. *Men from Earth.* 2nd ed. New York: Bantam Falcon Books, 1991.

Aldrin, Edwin E. Jr. with Wayne Warga. *Return to Earth.* New York: Random House, 1973.

Allday, Jonathan. *Apollo in Perspective: Spaceflight Then and Now.* Bristol and Philadelphia: Institute of Physics Publishing, 2000.

Armstrong, Robert Bruce. *The History of Liddesdale.* Vol. I. Edinburgh, 1883.

Arnold, H. J. P., ed. *Man in Space: An Illustrated History of Spaceflight.* New York: Smithmark, 1993.

Baker, David. *The History of Manned Spaceflight.* New Cavendish Books, 1981. Reprint. New York: Crown Publishers, 1982.

Ball, John. *Edwards: Flight Test Center of the USAF.* New York: Duell, Sloan, and Pearce, 1962.

Barbour, John. *Footprints on the Moon.* New York: Associated Press, 1969.

Bean, Alan. *Apollo: An Eyewitness Account by Astronaut/Explorer Artist/Moonwalker Alan Bean.* Shelton, CT: Greenwich Workshop, Inc., 1998.

Berg, A. Scott. *Lindbergh.* New York: G. P. Putnam's Sons, 1998.

Bilstein, Roger. *Stages to Saturn: A Technological History of the Apollo/Saturn Launch Vehicles.* Washington, DC: NASA SP-4206, 1980.

——. *Orders of Magnitude: A History of the NACA and NASA, 1915–1990.* Washington, DC: NASA SP-4406, 1989.

Borman, Frank, with Robert J. Serling. *Countdown.* New York: Morrow, 1988.

Bowman, Martin W. *Lockheed F-104 Starfighter.* London: Crowood Press, 2001.

Boyne, Walter J., and Donald S. Lopez. *The Jet Age: Forty Years of Jet Aviation.* Washington, DC: Smithsonian Institution Press, 1979.

Brooks, Courtney G., James M. Grimwood, and Loyd S. Swenson Jr. *Chariots for Apollo: A History of Manned Lunar Spacecraft.* Washington, DC: NASA SP-4205, 1979.

Buckbee, Ed, with Wally Schirra. *The Real Space Cowboys*. Burlington, Ontario: Apogee Books, 2005.

Burgess, Colin. *Fallen Astronauts: Heroes Who Died Reaching the Moon*. Lincoln: University of Nebraska Press, 2003.

Burrows, William E. *This New Ocean: The Story of the First Space Age.* New York: Modern Library, 1999.

Carpenter, M. Scott, Gordon L. Cooper Jr., John H. Glenn Jr., Virgil I. Grissom, Walter M. Schirra Jr., Alan B. Shepard, and Donald K. Slayton. *We Seven.* New York: Simon & Schuster, 1962.

Cayton, Andrew R. L. *Ohio: The History of a People.* Columbus: Ohio State University Press, 2002.

Cernan, Eugene, with Don Davis. *The Last Man on the Moon: Astronaut Gene Cernan and America's Race in Space.* New York: St. Martin's Griffin, 1999.

Chaikin, Andrew. *A Man on the Moon.* New York and London: Penguin Group, 1994.

——. *A Man on the Moon.* 3 vols. (I: *One Giant Leap;* II: *The*

Odyssey Continues; III: *Lunar Explorers*). Alexandria, VA: Time-Life Books, 1999.

Collins, Michael. *Carrying the Fire: An Astronaut's Journeys.* New York: Farrar, Straus and Giroux, 1974.

——. *Liftoff: The Story of America's Adventure in Space.* New York: Grove Press, 1988.

Compton, W. David. *Where No Man Has Gone Before: A History of the Apollo Lunar Exploration Missions.* Washington, DC: NASA SP-4214, 1989.

Conrad, Nancy, and Howard A. Klausner. *Rocketman: Astronaut Pete Conrad's Incredible Ride into the Unknown*. New York: New American Library, 2005.

Cooper, Gordon, with Bruce Henderson. *Leap of Faith: An Astronaut's Journey into the Unknown.* New York: HarperTorch, 2000.

Cooper, Henry S. F. Jr. *Apollo on the Moon.* New York: Dial, 1973.

——. *Moon Rocks.* New York: Dial, 1970.

Corn, Joseph J. *The Winged Gospel: America's Romance with Aviation, 1900–1950.* New York and Oxford: Oxford University Press, 1983.

Cortright, Edgar M., ed., *Apollo Expeditions to the Moon.* Washington, DC: NASA SP-350, 1975.

Cunningham, Walter, with Mickey Herskowitz. *The All-American Boys.* New York: Macmillan, 1977.

Dawson, Virginia P. *Engines and Innovation: Lewis Laboratory and American Propulsion Technology.* Washington, DC: NASA SP-4306, 1991.

Dethloff, Henry C. *Suddenly, Tomorrow Came . . . : A History of the Johnson Space Center.* Washington, DC: NASA SP-4307, 1993.

Dick, Steven J., ed. *NASA's First 50 Years: Historical Perspectives*. Washington, DC: NASA SP-2010-4704, 2009.

Dick, Steven J. and Roger D. Launius, *Critical Issues in the History of Spaceflight*. Washington, DC: NASA SP-2006-4702, 2006.

——. *Societal Impact of Spaceflight*. Washington, DC: NASA SP-2007-4801, 2007.

Duke, Charlie and Dotty. *Moonwalker.* Nashville, TN: Oliver-Nelson Books, 1990.

Emme, Eugene M. *Two Hundred Years of Flight in America: A Bi-*

centennial Survey. San Diego, CA: American Astronautical Society, 1977.

Engen, Donald. *Wings and Warriors: Life as a Naval Aviator.* Washington and London: Smithsonian Institution Press, 1997.

Evans, Michelle. *The X-15 Rocket Plane: Flying the First Wings into Space*. Lincoln and London: University of Nebraska Press, 2013.

Farmer, Gene, and Dora Jane Hamblin. *First on the Moon.* New York: Little, Brown, and Co., 1969.

Fraser, George MacDonald. *The Steel Bonnets: The Story of the Anglo-Saxon Border Reivers.* London: Collins Harvill, 1989.

French, Francis, and Colin Burgess. *In the Shadow of the Moon: A Challenging Journey to Tranquility, 1965-1969*. Lincoln: University of Nebraska Press, 2007.

Fries, Sylvia Doughty. *NASA Engineers and the Age of Apollo.* Washington, DC: NASA SP-4104, 1992.

Gainor, Chris. *Arrows to the Moon: Avro's Engineers and the Space Race.* Burlington, Ontario: Apogee Books, 2001.

Garber, Stephen J., ed. *Looking Backward, Looking Forward: Forty Years of U.S. Human Spaceflight Symposium, 8 May 2001.* Washington, DC: NASA SP-2002-4107, 2002.

Glenn, John, with Nick Taylor. *John Glenn: A Memoir.* New York and Toronto: Bantam Books, 1999.

Goldstein, Laurence. *The Flying Machine and Modern Literature.* Bloomington: Indiana University Press, 1986.

Gorn, Michael H. *Expanding the Envelope: Flight Research at NACA and NASA.* Lexington: University Press of Kentucky, 2001.

Grandt, A. F. Jr., W. A. Gustafson, and L. T. Cargnino. *One Small Step: The History of Aerospace Engineering at Purdue University.* West Lafayette: School of Aeronautics and Astronautics, Purdue University, 1996.

Gray, George W. *Frontiers of Flight: The Story of NACA Research.* New York: Knopf, 1948.

Gray, Mike. *Angle of Attack: Harrison Storms and the Race to the Moon.* New York: W. W. Norton & Co., 1992.

Gunston, Bill. *Attack Aircraft of the West.* London: Ian Allan, 1974.

Hacker, Barton C., and James M. Grimwood. *On the Shoulders of Titans: A History of Project Gemini.* Washington, DC: NASA SP-4203, 1977.

Hallion, Richard P. *On the Frontier: Flight Research at Dryden, 1946–1981.* Washington, D.C.: NASA SP-4303, 1988.

——. *Supersonic Flight: Breaking the Sound Barrier and Beyond,* rev. ed. Washington, DC: Brassey's, 1997.

——. *Test Pilots: The Frontiersmen of Flight,* rev. ed. Washington and London: Smithsonian Institution Press, 1988.

——. *The Naval Air War in Korea.* New York: The Nautical & Aviation Publishing Co. of America, 1986.

Hansen, James R. *The Bird Is on the Wing: Aerodynamics and the Progress of the American Airplane.* College Station: Texas A&M University Press, 2003.

——. *Engineer in Charge: A History of the Langley Aeronautical Laboratory, 1917–1958.* Washington, DC: NASA SP-4305, 1987.

——. *Spaceflight Revolution: NASA Langley From Sputnik to Apollo.* Washington, DC: NASA SP-4308, 1995.

Harland, David M. *How NASA Learned to Fly in Space: An Exciting Account of the Gemini Missions.* Burlington, Ontario: Apogee Books, 2004.

——. *The First Men on the Moon: The Story of Apollo 11.* Chichester, UK: Praxis Publishing, 2007.

Henes, Donna. *The Moon Watcher's Companion.* New York: Marlowe & Company, 2002.

Heppenheimer, T.A. *Countdown: A History of Space Flight.* New York: John Wiley & Sons, 1997.

Hersch, Matthew. *Inventing the American Astronaut*. New York: Palgrave Macmillan, 2012.

Hurt, Douglas R. *The Ohio Frontier: Crucible of the Old Northwest, 1720–1830.* Bloomington: Indiana University Press, 1996.

Illiff, Kenneth W., and Curtiss L. Peebles. *From Runway to Orbit: Reflections of a NASA Engineer.* Washington, DC: NASA SP-2004-4109, 2004.

Irwin, James B., with William A. Emerson Jr. *To Rule the Night.* Philadelphia: Holman (Lippincott), 1973.

Irwin, Mary, with Madelene Harris. *The Moon Is Not Enough.* Grand Rapids: Zondervan Corporation, 1978.

Jenkins, Dennis. *Hypersonics Before the Shuttle: A Concise History of the X-15 Research Airplane.* Monographs in Aerospace

History No. 18. Washington, DC: NASA SP-2000-4518, June 2000.

Jones, Robert Leslie. *The History of Agriculture in Ohio to 1880.* Kent, OH: Kent State University Press, 1983.

Kelly, Thomas J. *Moon Lander: How We Developed the Apollo Lunar Module.* Washington and London: Smithsonian Institution Press, 2001.

King, Elbert A. *Moon Trip: A Personal Account of the Apollo Program and Its Science.* Houston: University of Houston Press, 1989.

Knepper, George. *Ohio and Its People.* Kent, OH, and London, England: Kent State University Press, 1989.

Knott, Richard C. *A Heritage of Wings: An Illustrated History of Naval Aviation.* Annapolis: Naval Institute Press, 1997.

Koppel, Lily. *The First Wives Club: A True Story*. New York and Boston: Grand Central Publishing, 2013.

Kraft, Chris. *Flight: My Life in Mission Control.* New York and London: Plume Books, 2001.

Kranz, Gene. *Failure Is Not an Option: Mission Control from Mercury to Apollo 13 and Beyond.* New York and London: Simon & Schuster, 2000.

Lambright, W. Henry. *Powering Apollo: James E. Webb of NASA.* Baltimore and London: Johns Hopkins University Press, 1995.

Launius, Roger D. *Apollo: A Retrospective Analysis.* Monographs in Aerospace History No. 3. Washington, DC: Government Printing Office, July 1994.

Lay, Beirne Jr. *Earthbound Astronauts: The Builders of Apollo-Saturn.* Englewood Cliffs, NJ: Prentice-Hall, 1971.

Leckey, Howard L. *The Tenmile Country and Its Pioneer Families: A Genealogical History of the Upper Monogahela Valley.* Salem, MA: Higginson Book Co., 1950.

Leopold, George. *Calculated Risk: The Supersonic Life and Times of Gus Grissom*. West Lafayette, IN: Purdue University Press, 2016.

Levine, Arnold S. *Managing NASA in the Apollo Era.* Washington, DC: NASA SP-4102, 1982.

Lewis, Richard S. *Appointment on the Moon.* New York: Ballantine, 1969.

——. *The Voyages of Apollo: The Exploration of the Moon.* New York: Times Book Company, 1974.

Life, Special Issue, "Man in Space: An Illustrated History from Sputnik to Columbia," March 17, 2003.

Light, Michael. *Full Moon*. New York: Knopf, 1989.

Loftin, Laurence K. Jr. *Quest for Performance: The Evolution of Modern Aircraft.* Washington, DC: NASA SP-468, 1985.

Logsdon, John M. *John F. Kennedy and the Race to the Moon*. New York: Palgrave Macmillan, 2010.

Lovell, Jim, and Jeffrey Kluger. *Lost Moon: The Perilous Voyage of Apollo 13.* Boston and New York: Houghton Mifflin, 1994.

Mack, Pamela E., ed. *From Engineering Science to Big Science: The NACA and NASA Collier Trophy Research Project Winners.* Washington, DC: NASA SP-4219, 1998.

MacKinnon, Douglas, and Joseph Baldanza. *Footprints.* Illustrated by Alan Bean. Washington, DC: Acropolis Books, 1989.

Mailer, Norman. *Of a Fire on the Moon.* New York: Little, Brown and Co., 1969.

Masursky, Harold, G. William Colton, and Farouk El-Baz, eds. *Apollo Over the Moon: A View from Orbit.* Washington, DC: NASA SP-362, 1978.

McCurdy, Howard E. *Space and the American Imagination.* Washington and London: Smithsonian Institution Press, 1997.

McDonald, Allan J., with James R. Hansen, *Truth, Lies, and O-Rings: Inside the Space Shuttle Challenger Disaster*. Gainesville: University Press of Florida, 2009.

McDougall, Walter A. *The Heavens and the Earth: A Political History of the Space Age.* New York: Basic Books, 1985.

Michener, James A. *The Bridges at Toko-Ri.* New York: Random House, 1953.

Miller, Ronald, and David Sawers. *The Technical Development of Modern Aviation.* New York: Praeger, 1970.

Mindell, David. *Digital Apollo: Human and Machine in Spaceflight*. Cambridge, MA: MIT Press, 2008.

Mitchell, Edgar, and Dwight Williams. *The Way of the Explorer: An Apollo Astronaut's Journey Through the Material and*

Mystical Worlds. Franklin Lakes, NJ: New Page Books, 2008.

Moore, John. *The Wrong Stuff: Flying on the Edge of Disaster.* North Branch, MN: Specialty Press, 1997.

Murray, Charles, and Catherine Bly Cox. *Apollo: The Race to the Moon.* New York: Simon & Schuster, 1989.

Mutch, Thomas A. *A Geology of the Moon: A Stratigraphic View.* Princeton: Princeton University Press, 1970.

NASA. *Managing the Moon Program: Lessons Learned from Project Apollo.* Monographs in Aerospace History No. 14. Washington, DC: Government Printing Office, July 1999.

NASA. *Proceedings of the X-15 First Flight 30th Anniversary Celebration.* Washington, DC: NASA Conference Publication 3105, 1991.

Newell, Homer E. *Beyond the Atmosphere: Early Years of Space Science.* Washington, DC: NASA SP-4211, 1980.

Newton, Wesley P., and Robert R. Rea. *Wings of Gold: An Account of Naval Aviation Training in World War II.* Tuscaloosa: University of Alabama Press, 1987.

Norberg, John. *Wings of Their Dreams: Purdue in Flight*. West Lafayette, IN: Purdue University Press, 2003.

Oberg, James E. *Red Star in Orbit.* New York: Random House, 1981.

Peebles, Curtis. *The Spoken Word: Recollections of Dryden History, the Early Years.* Washington, DC: NASA SP-2003-4530, 2003.

Pellegrino, Charles R., and Joshua Stoff. *Chariots for Apollo: The Making of the Lunar Module.* New York: Atheneum, 1985.

Petroski, Henry. *To Engineer Is Human: The Role of Failure in Successful Design.* New York: Vintage Books, 1992.

Pizzitola, Anthony. *Neil Armstrong: The Quest for His Autograph*. CreateSpace Independent Publishing Platform, 2011.

Pyle, Rod. *Destination Moon: The Apollo Missions in the Astronauts' Own Words*. New York: Collins, 2005.

Rahman, Tahir. *We Came in Peace for All Mankind: The Untold Story of the Apollo 11 Silicon Disc*. Overland Park, KS: Leathers Publishing, 2008.

Reeder, Charles Wells. *The Interurbans of Ohio.* Columbus: Ohio State University Press, 1906.

Reid, Robert L. *Always a River: The Ohio River and the American Experience.* Bloomington: Indiana University Press, 1991.

Reynolds, David. *Apollo: The Epic Journey to the Moon.* New York and San Diego: Harcourt, 2002.

Roland, Alex. *Model Research: The National Advisory Committee for Aeronautics.* 2 vols. Washington, DC: NASA SP-4103, 1985.

Rosof, Barbara D. *The Worst Loss: How Families Heal from the Death of a Child.* New York: Henry Holt and Company, 1994.

Saltzman, Edwin J., and Theodore G. Ayers. *Selected Examples of NACA/NASA Supersonic Research.* Dryden Flight Research Center, Edwards AFB, CA: NASA SP-513, 1995.

Schirra, Walter M. Jr., with Richard N. Billings, *Schirra's Space.* Boston: Quinlan Press, 1988.

Schmitt, Harrison H. *Return to the Moon*. New York: Copernicus Books and Praxis Publishing, Ltd., 2006.

Scott, David, and Alexei Leonov. *Two Sides of the Moon.* New York: Thomas Dunne Books (St. Martin's Press), 2004.

Scott, Walter. *Minstrelsy of the Scottish Border.* 3 vols. London, 1869; Singing Tree, 1967.

Seamans, Robert C. Jr. *Aiming at Targets: The Autobiography of Robert C. Seamans, Jr.* Washington, DC: NASA SP-4106, 1996.

Siddiqi, Asif A. *Challenge to Apollo: The Soviet Union and the Space Race, 1945–1974.* Washington, DC: NASA SP-2000-4408, 2000.

Slayton, Donald K., with Michael Cassutt. *Deke! U.S. Manned Space: From Mercury to Shuttle.* New York: Forge, 1994.

Spudis, Paul D. *The Once and Future Moon.* Washington and London: Smithsonian Institution Press, 1996.

Stafford, Tom, with Michael Cassutt. *We Have Capture.* Washington and London: Smithsonian Institution Press, 2002.

Sullivan, Scott P. *Virtual Apollo: A Pictorial Essay of the Engineering and Construction of the Apollo Command and Service Modules.* Burlington, Ontario: Apogee Books, 2003.

——. *Virtual LM: A Pictorial Essay of the Engineering and Construction of the Apollo Lunar Module.* Burlington, Ontario: Apogee Books, 2004.

Taylor, Stuart Ross. *Lunar Science: A Post-Apollo View.* New York: Pergamon, 1975.

Thompson, Milton O. *At the Edge of Space: The X-15 Flight Program.* Washington and London: Smithsonian Institution Press, 1992.

Thompson, Neal. *Light This Candle: The Life and Times of Alan Shepard, America's First Spaceman.* New York: Crown Publishers, 2004.

Thruelson, Richard. *The Grumman Story.* New York: Praeger, 1976.

Trento, Joseph J. *Prescription for Disaster: From the Glory of Apollo to the Betrayal of the Shuttle.* New York: Crown Publishers, 1987.

Upton, Jim. *Lockheed F-104 Starfighter.* Minneapolis: Specialty Press, 2003.

Vaughn, Diane. *The Challenger Launch Decision: Risky Technology, Culture, and Deviance at NASA.* Chicago: University of Chicago Press, 1997.

Vincenti, Walter G. *What Engineers Know and How They Know It: Analytical Studies from Aeronautical History.* Baltimore and London: The Johns Hopkins University Press, 1990.

Wachhorst, Wyn. *The Dream of Spaceflight: Essays on the Near Edge of Infinity.* New York: Basic Books, 2000.

Wallace, Harold D. *Wallops Station and the Creation of an American Space Program.* Washington, DC: NASA SP-4311, 1997.

Wallace, Lane. *Flights of Discovery: 50 Years at the NASA Dryden Flight Research Center.* Washington, DC: NASA SP-4309, 1996.

Waltman, Gene L. *Black Magic and Gremlins: Analog Flight Simulations at NASA's Flight Research Center.* Monographs in Aerospace History No. 20. Washington, DC: NASA SP-2000-4250, 2000.

Wead, Doug. *All the Presidents' Children.* New York and London: Atria Books, 2003.

Wendt, Guenter, and Russell Still. *The Unbroken Chain.* Burlington, Ontario: Apogee Books, 2001.

Wiley, Samuel T., ed. *Biographical and Historical Cyclopedia of Indiana and Armstrong Counties, Pennsylvania.* Philadelphia: Gresham & Co., 1891.

Wilford, John Noble. *We Reach the Moon.* New York: Bantam Books, 1969.

Wilhelms, Don E. *The Geologic History of the Moon.* Washington, DC: U.S. Geological Survey Professional Paper 1348, 1987.

——. *To a Rocky Moon: A Geologist's History of Lunar Exploration.* Tucson and London: University of Arizona Press, 1993.

Wolfe, Tom. *The Right Stuff.* New York: Farrar, Straus and Giroux, 1979.

Worden, Al, and Francis French. *Falling to Earth: An Apollo 15 Astronaut's Journey to the Moon.* Washington, DC: Smithsonian Books, 2011.

Yeager, Chuck, and Leo Janos. *Yeager: An Autobiography.* Toronto and New York: Bantam Books, 1985.

Young, James O. *Meeting the Challenge of Supersonic Flight.* Edwards AFB, CA: U.S. Air Force Flight Test Center History Office, 1997.

Young, John, and James R. Hansen, *Forever Young: A Life of Adventure in Air and Space*. Gainesville, FL: University Press of Florida, 2012.

ARTICLES

Asher, Gerald. "Of Jets and Straight Decks: USS *Essex* and Her Air Wings, 1951–1953," *Airpower* 32 (Nov. 2002): 26–40.

Brinkley, Douglas. "The Man and the Moon," *American History* 39: 26–37, 78–79.

Gates, Thomas F. "The Screaming Eagles in Korea, 1950–1953: Fighting 51, Part II," *The Hook* 24 (Winter 1996): 19–31.

Gray, Paul N. "The Bridges at Toko-Ri: The Real Story," *Shipmate* (July–Aug. 1997).

Home-Douglas, Pierre. "An Engineer First," *Prism* 13: 42–45.

Honegger, Barbara, and USAF Lt. Col. (Ret.) Hank Brandli. "Saving Apollo 11," *Aviation Week and Space Technology* (Dec. 13, 2004): 78–80.

Kaufman, Richard F. "Behind the Bridges at Toko-Ri," *Naval Aviation News* 84 (Mar.–Apr. 2002): 18–23.

Michener, James A. "The Forgotten Heroes of Korea," *The Saturday Evening Post,* May 10, 1952, 19–21 and 124–28.

Reilly, John. "The Carriers Hold the Line," *Naval Aviation News* 84 (May–June 2002): 18–23.

Thompson, Warren E. "The Reality Behind Toko-Ri," *Military Officer* 1 (June 2003): 54–59.

PROFILES OF ARMSTRONG

Abramson, Rudy. "A Year Later: Armstrong Still Uneasy in Hero Role," *Los Angeles Times,* July 19, 1970.

Ambrose, Stephen E., and Douglas Brinkley. "NASA Johnson Space Center Oral History Project Oral History Transcript: Neil A. Armstrong," *Quest: The History of Spaceflight Quarterly* 10 (2003). It is also available online at www.jsc.nasa.gov/oral_histories.

Andry, Al. "America's Enigmatic Pioneer," *Cincinnati Post,* July 20, 1989.

"Armstrong Aimed at Moon Walk," *Dayton Journal Herald,* July 10, 1969.

"Armstrong Still the Same Old Neil," *Lincoln* [NE] *Journal,* July 20, 1978.

"Astronaut Neil Armstrong 'Embodied Our Dreams,'" Aug. 27, 2012, National Public Radio (interview with Neil deGrasse Tyson), accessed at http://www.npr.org/2012/08/27/160095721/remembering-astronaut-neil-armstrong.

Babcock, Charles. "Moon Was Dream to Shy Armstrong," *Dayton Journal Herald,* July 11, 1969.

Bebbington, Jim. "Armstrong Remembers Landing, Delights Auglaize Show Crowd," *Dayton Daily News,* July 18, 1994.

Benedict, Howard. "Ten Men on the Moon," *Florida Today,* Dec. 3, 1972.

Berkow, Ira. "Neil Armstrong Stays Alone in His Private Orbit," *Rocky Mount* [NC] *Telegram,* Dec. 15, 1975.

——. "Cincinnati's Invisible Hero," *Cincinnati Post,* Jan. 17, 1976.

Brinkley, Douglas. "The Man on the Moon," *American History* 39 (Aug. 2004): 26–37, 78.

Chriss, Nicholas C. "After Tranquility, Astronauts Lives Were Anything but Tranquil," *Houston Chronicle,* July 16, 1989.

Cohen, Douglas. "Private Man in Public Eye," *Florida Today,* July 20, 1989.

Conte, Andrew. "The Silent Spaceman: 30 Years After Moon Landing, Armstrong Still Shuns Spotlight," *Cincinnati Post,* July 17, 1999.

Cromie, William. "Armstrong Plays Down His Mark on History," *Washington Sunday Star,* July 13, 1969.

Day, Dwayne. "Last thoughts about working with the First Man," *The Space Review*, Dec. 31, 2012, accessed at http://www.thespacereview.com/article/2209/1.

Dillon, Marilyn. "Moon Walk Remains a Thrill," *Cincinnati Enquirer,* June 12, 1979.

Domeier, Douglas. "From Wapakoneta to the Moon," *Dallas Morning News,* June 21, 1969.

Dordain, Jean-Jacques. "Personal Tribute to Neil Armstrong," esa.com, Aug. 26, 2012.

Dunn, Marcia. "Neil Armstrong, 30 Years Later: Still Reticent After All These Years," Associated Press story, July 20, 1999, accessed at ABCNEWS.com.

Earley, Sandra. "In Search of Neil Armstrong," *Atlanta Journal and Constitution Magazine,* May 20, 1979.

Furlong, William (World Book Science Service). "Bluntly, He Places Ideas Above People," *Lima News,* June 13, 1969.

Galewitz, Phil. "Astronaut's Museum Speaks for Him," *Palm Beach Post,* Feb. 16, 2003.

Graham, Tim. "A Rare Talk with the Man from the Moon," *Cincinnati Post,* Mar. 3, 1979.

Greene, Bob. "Neil Armstrong Down to Earth," *St. Louis Post-Dispatch,* May 10, 1979.

——. "A Small Town and a Big Dream," *Cincinnati Post,* Oct. 24, 1992.

Hansen, James R. "The Truth about Neil Armstrong," Space.com, Aug. 23, 2013.

——. "The 11 biggest myths about Neil Armstrong, first man on the Moon," cbsnews.com, July 18, 2014.

Harvey, Paul. "Neil Called Semi-Recluse," *Cincinnati Enquirer,* May 13, 1981.

Hatton, Jim. "Neil Says Feet Firmly on Terra Firma," *Cincinnati Enquirer,* Dec. 2, 1974.

Hersch, Matthew. "Neil A. Armstrong, 5 August 1930–25 August 2012," in *Proceedings of the American Philosophical Society* 157 (Sept. 2013): 347–51.

Home-Douglas, Pierre. "An Engineer First," *Prism* 13 (summer 2004): 42–45.

Johnston, John, Saundra Amrhein, and Richelle Thompson. "Neil Armstrong, Reluctant Hero," *Cincinnati Enquirer,* July 18, 1999.

Kent, Fraser. "'Good, Gray Men' Fly to Moon," Cleveland *Plain Dealer,* July 15, 1969.

Knight News Service. "Armstrong the Star Sailor Born to High Flight," *Cincinnati Enquirer,* July 20, 1979.

Lawson, Fred. "Hero Seeks Privacy After Moon Walk," *Dayton Daily News,* July 15, 1984.

Lyon, David. "Moon's Armstrong Just Guy Next Door to Neighbors," *Dayton Daily News,* Dec. 7, 1972.

Martin, Chuck. "Lebanon's Code of Silence Shields Armstrong," *Cincinnati Enquirer,* July 18, 1999.

Mason, Howard. "After the Moon: What Does an Astronaut Do?" *New York Times Magazine,* Dec. 3, 1972.

Mosher, Lawrence. "Neil Armstrong: From the Start He Aimed for the Moon," *National Observer,* July 7, 1969.

Motsinger, Carol. "Welcome back to Cincinnati, Neil Armstrong," *Cincinnati Enquirer,* July 8, 2016.

"Neil Armstrong, Man for the Moon," *The National Observer,* July 7, 1969.

Purdy, Matthew. "In Rural Ohio, Armstrong Quietly Lives on His Own Dark Side of the Moon," *New York Times,* July 20, 1994.

Reardon, Patrick. "A Quiet Hero Speaks: Neil Armstrong Finally Opens Up," *Chicago Tribune,* Sept. 27, 2002.

Recer, Paul. "U.S. Moonmen Returned to Earth Changed Men," *Cincinnati Enquirer,* July 30, 1972.

Ronberg, Gary. "A Private Lifetime on Earth," *Philadelphia Enquirer,* July 18, 1979.

Rosensweig, Brahm. "Whatever Happened to Neil Armstrong?" Discovery Channel, accessed at www.exn.com, July 6, 1999.

Salvato, Al. "In Search of the Man on the Moon," *Cincinnati Post,* July 16, 1994.

Sator, Darwin. "Astronaut Armstrong Firmly Planted on Earth," *Dayton Daily News,* May 8, 1975.

Sawyer, Kathy. "Neil Armstrong's Hard Bargain with Fame," *Washington Post Magazine,* July 11, 1999.

Sell, Mark. "Armstrong: 'It's Over; and I'd Like to Forget It,' " *Florida Today,* Oct. 1, 1978.

Shepherd, Shirley. "On Wapakoneta, Astronaut Neil Armstrong and a Reporter's Woes," *Muncie* [IN] *Star,* July 1, 1969.

Snider, Arthur J. "Neil Armstrong Proves to Be Very Much an Earthling," *Chicago Daily News,* Aug. 11, 1977.

Stanford, Neal. "Pride in Achievement: NASA Hails Apollo Program as 'Triumph of the Squares,' " *Christian Science Monitor,* July 16, 1969.

Stevens, William K. "The Crew: What Kind of Men Are They?" *New York Times,* July 17, 1969.

Van Sant, Rick. "Nine Years Later, Moon-Walker Still Not Star-Struck," *Cincinnati Post,* July 20, 1978.

Wheeler, Lonnie. "The Search Goes On," *Cincinnati Enquirer,* Mar. 4, 1979.

Wilford, John Noble. "Three Voyages to the Moon: Life After Making History on TV," *New York Times,* July 17, 1994.

Wolfe, Christine. "Just Professor, Not Spaceman," *Cincinnati Enquirer,* June 19, 1988.

Wright, Lawrence. "Ten Years Later: The Moonwalkers," *Look* (July 1979): 19–32.

REFERENCE SOURCES

Angelo, Joseph A. Jr. *The Dictionary of Space Technology.* New York: Facts on File, Inc., 1982.

Cassutt, Michael. *Who's Who in Space: The First 25 Years.* Boston: G. K. Hall & Co., 1987.

Hawthorne, Douglas B. *Men and Women of Space.* San Diego: Univelt, Inc., 1992.

Heiken, Grant, David Vaniman, and Bevan M. French. *Lunar Sourcebook: A User's Guide to the Moon.* New York: Cambridge University Press, 1991.

Jenkins, Dennis, Tony Landis, and Jay Miller. *American X-Vehicles: An Inventory: X-1 to X-50.* Monographs in

Aerospace History No. 31. Washington, DC: NASA SP-2003–4531, June 2003.

Launius, Roger D., and J. D. Hunley. *An Annotated Bibliography of the Apollo Program.* Monographs in Aerospace History No. 2. Washington, DC: Government Printing Office, July 1994.

Orloff, Richard W. *Apollo by the Numbers: A Statistical Reference.* Washington, DC: NASA SP-2000-4029, 2000.

Portree, David S. F., and Robert C. Trevino. *Walking to Olympus: An EVA Chronology.* Washington, DC: NASA Monographs in Aerospace History Series No. 7, Oct. 1997.

Stillwell, Wendell H. *X-15 Research Results.* Washington, DC: NASA SP-60, 1965.

Surveyor Program [Office]. *Surveyor Program Results.* Washington, DC: NASA SP-184, 1969.

The Apollo Spacecraft: A Chronology. Four vols: Vol. I: "Through November 7, 1962," Ivan D. Irtel and Mary Louise Morse, eds.; Vol. II; "November 8, 1962–September 30, 1964," Mary Louise Morse and Jean Kernahan Bays, eds.; Vol. III: "October 1, 1964–January 20, 1966," Courtney G. Brooks and Ivan D. Ertel, eds.; Vol. IV: "January 21, 1966–July 13, 1974," Ivan D. Ertel and Roland W. Newkirk with Courtney G. Brooks, eds. Washington, DC: NASA SP-4009. 1969, 1973, 1976, 1978.

Wells, Helen T., Susan H. Whitely, and Carrie E. Karegeannes. *Origins of NASA Names.* Washington, DC: NASA SP-4402, 1976.

SELECT ELECTRONIC MEDIA ON APOLLO 11

Remembering Apollo 11: The 30th Anniversary Data Archive CD-ROM. NASA SP-4601, 1999.

Remembering Apollo 11: The 35th Anniversary Data Archive CD-ROM. NASA SP-4601, 2004.

Apollo Moon Landing 1969 BBC Television Coverage CD. Pearl, 1994.

Apollo 11 HD Videos, accessed at https://www.nasa.gov/multimedia/hd/apollo11_hdpage.html.

Apollo Lunar Surface Journal, eds. Eric M. Jones and Ken Glover, accessed at https://www.hq.nasa.gov/alsj.

JUVENILE LITERATURE

Brown, Don. *One Giant Leap: The Story of Neil Armstrong.* Boston: Houghton Mifflin Company, 1998.

Connolly, Sean. *Neil Armstrong: An Unauthorized Biography.* Hong Kong: Heinemann Library, 1999.

Dunham, Montrew. *Neil Armstrong, Young Flyer.* New York: Aladdin Paperbacks (Simon & Schuster), 1996.

Edwards, Roberta. *Who Was Neil Armstrong?* New York: Grosset & Dunlap (Penguin Random House), 2008.

Kramer, Barbara. *Neil Armstrong, the First Man on the Moon.* Berkeley Heights, NJ: Enslow Publishers, Inc., 1997.

Rau, Dana Meachen, *Neil Armstrong.* New York: Children's Press (Scholastic, Rookie Biographies), 2003, 2014.

Smith, Jacob. *Neil Armstrong Biography for Kids Book: The Apollo 11 Moon Landing, With Fun Facts and Pictures on Neil Armstrong*. Amazon Digital Services, LLC, 2017.

Westman, Paul. *Neil Armstrong, Space Pioneer.* Minneapolis: Lerner Publications Company, 1980.

Zemlicka, Shannon. *Neil Armstrong*. Minneapolis: Lerner Publications Company, 2002.

Note on Sources

Readers interested in seeing all the source notes for *First Man* may do so by consulting the notes section (64 pages long) at the back of either of the first two editions of this book.

In addition, the notes for this 2018 edition are now available on a website that the author has created: www.simonandschusterpublishing.com/firstman.

Index

About the Author

James R. Hansen is Professor Emeritus of History at Auburn University in Alabama. An expert in aerospace history and the history of science and technology, Hansen has published a dozen books and numerous articles covering a wide variety of topics, including the early days of aviation, the history of aerospace engineering, NASA, the Moon landings, the space shuttle program, and China's role in space. A Universal Studios film adaptation of *First Man* hits the silver screen in October 2018, with a script written by Academy award-winning screenwriter Josh Singer (*Spotlight*). Academy Award winner Damien Chazelle (*La La Land*) is directing the film, which stars Ryan Gosling as Neil Armstrong. Hansen lives in Auburn, Alabama, with his wife, Peggy. He has two children, Nathaniel and Jennifer (Gray), and three grandchildren, Isabelle (Gray), Mason (Gray), and Luke (Hansen).